SCALE TRANSITIONS AS FOUNDATIONS OF PHYSICS

SCALE TRANSITIONS AS FOUNDATIONS OF PHYSICS

Nicolae Mazilu

University of Akron, Ohio, USA

Maricel Agop

Asachi Technical University, Iasi, Romania

Ioan Merches

Al. I. Cuza University, Iasi, Romania

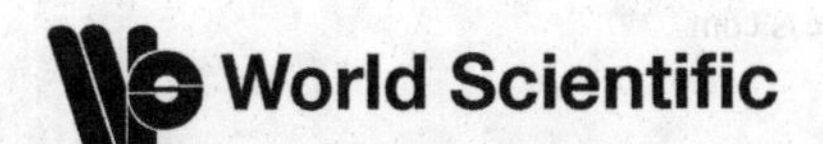

NEW JERSEY • LONDON • SINGAPORE • BEIJING • SHANGHAI • HONG KONG • TAIPEI • CHENNAI • TOKYO

Published by

World Scientific Publishing Co. Pte. Ltd.
5 Toh Tuck Link, Singapore 596224
USA office: 27 Warren Street, Suite 401-402, Hackensack, NJ 07601
UK office: 57 Shelton Street, Covent Garden, London WC2H 9HE

British Library Cataloguing-in-Publication Data
A catalogue record for this book is available from the British Library.

SCALE TRANSITIONS AS FOUNDATIONS OF PHYSICS

ISBN 978-981-123-186-5 (hardcover)
ISBN 978-981-123-187-2 (ebook for institutions)
ISBN 978-981-123-188-9 (ebook for individuals)

For any available supplementary material, please visit
https://www.worldscientific.com/worldscibooks/10.1142/12151#t=suppl

Desk Editor: Nur Syarfeena Binte Mohd Fauzi

Typeset by Stallion Press
Email: enquiries@stallionpress.com

Printed in Singapore

"If I were again beginning my studies, I would follow the advice of Plato and start with mathematics"

Galileo Galilei

Contents

Foreword

This book is about the structure of a universe in general, as it can be penetrated within the limits of today's human knowledge. Of course, one cannot understand by this the *whole* human knowledge, but only that part of it proper to applications in building what can be called a model of the universe. First, we concentrate on arguing why such a model has to be cosmological, and this explains by itself why the part of human knowledge involved in such a task has to be mainly physics and mathematics. Fact is that the birth of wave and quantum mechanics at the beginning of the last century led to the explicit consideration of the *concept of interpretation*, meant to solve the essential discrepancy between the mathematical description of a model of reality and the physical *possibility of existence* and *human authentication* of that reality. This concept was described by us in a previous work (Mazilu *et al.*, 2019), whose main upshot is a specific image of the fundamental physical unit of a universe, satisfying the criterion of time and space scale transitions: the planetary model of hydrogen atom, with spatially extended physical components for the central matter of nucleus and the revolving matter of electron.

Regarding the possibility of existence, the physical components of the model of planetary model, taken separately, can each count as fundamental particles in the common connotation in physics today, except, perhaps, for a hint regarding the topology of space containing their matter. To wit, what we call electron has a toroidal topology in the model, and we are certainly required to decide what is its general status as an independent physical particle (Williamson and van der Mark, 1997). In other words, we are required to account for the relationship between the internal structure of the matter inside a particle, and the topological form of its physical occurences. Then, that internal structure must be described cosmologically, and we do this by a general, specifically mathematical, concept of confinement, as suggested in a methodical analysis of the classical dynamical treatment of the Kepler problem. The main idea here, of a cosmological origin, is that the matter and the space enter in a reciprocal relation with equal ranks. Specifically, if one can talk of the matter in space, one has also to talk of the space in matter: at least from physical point of view they should be necessarily described the very same way. More to the point, talking of the *space inside matter*, in the sense of Newton, compels us to assuming that the matter should be considered as *matter inside space* as well. Both these categories are therefore topologically finite, and this is the general characteristic of a cosmology in the Einsteinian connotation. As to a general connotation, the space *per se* and the matter *per se* are two categories mutually exclusive whose existence is only hypothetical.

The space and time scales' transition criterion is obtained by a generalization of the idea of adiabatic invariance, and is built based on the concept of group invariance. The notion derived from Sir Michael Berry's unearthing of the adiabatic phase changes hidden in the usual wave mechanical treatment of quantum systems (Berry, 1984), can be extended to an exact invariance of the Newtonian force fields, with respect to a gauging procedure which involves time and space scales (Berry and Klein, 1984). We do this by a group action involving the $\mathrm{SL}(2,\mathrm{R})$ algebraic structure, which is implicit in the Berry–Klein gauging definition. Starting from this point, everything follows by mathematical rules: there is an internal world of physical particles, inaccessible to human means, where the constitutive Hertz particles of the matter in the *per se* condition — *i.e.* matter that does not contain space inside — are in equilibrium. In other words, the interpretation of the matter *per se* has to be done by ensembles of Hertz particles in static equilibrium (Mazilu *et al.*, 2019).

If we consider the space as an essential possibility of motion, then inside the matter *per se* there is no motion, and this is why we can interpret it by ensembles of Hertz particles in equilibrium.

However, at some point of our presentation, the possibility of human authentication becomes critical and requires a process reciprocal to interpretation: we called this a *reverse interpretation*. The name is intended to suggest the fact that, when it comes to human perception, which is usually epitomized by the idea of motion, we usually have to do either with isolated particles, or with finite ensembles of particles at most, and we have to associate a continuum to these. However, a continuum like this has to be considered as matter *per se*, and in the matter *per se* there is no motion. The groupal method of scale transition insures an invariant method though for associating a continuum to the perceived motion, with the outcome of a general idea of *physical coordinate system* in the sense defined by Bartolomé Coll, that also allows for a general concept of *physical reference frame* (Coll, 1985).

By and large, the idea is that the scale transition is actually the engine of the whole natural philosophy, which always adapted, sometimes even completely reconstructed, its main modern instrument — the theoretical physics — in order to cope with it. The first three chapters of this work render, from the perspective of scale transition, significant historical moments of theoretical physics, followed, in the remaining chapters, by suitable descriptions of the mathematical tools to go with this idea.

The core of mathematical philosophy to follow suit, is the Berry–Klein theory of invariance of the field of forces to scale transition, specifically, its close connection with the Ermakov–Pinney equation, which is thus bound to occupy a good portion of our presentation. This equation allows us to talk sense in making the difference between distance and length, and between kinematics and dynamics outside the concept of force. The basis of this whole mathematical philosophy is the concept that wave mechanics was forced to adopt with the advent of quantization, namely the concept of interpretation. This is for physics what for mathematics is the continuum hypothesis, or the axiom of choice, with which, in fact, the concept of interpretation has a direct parentage. However, what we want expressly to stress again and again, is the one point of view never taken into consideration by the natural philosophy — forgotten at its very birth, if we may say so — and which was aroused, the first and only time, by Heinrich Hertz in his *Principles of Mechanics* (Hertz, 2003). Probably, even Hertz himself did not know what to do with the concept he defined — the course of his beautiful work follows only the mathematical line of presenting the principles, 'in a new form', as he says — but the truth is that only the wave and quantum mechanics unveiled the true gnoseological capabilities of this concept, mostly by the definition of interpretation due to Charles Galton Darwin. The best illustrative example is the discussion around the cosmological problem started by Einstein in 1917 and reported by us here in Chapter 1. It revealed the important position in the natural philosophy of the Einsteinian point of view, and the clear difference between this and the old classical Newtonian point of view. It also revealed that the two points of view can be reconciled, if we may say so, into a general, apparently more realistic, natural philosophy, whereby the wave mechanics plays an essential part. Inasmuch as here we just have to pinpoint that part, it seems better to start from the Hertz's concept which we feel worth reproducing again and again.

For now, we reproduce, and discuss, only the necessary original definitions and commentaries (Hertz, 2003, pp. 45–46), keeping in store our understanding, to be revealed gradually. Therefore, quoting:

> **Definition 1**. A *material particle* is a characteristic by which we associate without ambiguity a given point in space at a given time with a given point in space at any other time.
>
> Every material particle is *invariable and indestructible.* The points in space which are denoted at two different times by the same material particle, coincide when the times coincide. Rightly understood the definition implies this.
>
> **Definition 2**. The number of material particles in *any space*, compared with the number of material particles *in some chosen space at a fixed time*, is called *mass* contained in the first space.
>
> We may and shall consider the number of material particles in the space chosen for comparison to be infinitely great. The *mass of the separate material particles* will therefore, by the definition, be *infinitely small.* The mass in any given space may therefore have any rational or irrational value.

> **Definition 3**. A *finite* or *infinitely small mass*, conceived as being *contained in an infinitely small space*, is called a *material point.*
>
> A material point therefore consists of any number of material particles *connected with each other*. This number is always to be infinitely great: this we attain by supposing the material particles to be of a *higher order of infinitesimals* than those material points which are regarded as being of infinitely small mass. The masses of material points, and especially the masses of infinitely small material points, may therefore bear to one another any rational or irrational ratio. (*our emphasis, n/a*).

The trend of progressing of his Mechanics does not seem to indicate that Hertz followed a program as outlined in this list of definitions, at least not from the points of view later revealed in physics. It is quite normal: the reasons listed by Hertz in his Preface to the treatise show that he mainly followed the soft spots of the concept of force at that time. This is perhaps the reason that the treatise does not play today the foundational part it deserves in our physical knowledge. However, an early analysis by Poincaré suggests that Hertz's work has to be taken more seriously, even as it is, for, even as such it touches fundamental issues of the human knowledge. Quoting:

> I insisted on this discussion longer than Hertz himself; I meant to show though that Hertz didn't simply look for quarrel with Galilei and Newton; we must agree to the conclusion that in the framework of the classical system *it is impossible to give a satisfactory idea for force and mass.* [(Poincaré, 1897); *original emphasis*]

The exquisite analysis of the great scholar, and everyone else's ever interested in the Mechanics of Hertz, for that matter, do not appear to take due notice of the definitions and comments excerpted by us above. Fact is that the excerpt, which apparently is referring only to mass, touches actually both of the two fundamental ideas mentioned by Poincaré, and with them the objective reasons of the subsequent general relativity and wave mechanics. In this respect, two things are worth noticing right away, bearing directly on our subject matter here.

The first thing to be noticed is the definition of mass. With the benefit of more than a century of physics, we consider it in order to show what it occasioned, and what actually Hertz's definitions targeted. The mass to Newton is, apparently, the quantity of matter, whose definition is (Newton, 1974):

> **Definition I**. The Quantity of matter is the measure of the same arising from its density and bulk conjunctly.

Seems like a vicious cycle: assuming that we know what the 'bulk' is, specifically by identifying it with the 'volume' — as the course of *Principia* seems to entitle us — in order to know what quantity of matter is, again, specifically identifying it with the mass, we need to know what the density is. On the other hand, in order to know what the density is we need to know what the mass is, because in these conditions, the density itself also 'arises from mass and bulk conjunctly'. But, we should not take things the hard way! Fact is that in the times of Newton the density was much closer to the natural philosophy than it became later when the man started asserting a better understanding of the *structure of matter*. The density was then an attribute that could be 'felt', so to speak, as it addresses ultimately to our senses. We can feel that the water is denser than the air, that a stone is denser than the water, etc., and we can even create an approximate, or even exact, scale of densities thus based only on our senses, by quantitatively comparing the densities. In fact, the Archimedes' principle has already had been known and used for almost two millennia at the times of Newton, thereby providing a method of measuring the densities, at least for solids. Even today in engineering fluid mechanics, one uses the old scale of comparative densities, which takes the density of water as unity. Seen in this light, and conceiving the mass as 'quantity of matter' and the bulk as 'volume', like we said, the definition of Newton is perfect, and was used by him as such, to lead to some very important consequences, especially technological.

However, from the point of view of the natural philosophy at large, this definition was not satisfactory, inasmuch as it leaves the relation between matter and space undecided. Newton himself was compelled into explaining this relation further, within the limits of the concept of density. Quoting:

> **COR. III**. *All spaces are not equally full*; for if all spaces were equally full, then the specific gravity of the fluid which fills the region of the air, on account of the extreme density of the matter, would fall nothing short of the specific gravity of quicksilver, or gold, or any other *the most dense* body; and, therefore, neither gold, nor any other body, could descend in air; for bodies do not descend in fluids, unless they are specifically heavier than the fluids. And if *the quantity of matter in a given space* can, by any rarefaction, be diminished, what *should hinder a diminution to infinity*?
> **COR. IV**. If all the solid particles of all bodies are of the same density, and cannot be rarefied without pores, then a *void, space, or vacuum must be granted. By bodies of the same density, I mean those whose inertias are in the proportion of their bulks.* [(Newton, 1974, Volume II, p. 414); *our italics*]

Notice the *reference of density to the inertia* in the last phrase here. One can say that Newton must have felt that the concept of density, as it was defined classically, is insufficient for the ranking of densities in a natural philosophy. Indeed, in order to define a scale of densities, one needs to define first what 'equal density' means, and that requires the characterization of 'filling of a space'. Thus, the concept of force, allowing to the solid particles of bodies a certain 'degree of filling' of a space, must add some further differentiate to the category of matter. The last phrase, 'inertias in proportion of bulks' in the above excerpt, explains the whole Newton's philosophy of definition of forces by a procedure of measurement respecting the same principle of comparison as the measurement of density [see Corollary III of Proposition VII, in the Volume I of *Principia* (Newton, 1974)].

Now, one can say that the most notable consequences of this definition of the density is the fact that it occasioned a definition of the force as a vector field, *i.e.* as a *state of space containing matter*. This fact has been fruitfully exploited afterwards by Maxwell, for instance, in creating his electromagnetic theory of light. The basic modern philosophy here, logically derived from the geometrical analysis, is that the force in a portion of space is related to the density of matter from that portion by the so-called *Poisson equation.* This way the density of matter is directly related to the forces, which can be ascribed to it. Of course, this formalism has drawbacks, like any human endeavor for that matter. For instance, it shows that the Newtonian force, of magnitude going inversely with the square of distance can be considered as universal *only in space ranges where the density of matter is zero.* However, we need to utter it, these space ranges are amongst those spaces that must be granted in matter, according to Newton's phrase above, spaces internal to matter, just as 'internal' as the matter is 'internal' to space. These are those kind of coordinate spaces for which the Schrödinger equation was initially written, as Darwin would say (Darwin, 1927). In other words, the relation between space and matter, taken as concepts, is reciprocal, and this fact must be reflected in their description.

Indeed, the way our senses usually present them to reason, *viz.* as intuitions, if it is to use a Kantian assessment in this natural philosophy, the matter is contained in space, therefore it is 'internal' to space. In fact the Newtonian force is the *only* force compatible with a zero density of the matter, capable to explain the presence of matter in a space. The way these conclusions came to light can be seen, first of all, in the modern cosmology. Indeed, from astrophysical and astronomical observations it is inferred that the matter in universe has a *very small* mean density (around $1.0 \times 10^{-27}\,\text{kg/m}^3$). This was deduced by considerations independent of force, from the region of the universe accessible to our observations. On the other hand, we need to recall the same space behavior for the forces between the particles representing electricity. This fact was discovered by Coulomb. The electric particles, again, enter the structure of any known continuum, but in an electrically neutral continuum the charge density is exactly zero, not approximately. Thus, the figure above for the density might mean plainly zero, if we count on our experience with electricity and do not use and abuse of extremely small numbers. Or, if we choose to work with any extreme numbers at all, we may also choose to accept variations of the electron charge, for instance, in order to explain the matter density as it is. However, the most important predictions of the classical theory are that the only forces compatible with a *constant* density, according to Poisson stand in mechanics, are the *elastic* forces, and a neat zero density is *always* and *unconditionally* compatible with some *noncentral forces*. These two specific consequences of Newtonian theory of forces, properly combined led in time directly to the theory of light, even within the framework of the classical mechanics. As it happened though, the theory of light took a wide turn, apparently with the specific purpose of avoiding the Newtonian dynamics.

Now, there are a lot of consequences of Newtonian dynamics, all of them being certainties revealed by mathematics, as Newton conceived it. One of these consequences endorsed by mathematics, compelled science to exclusively consider the density as a primitive quantity, in the precise sense that it controls the forces via Poisson equation. This theoretical consequence of Newtonian dynamics, brought the concept of density of matter to a critical point by the beginning of 20th century. The point was that the cosmological density depends on the space scale we are considering when describing the universe around us. For once, this gave birth to the theory of general relativity, whose main theme was to add a new differentia to the concept of density, namely the counting: if the universe is a physical structure, its description is pending on the matter density, according to the very classical precepts. And the matter density in a universe cannot be decided only locally in the Archimedean manner: one also has to count the fundamental matter formations, in order to find a numerical density of them.

The above excerpt from Hertz's *Mechanics* is referring to this new differentia of the density, apparently impossible within Newtonian mechanics. One can see, indeed, that this new concept of density may be fundamental for general relativity or for wave mechanics, but not for Newtonian mechanics. And while these branches of physics took occasionally notice of the concept as such, it was not used in that guise for any foundation of the physics. In hindsight we can even state the reason: the Hertz's definitions have never been considered as significant. And this was, and still is, utterly necessary, for those definitions are the only ones that make the problem of connection between space and matter just as essential for our knowledge, as the fundamentals of mathematics expressed in the set theory. Just like in the set theory, Hertz reveals the importance of the continuum hypothesis and of the axiom of choice, and the necessity of their connection with a theory of measure.

Remaining within the realm of physics for the moment, we have to mention that there were important episodes of physics pointing out, almost explicitly we should say, toward a necessity of 'Hertzian approach', as it were. However, they were 'diverted' in the direction of what one considers today as some very significant achievement of the physics itself. One of these was the concept of 'application of the field on charge and vice versa', advanced by Louis de Broglie, and mentioned by us in Chapter 2 here. But we can make another, more significant case, insofar as it is explicitly referring to the notion of manifold representing a physical quantity according to an observation of Bernhard Riemann on the theory of quantities, and connect it with the more modern set theory. This episode is all about the theory of colors, and involves the central personality of the modern physics, Erwin Schrödinger. From the point of view of the idea of application, one can say that, by analogy with the de Broglie's charge, this case can be presented as an 'application of the color to matter'. It can even be presented in the same spirit, but with three 'application functions', in the spirit of the classical Helmholtz theory of colors which involves a three-dimensional color manifold (Mazilu *et al.*, 2019). Quoting:

> The *manifold of lights* has a higher power than the *power of the continuum*, namely that of a space of functions; and hence an indefinitely large number of dimensions. A priori it would seem possible that this could also hold true for the *manifold of color qualities*, or at least that it could have had a very large number of dimensions; as in the case of manifold of combined tones, since the ear acts to some extent as a harmonic analyzer. *That is not the case here.* Rather, according to *the principle of matching appearance on adjacent fields*, the lights arrange themselves into large groups — *each one of the power of the function space* — and the manifold of these groups of the same appearance is, for normal color-perception persons, *of the dimension three* — the highest ever observed. This dimensionality *is a fundamental fact of basic colorimetry*, and its derivation from experience will be our closer concern here.

[(Schrödinger, 1920); *our translation and Italics; see also* (Niall, 2017), *Chapter II, first communication*]

The Schrödinger work can be labeled, as we said, as an apprenticeship into the general theory of quantities, in order to train himself enough to be able to create six years later, the future wave mechanics as an application of the matter on space. For, such can be branded the wave mechanics in the connotation of Schrödinger, insofar as it is referring to the fundamental structure of the matter, the hydrogen atom.

In hindsight, one can say that the *quantification*, as a general procedure of associate a manifold to physical quantities, deviated into *quantization* and, further on, into *application*, due to the fact that the physics has not followed the path indicated by Hertz's program. For once, the essence of de Broglie's and Schrödinger's 'application of matter to space' and Schrödinger's 'application of colors to space', just showed that a probabilistic point of view has to be involved here at any rate. However, the theory of probabilities was not, at the time, and we should add that it is not even today, capable of undertaking the task the way these cases — especially the theory of colors — presented it, if it does not explicitly account for the concept of scale transition. For, the application of the lights to colors, with its weight function assuming negative values (Schrödinger, 1920), would have to be considered a necessary step toward the wave function, which is known to have led to negative probabilities both directly as well as indirectly, via a density function. As we see it, only late in the last century, in an approach devised by Nicholas Georgescu-Roegen would make the theory of probabilities capable of such a job (Georgescu-Roegen, 1971).

Now, the mathematical apparatus we selected to serve the purpose of the idea of scale transition, can be branded as a tool coping with our experience at large. Like in the case of Earth, whose innermost region is unreachable by human means, even though the humans are living on its crust, a physical particle can be assumed as structured by a *central nucleus* of pure matter, where *the space does not have access*, and a physical structure that represents the *rest of the universe* relative to that physical particle. Only in this last part of the particle a physical structure can occur, for only here the space can exist. The pure matter of the central nucleus — the matter *per se*, as we have called it every now and then — is a continuum which can only be interpreted as a Madelung fluid of Hertz material particles. This interpretation can be accomplished in the manner of Schrödinger, by ensembles of material particles having the powers of continua. Ensembles of these particles, but *at rest with respect to each other*, constitute Hertz material points. As according to our experience, the rest state of any ensemble of material points does not seem to last a finite time, a material point should be always described in the infrafinite scale of time. Thus, a Hertz material point can be taken as an equilibrium ensemble of material particles, but then we ought to explain how those constitutive material particles can be at relative rest with respect to each other, while in motion of ensemble, with respect to other like ensembles. The explanation can be done actually, if we accept forces between the material particles needed in the interpretation of the continua, for these forces have to be of a special type.

Indeed, a Hertz material particle is simply a position endowed with physical properties. These physical properties should be of the kind that generate those special forces satisfying the condition of equilibrium. First, at any scale such forces must obey the third principle of dynamics, because a material particle acted upon by them cannot be in equilibrium but only if they act as vectors upon it and, moreover, are in equilibrium in any direction in matter. Insofar as the 'direction' has its origin in our space perception of the world, involving, as it were, a certain subjectivity in its definition, in this argument is, therefore, assumed that the *concept of direction* transcends between space and matter, as we shall show in Chapter 5 here. This fact is of tremendous importance for physics.

In order to define a particle in the static instance one must admit that it is acted upon by forces invariant with respect to the matter extension. According to the Berry–Klein theory, these should be Newtonian type forces. In this case, the only physical properties a particle can support, are the gravitational mass and the two charges:

electric and magnetic. However, because the definition of statics needs a Wigner dynamic principle, these forces define a velocity field. Further on, because a statics defines a Hertz material point, in this instance the material point is defined as a reference frame. All these conclusions are presented by us here, based on the Berry–Klein theory, as consequences of the Ermakov–Pinney theorem, via continuous group theory.

Speaking of the coordinate space, we should say a few words regarding the concept of reference frame. Rarely, if ever, is the theoretical physics concerned with the definition of coordinates: the usual consent is that the coordinates exist, they do have the meaning we happen to assign them, and the theoretical results can be expressed in such a way that, when it comes to verification, there is always a correspondence with the reality of things described by those coordinates. It is on this state of the case that the modern global positioning on Earth came to remind us that the special relativity, with all its specific requirements, actually represents a modality of defining the coordinates physically, by a condition of equivalence between them. This condition is always tacitly assumed in the classical cases, but never explicitly stated: the space coordinates are defined geometrically, first and foremost, by coordinate lines, along which we need to pinpoint some values uniquely corresponding to a position in space. However, for a holistic theory of the universe such a definition of the coordinates is not enough.

First of all, if it is that some analogy with the radiation defining the relativistic coordinate systems be involved here, one needs a description of radiation from thermodynamical point of view, inasmuch as this point of view involves an enclosure containing the radiation. Then we need to choose that description of the thermal radiation which proved to be invariant to the changing in dimensions of the enclosure, and imposed the modern cosmology. Historically, such an addition to the reference frame was realized in the form of Wien–Lummer enclosure for radiation studies (Wien and Lummer, 1895). This device had a crucial role in establishing the radiation laws, and their concordance with the observed properties of the radiation at large. That concordance, in the form of

Wien's displacement law, contributed in establishing the quantum theory as a natural theory of light as we know it today.

It is along this line of historical development of physics, that the theories of the light ray have been improved, until they have reached the discovery of a new phenomenon, namely the *holography*, to be added to classical ones: reflection, refraction and diffraction. The holography is then the general phenomenon to be described by interpretation according to Schrödinger equation. The Wien–Lummer enclosure is, in fact, such a reference frame.

Let us say a few words on the concept of coordinate system. In order to describe the spatial position in space, we have to use a certain reference frame, and consequently one always needs such a concept, usually connected to a particular geometry that offers the meaning of coordinates. This is the typical case in physics, and it even became cursory, if we may say so, to the point where sometimes the coordinates are only mentioned, with no precaution of defining them in a way or another. It is the modern concept of global positioning that came to impose a closer consideration of the concept of *space position* itself, in the definition of which *the idea of light ray* needs to be taken as a fundamental concept.

The best concept of a definition of a system of coordinates, in our opinion, is that of Bartolomé Coll who aims at defining physically the coordinate lines, and actually builds a general natural philosophy, as it were, to be followed in such a construction in any universe (Coll, 1985). We shall use Coll's philosophy, but only for the three-dimensional case. In order to make this philosophy into an appropriate one for the physics in general, we reproduce here three essential endnotes from the Coll's work just cited. First, comes the idea of *lines of coordinates*:

> Typically, the definition domains of such systems correspond to *world tubes* obtained by evolution of *space-like tetrahedral figures*, over whose four faces, at every instant, *light beams fall on.* (Coll, 1985, Endnote [6], *our emphasis*)

In the present work we use 'tetrahedral figures' in Euclidean reference frames where only one of the faces of a tetrahedron is to be taken into consideration. This way eight tubes will be constructed for each and every reference frame as eight light beams. The structure of these light beam should be, and it is indeed, as 'physical' as possible. For us it is the classical light ray, completed with a physical interpretation due to Louis de Broglie (de Broglie, 1927b,c). The word 'interpretation' is taken here in the precise connotation of this concept necessary for the construction of wave mechanics (Darwin, 1927). Only, de Broglie's 'wave phenomenon called material point' gets here baptized by Bartolomé Coll as 'luxon', a name that we should assume without any reserve, in view of the vagueness of the light particle. Quoting, again:

> Here we consider light in the *geometric optics* approximation, that is to say, *as a fluid* of point like "luxons" (Coll, 1985, Endnote [7], *our emphasis*).

Therefore, along a world tube representing a light beam, particularly a light ray, the luxons are traveling, and the light wave is laterally limited by the tube. Louis de Broglie has defined the world tube as a *capillary tube*, with the simultaneous luxons defining, in turn, a surface evolving along the tube, just as Newton did for his definition of the light ray (Newton 1952, p. 1), thereby generating the whole physics of light. Among other things it becomes possible to define a physical coordinate line with respect to this capillary tube: it is the line normal to that surface determined by simultaneously traveling luxons. Quoting, again from Bartolomé Coll:

> ... the theoretical or experimental conclusion that a specific physical field, under particular conditions, depends only on the variable r is void or, at least, confuse, if the nature of the coordinate line "r variable" is not precised (radial, cylindrical, angular or other). And, from the experimental point of view, we need to describe *the physical procedure for its construction.* It is this sense that has, for us here, the word *operational* (Coll, 1985, Endnote [4], *our emphasis*).

This is the meaning of the word 'operational' for us too, all along the present work. The ray is for us a de Broglie capillary tube along which some luxons are circulating, as shown in Chapter 5 in connection with the Newtonian forces. Only, the luxons must be particularly defined, and this is the point where the cosmology enters the stage. For, we carry out this definition in relation with the physical structure of the universe. The whole mathematics beyond this construction was presented by us under the sweeping idea of three-dimensional analyticity, in Chapters 3 and 4. The concept under which a physics of matter and space alike, can be imagined mathematically, is then presented by us in Chapter 8 under the just as sweeping name of *apolarity*. In a word, the connection between matter and a reference frame, just like the connection between space and a reference frame is mathematically provided by the concept of apolarity

A confession is necessary from our part before starting our narrative: everything in this work is hardly original! Fact is that, as we went on with the elaboration on contributions to natural philosophy like those of Earnshaw, Maxwell, Poincaré, Lorentz and some other remarkable personalities of physics, two things became clear to us: first, the concepts we wanted to elaborate on, and which we shall exhibit here in due time, were touched, sometimes even long ago, at some historical moments in our evolution, in a way or another; secondly, however, it seems to us that physics is not yet aware of these spiritual encounters, to the point of a conceptual attitude. So, if anything, our originality, as it were, is that in order to raise that essential awareness, we considered our duty to elaborate on those concepts theoretically, and chose the historical presentation of facts with due theoretical conclusions, of course, as the best way of exhibiting the principles. The work hopefully gained in clarity of expounding the principles, but unfortunately lost in technical realizations, to which we plan to dedicate a separate work.

Another point we need to highlight from the very beginning, is one of a 'philosophical terminology', so to speak. We can hardly assume for ourselves a high proficiency in English verbalization, but we were literally forced to an attitude: that of using the word 'differentia' instead of 'determination', for the specific attributes defining a concept. The expression 'determination' has, mostly for the category of readers to whom our present work is dedicated, the connotations of 'willpower, 'fortitude', 'resolve' and such like.

When it comes to locating a concept, we need to define it negatively, by a difference with respect to other concepts. This connotation is indeed assumed, in English, by the word 'determination', but for a restricted category of individuals, as it seems, namely the philosophers. For instance, the translations of great philosophical works of Kant and Hegel, liberally use the word 'determination' with no danger of misunderstanding among philosophers. We too use liberally here the words 'idea' and 'concept', unavoidable by the nature of the present work but, also unavoidably, we need to define them specifically. It is for this definition of a concept or idea, that we use, instead of the philosophical term 'determination', the more suggestive term 'differentia' — with the plural 'differentiae' — appropriated in English from Latin, and used to define a concept 'negatively', as it were. A notable example may help in elucidating the issue.

One can safely assume that for the common reader the 'space' and 'time "are a priori intuitions", not 'concepts', to use the words of the illustrious Immanuel Kant. This typical philosophy is plainly illustrated by the words of Karl May, for instance, containing, in our opinion, the universal reason of that *a priori* in the matters of intuition of space and time and, therefore, the necessity of spiritual evolution of society, whose basis is the space and time, taken on this occasion as concepts:

> I don't agree, however I don't dispute, the theory according to which the todays' Indians would not be the descendants of those noble people of old; nevertheless, even if it would be so indeed, I cannot find a reasonable ground for the statement that the Indians would be unapt for a spiritual evolution. Obviously, by *robbing them of space and time, you coerce them into downfall and obliteration.* (Karl May, *Winnetou*)

The last emphasized words illustrate the modern principle of incarcerating, in order to eliminate the antisocial people from social life. The point is that only starting from this fundamental state of perceiving the space and time, is the man able to ascend to a concept with the categories of space and time. For once, this is the playground of physics, and therefore the reason to take into consideration concepts, not intuitions. For the category of space we have the example of the concept of metric space, best illustrated by the Riemann's famous Dissertation which generated the modern metric differential geometry (Riemann, 1867). However, more illustrative turns out to be the category of time: classically, one can say that the first concept pertaining to this category, whose 'differentia' is the *uniformity*, occurred to Newton, who used the concept in the natural philosophy in order to characterize the forces. With the birth of electrodynamics, it became apparent that a new 'differentia' of the concept of time is necessary, namely the *sequencing property*. This was formally instituted by Einstein in 1905, by building the special relativity, which is mathematically based on electrodynamics, whereby the classical concepts of space position and time moment are combined into the concept of event (Einstein, 1905). Thus, the concept of time as we see it, has two 'differentiae' at this stage of theoretical physics: the *uniformity*, generating the mathematics of continuity necessary to physics. This was initiated by Newton as the calculus of fluxions. Then, there is the *sequencing* property, whereby the time is mainly understood by the idea of ensemble of common moments of simultaneous events. This is how the time is *conceived* in the present work. Now, let us start showing how the concept works!

Chapter 1

Classical Moments of Interpretation

At the end of a previous book (Mazilu *et al.*, 2019), we have summarized the points of its continuation, 'in no particular order', as we expressed it there. Here, however, we need to specify the choice of a certain order, and the manner to follow it in this part of the work for, obviously, otherwise we cannot properly accomplish the work. And in order to achieve this goal we need to expound its very basis. This is the point where the planetary model with space extended physical components comes in handy for an intuitive regulation of our line of thoughts and ideas.

This model is here understood as having just two physical components, as in the classical treatment of hydrogen atom, except that these components *are extended in space*, viz. they are not material points in the classical sense of the word, but in Hertz's acceptance. To wit, the model is constituted of two physical components: one central, spatially extended particle creating the force field which defines the dynamical model as a physical structure, the other particle, still spatially extended, revolving around the central one in a motion describable, in a first instance, as a Kepler motion. However, as the real physical structures of the universe seem to display fundamental physical components involving more than one revolving particle — the planets around Sun, the different kinds of atoms representing chemical elements, the molecules representing substances in general, *etc.* — we are compelled to assume that the hydrogen-like planetary structure is, in fact, involved in a kind of 'construction' of these physical components at any scale. It will be, therefore, a task of our theory to describe these real physical structures as 'syntheses', so to speak, based on such planetary hydrogen-like fundamental units.

As to the two physical components of such a unit — the extended physical particles or, simply, *physical particles*, being understood that these particles are, by definition, always spatially extended — they will be conceived as having each a *nucleus of pure matter* — the matter *per se* in what follows — where *the space does not have access.* Outside this nucleus we have the rest of particle's universe, characterized by the fact that here the space penetrates the matter *per se*, thus offering support for a proper description of motion with respect to the particle. This approach allows, in turn, for the description of the universe as a physical structure. At least this is how the universe was thought thus far in physics, as well as everywhere in fact, and we are not here to argue against the view. Two or three classical examples discussed in the current opening to our work, will hopefully be able to show the reader just what is the essence of this line of thinking and, furthermore, what we think as essential for a proper development of the natural philosophy. It will appear thus that the concept of *interpretation of a continuum*, whose definition was enforced upon physics a century ago by the necessities of wave mechanics with its quantal amendment, is essential even for a description of the matter *per se*. However, as it turns out, the interpretation is not sufficient, and needs to be further completed: at points where the matter is discretely conceives as penetrated by space, *it needs to be 'reversely' interpreted, as it were, by a continuum.*

We can hardly raise here a claim of originality for such a model of a physical particle, anyways not for the outer physical structure where the matter is penetrated by space. As far as we are aware, and concerned actually, it occurred as a model of universe to Albert Einstein, who apparently found it to Henri Poincaré, and probably it is of an even earlier origin. Quoting:

> We now come to our concepts and judgments of space. It is essential here also to pay strict attention to the relation of experience to our concepts. It seems to me that Poincaré clearly recognized the truth in the account he gave in his book, La Science et l'Hypothèse.

> Among all the changes which we can perceive in a rigid body those are marked by their simplicity which can be made reversibly by a voluntary motion of the body; Poincaré calls these changes in position. By means of simple changes in position we can bring two bodies into contact. The theorems of congruence, fundamental in geometry, have to do with the laws that govern such changes in position. For the concept of space the following seems essential. We can form new bodies by bringing bodies B, C, ... up to body A; we say that we *continue* body A. We can continue body A in such a way that it comes into contact with any other body, X. The ensemble of all continuations of body A we can designate as the 'space of the body A'. Then it is true that all bodies are in the 'space of the (arbitrarily chosen) body A'. In this sense we cannot speak of space in the abstract, but only of the 'space belonging to a body A'. The earth's crust plays such a dominant rôle in our daily life in judging the relative positions of bodies that it has led to an abstract conception of space which certainly cannot be defended. In order to free ourselves from this fatal error we shall speak only of 'bodies of reference', or 'space of reference'. It was only through the theory of general relativity that refinement of these concepts became necessary, as we shall see later. [(Einstein, 2004, p. 3); *original Italics*]

If such an image is to be taken as a model for the universe at large, one can see that we have here a genuine *physical structure*: the essential constitutive unit of this universe is 'the body' but *in motion*, even in 'voluntary motion'. As far as the classical natural philosophy is concerned, this concept assumes the space as its essential possibility, inasmuch as only the space can confer *possibility* to motion. Such a physical structure is absolutely unique — the Einstein's body is described as 'arbitrarily chosen' — which excludes a multiple space scale from the very beginning.

Now, if there is an ingenuity from our part, it can be first of all recognized in the idea of physical particle. This, in turn, should be related to a 'individual universe' as it were, involving the matter *per se* as an essential ingredient, and a suggestion of multiplicity of the universes, *as related to physical particles.* In an earlier work (Mazilu and Porumbreanu, 2018) we described the Earth as the epitome of such a physical particle, along the idea that the 'dominant rôle of the Earth crust', as Einstein puts it, has an even more fundamental meaning, if we may say so, than usually recognized (Mazilu *et al.*, 2019). We envision here the physical particle as having a surface of separation between the matter *per se* and the physical structure of the associated universe, after the manner in which this surface is defined in the membrane model of the blackholes (Price and Thorne, 1986, 1988). The fact that the matter *per se* does not admit space inside it, is the counterpart of the tacitly accepted property of space *per se* of not containing matter 'inside it'. That is to say that the space is always understood as '*per se*' in physics, nevertheless, without stating it explicitly. Even more so, if the Mach's principle is somehow involved in our judgment, one needs to consider some equivalence between space and matter (see Einstein, 2004, pp. 57–58), and this equivalence cannot be reduced to action only: the matter *per se* and the space *per se* are then even formally equivalent to one another, in the sense that they 'repudiate' each other.

This description should be, in our opinion, essential for physics. The two categories are not physical structures, but are open only to interpretation: the *matter of the nucleus of a physical particle is conceived as a continuum* in need of interpretation, just as well as the space. Only, their interpretative ensembles are then different, and what we assume here is that the interpretation of matter should be done by ensembles of Hertz material particles, *i.e. positions in matter* endowed with *physical properties* in the classical connotation. Such physical properties are finite in number, like the properties of the color continuum (see Mazilu *et al.*, 2019, Chapter 5), with the only difference that they are established here by Newtonian forces. These forces need to be chosen as a constructive criterion of a geometry of physical properties, because they are *the only forces invariant with respect to scale changes*, even in the sense of Berry–Klein theory. Indeed, by Mariwalla's theorem (Mariwalla, 1982), the Kepler motions of Hertz's material particles are then universal, being invariant to time and space scale changes. Besides, the Newtonian forces are capable to offer us the possibility of interpreting the matter within the nucleus of a physical particle by *static ensembles*, in the sense that they satisfy the Wigner inertia principle, if they possess three physical characteristics according to our experience: *one gravitational mass* and *two charges, electric* and *magnetic*. Thus, the ensembles of *such* Hertz material particles are to be understood as static ensembles of identical Hertz material particles *in equilibrium*. This equilibrium is maintained *for infrafinite time intervals* by Newtonian forces between them, arising as a consequence of their charge and mass endowment. With their elements thus characterized, the equilibrium ensembles so contemplated are susceptible of a metric description, allowing a harmonic mapping between matter and the *fictitious space* we think it occupies.

These continua should be interpreted as Madelung fluids. More specifically no motion *per se* is valid here, as no space exists in matter, but the concept of motion should be replaced with the idea of a *stochastic process* as required by an SRT. This idea was suggested for the first time in a metric theory of spacetime by Carlton Frederick, who also advanced the thesis that the stochasticity component of a theory, which should necessarily replace the classical motion, is to be transferred to the components of the metric tensor of spacetime continuum containing matter (Frederick, 1976). We left aside mentioning here the manner of involvement of the space *per se*, which will be settled in due time, based on Nikolai Chernikov's mathematical theory (see Mazilu *et al.*, 2019, Chapter 12). This allows us to declare even by now that the principle of metric description goes for the matter just as well as it goes for the space. Then again, we need to take note of some complications that may occur with this kind of approach in the case of the spacetime, which may occur, just as well, in the case of space itself. The physics became aware of them not very long after the moment when the foundations of general relativity were laid on in the form of Einstein's natural philosophy, and even devised some genuine modalities to deal with them in special cases, mostly in cosmology (see, for instance, Barrow and Dabrowski, 1998). Quoting from a classical work:

> Every world-point is the origin of the double-cone of the *active future* and the *passive past*. Whereas in the special relativity these two portions are separated by an intervening region, it is certainly possible in the present case (*of general relativity, a/n*) for *the cone*

> *of the active future to overlap with that of the passive past*; so that, in principle, it is possible to experience events now that will in part be an effect of my future resolves and actions. Moreover, it is not impossible for a world-line (in particular, that of my body), also it has a time-like direction at every point, *to return to the neighborhood of a point which it has already once passed through.* The result would be a spectral image of the world more fearful than anything the weird fantasy of E. T. A. Hoffmann has ever conjured up. In *actual fact the very considerable fluctuations of the* g_{ik}'s that would be necessary to produce this effect *do not occur* in the region of the world in which we live. Nevertheless, there is *a certain amount of interest in speculating on these possibilities* inasmuch as they shed light on the *philosophical problem of cosmic and phenomenal time.* Although paradoxes of this kind appear, nowhere do we find any real contradiction to the facts directly presented to us in experience. [(Weyl, 1952, §34, p. 274); *our Italics a/n*]

Obviously, in cosmology we have to deal with the *sequencing differentia of the time concept*: that 'cosmic and phenomenal time' in this excerpt from the writing of the great mathematician and natural philosopher, leaves, indeed, no doubt as to which characteristic of time is involved here. As it turned out historically, though, that 'amount of interest' is not — and, in fact, we hold true that it should not be — quite as much 'philosophical', as it should be physical: it involves, as we shall show here in due time, the essential difference between the *compass of gravitation* and the *compass of inertia.*

Thus, inasmuch as the mathematical principle of metric description of the spacetime turned out to be, just like the principle of the metric description of space, that of *embedding*, the issues should be the same, only disguised under different concepts. As a consequence, the definition of both space and matter, in the *per se* condition, should be geometrical, whereby the geometry can be assumed from the beginning as Euclidean at any scale and dimension of space. Then considering, like Einstein once did, that at the infrafinite and finite scales the metric geometry is the same — specifically, a Euclidean one — the embedding leads to a non-Euclidean metric of the type involved in the initial conditions of the Kepler problem

(Mazilu *et al.*, 2019). This fact was, indeed, revealed by Albert Einstein in 1917, and then successfully used by Willem de Sitter in constructing a universe without matter, and by Felix Klein in theoretically describing the principles of a rational cosmology. All of a sudden this model revealed a breach between curvature and matter, brought about by the cosmological constant, which stays even today as it was at the moment of the introduction of this constant to our natural philosophical awareness. In hindsight, it is plainly illustrated by the existence of the Keplerian orbits of the planetary model, as we have to show in this very work. Let us, therefore, review, first, a few introductory historical cases of interpretation in the classical natural philosophy, by the way of introduction of this model. Details follow.

1.1 Classical Ether as an Implicit Case of Interpretation

The classical ether was the first model of matter, the way we understand it as existing in the nucleus of spatially extended particles: *untouchable* and *unfathomable*. It has been conceived as a kind of continuous medium even from the old times. However, only beginning with the 19th century it started being *systematically interpreted* in terms of 'ensembles of molecules', and thereby believed *to be a physical structure*: matter existing in space, therefore penetrated by space. Indeed, it is this last quality of the ether the one that has been brought along in the issues raised by the idea of 'dragging', which culminated with the special theory of relativity. One of the wittiest partisans of this kind of ether was Samuel Earnshaw, who even offered an exquisite interpretation of ether, *avant la lettre* as it were. We need to extract in quotation this idea of interpretation here, for it is completely formulated within the modern concept of interpretation but, predating the definition of Charles Galton Darwin by almost a century, it has the merit of exposing, for the first time, assumptions that, up to that date, were only implicit in the description of the concept of ether. Needless to say, they remained as such even to this date, but, being buried in our conscience under the label of 'historical', this obliged us to make them explicit. These assumptions can, therefore, also be taken as differentiae of the modern concept of interpretation:

> I. It is assumed that the ether *consists of detached particles*; each of which is in a position of equilibrium, and when slightly disturbed is capable of *vibrating in any direction.* (Many solid as well as aerial bodies transmit sound, which is generally supposed to imply the existence of the same properties in them as are here assumed to be true of the ether.)
>
> The most curious and perhaps least expected result of this assumption is, *that the molecular forces which regulate the vibrations of the ether do not vary according to Newton's law of universal gravitation*: and it is not a little remarkable, that a force, whether attractive or repulsive, varying according to this law, is *the only one which cannot possibly actuate* the particles of a *vibrating* medium.
>
> II. It is next assumed that the motion of a vibrating particle is more affected by the influence of the particles which are near to it than of those which are more remote. (This is certainly true of many other substances besides the ether.) The result which is sought to be derived from this assumption is, *that the molecular forces which regulate the vibration of the particles are* REPULSIVE, *and vary according to an inverse power of the distance greater than 2.*
>
> III. It is lastly assumed that the ether exists (or at least is capable of existing) as one mass held together by the *attraction* of its elementary molecules. This assumption is necessary, in order that the dispersion of the medium which would naturally result from the *repulsive* forces which regulate the vibration of its particles, may be thereby prevented.
>
> The result which is derived from this necessary assumption is that *each particle exerts* (in addition to the repulsive force before mentioned) *an attractive force, which varies according to Newton's Law of universal gravitation.*
>
> By reversing the problem, I have been able to shew, that though Newton's law is the only one which cannot enable the particles to vibrate, yet it is the only law of force which can enable them to constitute and maintain themselves a *permanent* medium, without endangering, or in any way affecting their *vibrating* or luminiferous property.
>
> I have on these grounds not hesitated to express my opinion, *that the particles of the luminiferous ether are each endued with*

> *two forces of distinct character and uses; one attractive, to preserve themselves a permanent medium, varying inversely as the square of the distance; and the other repulsive, to which is due their luminiferous property, varying in a higher inverse ratio of the distance than the square.* [(Earnshaw, 1842); *Italics and capitals in the original*]

Let us, first, take notice here of the penultimate paragraph, expressing what Earnshaw was 'able to shew'. Certainly, he did not show it the right way, elucidated only by the modern theory of Berry and Klein, expounded previously in Mazilu *et al.* (2019) as a 'cosmological' moment, which involves the modern notion of gauging. Still, Earnshaw's achievement represents for us a striking way of showing how the perception has to be guided by reason, and this very fact should be, in our opinion, taken face value; especially, if we desire to construct a proper theory of scale relativity!

For, indeed, the modern Berry–Klein Hamiltonian theory of interactions, as we presented it here, reveals the existence of an intricate, almost obscure we should say, connection between Newton's law of force and the elastic law of force. Such a connection testifies that this valuable theoretical discovery is in fact an expression of the old truth expressed in the right words of Earnshaw himself: the Newton's law of force is a law 'which can enable particles to constitute and maintain a *permanent* medium, without endangering, or in any way affecting their *vibrating* or luminiferous property'. The Berry–Klein presentation does not say quite explicitly 'the only law' but, as we shall see further in this work, this is precisely the case even when it comes to the concept of interpretation of a continuum in general. Moreover, that remark: 'enabling particles to constitute and maintain a permanent medium' points, in our view to the necessity of a reverse *interpretation — the interpretation of a discrete ensemble by a continuum*, as it were — to be detailed in due time in the present work. And again, this incident can be taken as an illustration of the fact that ideas have their own dynamics. However, it also entitles us to notice that one cannot nudge beyond the meaning of an idea in its capacity of truth, when such truth is uttered through the words of a genius. The only humanly freedom one can then take, is to try and discover that truth, and this takes time! Fact is, though, that this conclusion of Earnshaw did not mean too much for the modern physics, along the concept of truth: it concentrated instead on a direct result of the above thorough characterization of the ether *as a physical structure*, to wit, the known *Earnshaw theorem*. Roughly stated this theorem shows that the purely electrical physical structures of the matter are unstable. More to the point, *an ensemble of particles charged only with electricity cannot be in equilibrium in a static electric field alone* (Stratton, 1941), understanding by the state of equilibrium a *spatially confined physical structure*.

For once, this theorem can be taken — and we effectively take it — as establishing the ontological possibility of interpretation in general: *there is no static ensemble of Hertz material particles having a single physical property* in interpreting a continuum. For, in the realm of Newtonian forces, one single property would mean one single force between identical material particles necessary for interpretation, and as this force cannot be but either of attraction or of repulsion, such an ensemble of particles cannot exist according to our experience. On the other hand, surely our experience must be taken as a criterion of acceptance here, and the Newtonian forces — the essence of this experience — *are invariant to space and time scales*: whatever our experience reveals for them is valid at any space scale, be it infrafinite, finite or transfinite, if it is to use the hierarchy presented by Nicholas Georgescu-Roegen (Mazilu *et al.*, 2019). Such a structure, thought to be a physical structure, but a little more complete though — by the addition of magnetic properties, non-static in character, to electricity — is the Lorentz matter, analyzed in detail by the magnificent Henri Poincaré, who had a few tasks to accomplish with this analysis. As it turns out with the present occasion, one of these tasks is fundamental, insofar as, once accomplished, the other ones become simply just natural consequences. It is the introduction of gravitation as a property of the Lorentz matter. Let us, therefore, present the premises of the Lorentz theory of matter in a little more detail, at least as an issue of principle, as it helps us not only in rounding up the general concept of interpretation of a continuum, but mainly in establishing some theoretical principles for that reverse interpretation, as we have mentioned above. The best starting point, in our mind at least, is an explicit description of the interpretation as understood by Samuel Earnshaw, *i.e.* based on Newtonian forces, amended, though, on this very occasion, with the Hertz notions of material particle and material point.

1.2 Physical Description of a Hertz Material Point by Confinement

The equation of Newtonian dynamics for the classical Newtonian force is invariant with respect to a special transformation discovered by Mariwalla (1982). This invariance is not unconditional though: the transformation generating it involves simultaneously the position and the time of motion, but — and this is of a tremendous importance for us here — *at different scales for space and time*. Specifically, Mariwalla has shown that Newton's equations for planetary motion which describe dynamically the Kepler problem, written in vectorial form (Mazilu *et al.*, 2019) are invariant with respect to the simultaneous transformations of coordinates and time given by the formulas

$$\mathbf{R} = \frac{\mathbf{r}}{1+\mathbf{k}\cdot\mathbf{r}}; \quad d\tau = \frac{dt}{(1+\mathbf{k}\cdot\mathbf{r})^2}, \tag{1.1}$$

where $\mathbf{k}$ is, on this occasion, an arbitrary constant vector characterizing the transformation. As one can see, this is, indeed, a 'mixed' transformation, and creates the issue we want to emphasize here: it is referring, on the one hand, to the *finite space scale*, with the space represented in an Euclidean reference frame and, on the other hand, to the *infrafinite time scale*, for any meaningful time in that frame.

Before anything else, we are indebted to the reader with a proof of equation (1.1). It involves here only straight calculations, as follows: if, in general, we choose a scale function, λ say, to depend only on coordinates, and to scale both the time and coordinates in such a way that

$$\mathbf{r} = \lambda\mathbf{R}; \quad d\tau = \frac{dt}{\lambda^2}, \tag{1.2}$$

then we have:

$$\frac{d\mathbf{R}}{d\tau} = \lambda\frac{d\mathbf{r}}{dt} - \frac{d\lambda}{dt}\mathbf{r} \quad \therefore \quad \frac{d^2\mathbf{R}}{d\tau^2} = \lambda^2\left(\lambda\frac{d^2\mathbf{r}}{dt^2} - \frac{d^2\lambda}{dt^2}\mathbf{r}\right).$$

Using Newton's equations of motion (Mazilu *et al.*, 2019) the last equation here becomes

$$\frac{d^2\mathbf{R}}{d\tau^2} + \left(r^3\frac{d^2\lambda}{dt^2} + \kappa\lambda\right)\frac{\mathbf{R}}{R^3} = 0$$

so that, in cases where the dynamical problem involves *the very same Newtonian force* at different scales, we must have

$$r^3\frac{d^2\lambda}{dt^2} + \kappa\lambda = \kappa, \tag{1.3}$$

and the equations of motion deriving from the second principle of classical dynamics are indeed form-invariant. This means that a human being living for instance in a world where it establishes the event (**x**, **y**, **z**, **t**), uses the same equations to describe Kepler motion as a human being living in a world where it establishes the event (**X**, **Y**, **Z**, **T**). These two worlds are 'similar by a scale factor', as it were, both at finite scale in space and at infrafinite scale in time, in either one of the worlds. Now it is quite obvious that if in equation (1.3) we choose

$$\frac{d^2\lambda}{dt^2} = \mathbf{k}\cdot\frac{d^2\mathbf{r}}{dt^2}, \tag{1.4}$$

where **k** is an arbitrary constant vector, then we have

$$\lambda = 1 + \mathbf{k}\cdot\mathbf{r} \tag{1.5}$$

no matter of the constant κ. Therefore, the similarity factor between the two worlds is simply determined by the *projection of acceleration along any constant vector*. There are thus multiple Newtonian worlds, distinguished from one another by scale factors! However, once again, we need to take proper notice that this kind of scale factors — specifically *linear in the position in a reference frame* — are referring to the finite space scale and infrafinite time scale of those worlds.

As long as the classical dynamical equations of motion are involved in the description of such a world, this condition is inescapable.

This is, in fact, a kind of invariance discriminating in favor of Newtonian forces for, in general, we have

$$\frac{d^2\mathbf{R}}{d\tau^2} + \kappa\frac{\mathbf{R}}{R^3} = \lambda^3\left(\frac{d^2\mathbf{r}}{dt^2} + \kappa\frac{\mathbf{r}}{r^3}\right)$$

and any force *other than the Newtonian one* is liable to break the invariance of the equation of motion. By the same token, any equation of motion, *other than that involving acceleration as the second order derivative* is also liable to break the invariance. Mariwalla even noticed that the geometrical properties of Kepler motion (area law and the preservation of the plane of motion) are consequences of this type of invariance, which thus gives us all the invariants we need for the central motion under the *inverse square force.*

Another property comes with the observation that, in case *some other kinds of Newtonian forces* act between the bodies, so that instead of (1.3) we may have

$$r^3\frac{d^2\lambda}{dt^2} + \kappa\lambda \equiv \kappa'$$

with $\kappa' \neq \kappa$, then the scale function from equation (1.5) changes to

$$\lambda = \frac{\kappa'}{\kappa} + \mathbf{k}\cdot\mathbf{r}. \tag{1.6}$$

Thus, the Kepler problems for gravitational mass and charge, for instance, are not simply equivalent: *they are indeed formally identical, but for different scale factors* satisfying the description of Mariwalla's type of invariance. The forces though, involved in this kind of invariance, should be of the same family: central forces, of Newtonian character, *i.e. with their magnitude going inversely with the square of distance.* We are entitled then, to use these forces in constructing a Hertz material point according to a 'cosmological principle', as follows.

In the Newtonian stand, cosmology is indeed based on forces, and these forces are of a special type. As above, they should be central forces, but with magnitude depending exclusively on the distance between bodies, and in a quite specific way: inversely proportional with the square of that distance. Of course, in reality, such a structure of the universe is not possible, but only in cases where the structural units of this universe — the 'bodies' of Newtonian and Einsteinian natural philosophy — are so far apart from one another, *that their dimensions are negligible with respect to the distance between them.* In discovering the Newtonian forces — as we call them after an usual custom today — thus described, one can safely assume that this dimensional condition of their theoretical possibility should have already been a reality in the universe accessible to our knowledge. Indeed, otherwise we would not have the Kepler laws governing the motions of the celestial bodies, which led to the invention of such forces. However this dimensional geometrical condition is by no means sufficient for the task of building a cosmology. First of all, a physical structure of the very structural unit of the universe — the 'body' — is still needed. For, in time, the mankind became aware of the existence of still other Newtonian forces with the very same geometrical properties. The first instance of such forces in the quotidian world of our experience was *the electric force.* And this is how, in fact, we became aware of the necessity of a scale in characterization of the action of Newtonian forces. A phrase like that of Hermann Weyl, justifying the modern approach of cosmology, with a universe where...

> ...the electricity, which *obviously does not matter in the economy of cosmos, we now completely dismiss...* [(Weyl, 1923, §39, p. 290); *our rendering and Italics*]

just reflects such an awareness. Incidentally, we take the Weyl's approach to cosmology, as conveyed in the 39ff of the 5th and 6th editions of his celebrated *Space Time Matter*, as typifying the modern attitude in cosmology. It does not seem to exist any English translation of any of these editions; there is, however, a 1996 Russian translation of the 5th edition at "Janus" in Moscow.

The above excerpt from Weyl, shows that we are, at least formally, allowed to think of the presence of a certain space scale as being the expression of the dominance of actions of two *different forces* of the same *mathematical nature*, *i.e.* Newtonian forces. It is for these forces, as we said, that Michael Berry and George Klein proved their scale transcendence (Berry and Klein, 1984) as shown in (Mazilu *et al.* (2019, Chapter 9, equations (9.34)–(9.43))), and these are the only forces satisfying the invariance of the Mariwalla's transformation preserving the Newtonian dynamical description of the Kepler problem. This observation entices us to take notice of what seems to be obvious by itself: the very same property of these forces that allowed Hermann Weyl to produce the argument from the excerpt above, also allowed a long time ago to Charles Coulomb the experimental characterization of electric forces. Indeed, in the case of Coulomb, there is just a change in emphasis: in the 'economy' *of daily life*, it is the gravitation 'does not matter', therefore it can be dismissed, even though not quite by the same reasons the electricity at the cosmic level is dismissed. So, the apparent geometrical rigor, can be supplemented with this rigor of a physical nature, in judging the transition between universes at different space scales. Insofar as there is a big difference, however, on how that 'does not matter' can be described at the two space scales — the cosmos and daily life — the *physical characteristics* of the fundamental units of matter are thereby brought to the fore, in order to regulate the very transition of scales. It turns out that they allow for a mathematical characterization *independent of the dimensional conditions*, mandatory for a proper mathematical description of the dynamics of Kepler problem. More specifically, the dimensional condition defining the fundamental unit of matter in the physical structure of a certain universe, is capable of accommodating the existence of physical characteristics, and this fact can be formally described by a geometry of confinement.

First of all, in order to describe the fundamental units of matter in the same way in both cases — cosmos and daily life — the mass should be taken as the *gravitational mass.* Only thus can one declare that, for fundamental constitutive units possessing just gravitational mass and charge, in the case of 'cosmos' the Newtonian force of gravitational mass prevails quantitatively over that of electric charge, while in the case of 'daily life' the Newtonian force of electric charge — the Coulombian force, as it is usually called — prevails over the force of their gravitational mass. Therefore, if we are to describe *physically* a certain universe, its fundamental physical unit — the 'body', as we designated it every now and then, assuming a classical Newtonian standpoint — should exercise two kinds of Newtonian forces acting simultaneously *in any direction* in space, *at any distance*: gravitational and electric. The magnitude of the Newtonian force describing such a fundamental physical unit in a universe thus described, can be written as

$$G\frac{m^2}{r^2} - \frac{1}{4\pi\varepsilon}\frac{e^2}{r^2}. \tag{1.7}$$

Here another assumption appears as necessary in describing the universe, namely that the fundamental physical units of this universe are all identical, having the mass m and charge e, and the universe is represented as such in a space of gravitational constant G and electric permittivity ε . Also, equation (1.7) expresses the fact that, according to our experience, the two Newtonian forces act differently along the same direction: one is a force of attraction, the other of repulsion, and the two different signs of the monomials in (1.7) represent an algebraical writing with respect to the orientation of the direction along which the action is exerted. Thus, for instance, the universe at the cosmic scale can be characterized by a strong inequality:

$$Gm^2 - \frac{e^2}{4\pi\varepsilon} \gg 0 \tag{1.8}$$

so that, in the expression of Hermann Weyl, the 'electric force can be dismissed' *at any distance and in any direction*, while the universe at the daily scale can be characterized by

$$Gm^2 - \frac{e^2}{4\pi\varepsilon} \ll 0 \tag{1.9}$$

so that the 'gravitational force can be dismissed' *at any distance in any direction*. Certainly the 'distance' should have different quantitative meaning in the two cases, and perhaps the 'direction' too.

From the point of view of a scale relativity theory we would have then to decide what is the daily scale — for instance by declaring it the finite space scale, as the philosophy of Nicholas Georgescu-Roegen claims — in order to properly assign then, the correlated infrafinite and transfinite scales. However, as there is no obvious way of classically building a proper physical theory which incorporates such an idea — that is, other than the conditions imposed by Mariwalla's transformation — this problem is currently circumvented in physics by a specific uniqueness: the bodies are material points in a *unique* universe extended to infinity in the large and in the small. Practically though, in gaining our knowledge, we always had to recognize that the problem with the scale of distances still remains, and the science of physics dealt with it — and, actually, still does! — in a sort of 'case by case' approach. Indeed, we have to recognize that 'any distance' above is totally different between the case of cosmos and that of daily life. Even more so in the case of cosmos, by comparison with the microcosmos: one cannot define unambiguously a distance between the Earth, say, and an electron, it just does not make sense. In the modern cosmology, however, the conditions at infinity became critical and specifically related to a definite time sequence of events in the universe, which is thus conceived as a physical structure. However, with the SRT, the difference should be a matter of scale, and thus there are multiple universes decided by the hierarchy infrafinite-finite-transfinite. Whence, in our opinion, the necessity of describing simultaneous universes, to which we suggest now a positive possibility offered by the ideas of the absolute — or Cayleyan — geometry, based on the concept of Newtonian forces.

Start with the observation that the structure of a universe is always a hypothesis, so that the problem occurs: is there an ideal *static structure* of the universe, *formally the same at any scale*, that would be able to describe even the structure of the fundamental physical unit of the universe? Mathematically speaking this is always possible with Newtonian forces, insofar as, according to Berry and Klein's theory, the action of these forces transcends the time and space scales. Thus, for instance, a *static universe* dominated by the two Newtonian forces above, can be interpreted as an *ensemble of*

Hertz material particles at any scale, provided

$$Gm^2 - \frac{e^2}{4\pi\varepsilon} = 0. \tag{1.10}$$

This would mean a structure made of *identical Hertz particles*, each one of them endowed with gravitational mass and electric charge, in an ensemble in static equilibrium: the forces between particles are in equilibrium in any direction, at any distance, for any particle of the ensemble. Such an interpretation naturally restricts the ratio between electric charge and gravitational mass of the Hertz material particles.

However, there is another significant connotation of this approach to static equilibrium: we can construct an *absolute geometry* based on the static equilibrium of forces, using equation (1.10) as the equation of *an absolute* in the geometry of *physical attributes* of particles. First of all, contemplating some neatness of the mathematical theory — a 'mathematical beauty' that, according to Dirac any 'physical law should have' — let us arrange a uniform notation based on this equation, in order to simplify the algebra that follows. The terms in equation (1.10) are physically homogeneous, of units ($kg \cdot m^3 \cdot s^{-2}$). So, in order to make the notation uniform, we include the space-describing constants in definition of the physical properties of material particles, by the following symbol correspondence:

$$m\sqrt{G} \leftrightarrow m; \qquad \frac{e}{\sqrt{4\pi\varepsilon}} \leftrightarrow e.$$

This notation is intended to suggest that the first term in (1.10) is referring only to gravitational mass (m) in matter, while the second is referring only to the electric charge (e) in matter. This means that a universe is here interpreted by a static ensemble of identical Hertz material particles, each one of them having two physical characteristics: *mass* (gravitational) and *charge* (electric). Therefore, limiting ourselves, for the moment, to just gravitational mass and electric charge, the condition to be satisfied by an ensemble of identical Hertz material particles serving for the interpretation of a static universe independently of space scale should be written as

$$Q(m,e) \equiv m^2 - e^2 = 0. \tag{1.11}$$

The left-hand side of this identity symbolizes an algebraic 'quantic', which, as the right-hand side shows, is of second degree. Taken as absolute of a *geometry of the physical characteristics* of Hertz material particles, it divides the plane of these characteristics in two parts: the 'inside' part, for which we assume that the quadratic form is positive, and the 'outside' part, for which the quadratic form is negative. This convention is adopted so that the 'cosmos', characterized by equation (1.8) should be *inside the absolute*, very close to its center with respect to charge, while the 'microcosmos', described by equation (1.9), should be *outside of absolute*, very far away with respect to gravitational mass. The true measure of these degrees of 'closeness' is, nevertheless, offered by the values of quantity Q defined in equation (1.11). The positive part of this construction is that, once we give a metric for this geometry, we are always able to find space distributions of the quantities of gravitational mass and charge, using the *harmonic mappings*. The construction of these distributions may not be quite as easy as it sounds, but in some cases it may be indicative of the right path of our knowledge.

For instance, the Barbilian formula (Mazilu *et al.*, 2019) offers, indeed, a Cayleyan metric of the plane of the two physical characteristics above, which we take in the form

$$(ds)^2 = \left(\frac{mdm - ede}{m^2 - e^2}\right)^2 - \frac{(dm)^2 - (de)^2}{m^2 - e^2}.$$

The reason of this choice of sign for the Cayleyan metric becomes obvious by noticing that, after due calculations, this metric becomes a perfect square:

$$(ds)^2 = \left(\frac{mde - edm}{m^2 - e^2}\right)^2 \tag{1.12}$$

and a perfect square of a real quantity should be always positive. Thus, the interior of the absolute is characterized by a proper hyperbolic angle, ψ say, whose variation turns out to be our metric. Indeed we have:

$$(ds)^2 = (d\psi)^2; \quad \tanh\psi \equiv e/m.$$

The metric of physical characteristics of this universe depends on the ratio between charge and mass, a case well known in the history of physics. Only, we have to notice that here the mass is gravitational, while in the historical case the mass was inertial, as a consequence of dynamics used in describing the electron.

A problem surfaces when the charge 'splits', so to speak, *i.e.* there are two Newtonian forces of electric nature. This could be the case if the Hertz particles have *a third physical property*, for instance a *magnetic charge* and, attached to it, a magnetic force. As magnetic poles of the same name behave exactly like electric poles of the same charge, a static universe interpreted by ensembles of such Hertz material particles with three physical characteristics will be described by the quantity

$$Q(m, q_E, q_M) \equiv m^2 - q_E^2 - q_M^2 = 0; \quad e^2 \equiv q_E^2 + q_M^2 \tag{1.13}$$

instead of (1.11). In the second of these identities the charge **e** splits, 'Euclidean-wise' so to speak, into an electric charge q_E and a magnetic charge q_M. The identity above summarizes what we think of as one of the most beautiful pages of the modern natural philosophy ever written. This is why, we further think, it deserves a little more elaboration here, for the benefit of our reader and, in fact, of our knowledge at large.

According to this episode of natural philosophy, the second identity from equation (1.13) represents physically a specific invariance of the electromagnetic theory: the invariance with respect to what is today generally known as the *duality rotation*. Expressed simply, this rotation is a Euclidean rotation that leaves the *experimental* electric charge e of the 'split' defined by:

$$q_E = e \cdot \cos\theta; \quad q_M = e \cdot \sin\theta$$

invariant. Here θ is the angle variable describing the split among the possible experiments with the charge. The motivation then goes on to declare that the Maxwellian theory, privileging the electric charge monopoles and eliminating the magnetic ones, represents actually just one possible choice of the split angle — more to the point, $\theta = 0$ — among infinitely many others. The magnetic pole case is represented here by the choice $\theta = \pi/2$ for the experiments involving charges. Theoretically, the extra degree of freedom, represented concretely in physics by the existence of *duality rotation*, with respect to which the whole charge behaves invariantly, can be allocated to the known possibility of transition between *field description* by electric and magnetic intensities, and *Maxwell stresses* description (Katz, 1965). As to the natural philosophical reason of this possibility, it is indeed quite remarkable. Quoting:

> It is frequently pointed out that the crucial difference between electric and magnetic phenomena, which underlies this dissimilarity (*between Maxwell equation for the electric field and Maxwell equation for the magnetic field, n/a*), is that electric charges occur in nature as monopoles whereas magnetic charges do not so occur, but only as dipoles, and higher poles. This is demonstrated, for example, by breaking a permanent magnet in two. In so doing one *does not obtain free north and south poles*: each piece has again both polarities of equal magnitude. The mathematical formulation of this situation leads then to the equation div$B = 0$. On the other hand, electric charges can be obtained free, it is said.
>
> This reasoning is incomplete and deceptive. It is true that a permanent magnet *has equal and opposite magnetic charges near its ends*, and that by breaking the magnet in two and separating the parts and *inserting a chunk of empty space between them* new poles will appear on the new surfaces. But it is equally true that *electric charges occur only in equal pairs of opposite sign at opposite ends of a chunk of vacuum*, for example, by rubbing a rubber rod with a catskin and then separating the two. The vacuum between the rod and the catskin is analogous to the permanent magnet in that it has charges of equal magnitude and opposite sign at opposite ends. If we now *break the vacuum space between the two ends in two*, by *inserting, for example, an isolated conductor between them*, then charges are induced on the metal-vacuum interfaces such that each of the two chunks of vacuum carries again zero total charges at its ends. A well-known variation of this procedure is the so called ice-pail experiment of Faraday. One can pursue this reasoning further.

> The conclusion is that also electric charges occur only in pairs which can be looked at as the result of polarization. The only difference is that *magnetic poles appear as a result of polarization of a region of space filled with matter* (and so far no region of space filled with vacuum has yielded to polarization of this kind), whereas electric charges appear as *a result of polarization of a region of space filled with vacuum as well as at one filled with matter.*
>
> Logically and formally it is therefore possible to treat electricity and magnetism completely similarly, as long as *one is willing to treat a region of space filled with vacuum on the same footing as a region of space filled with matter.* [(Katz, 1965), *Italics ours a/n*]

This last sentence is, in our opinion, of considerable importance for the natural philosophy: rarely, if ever, is one willing to recognize in physics that while theoretically treating the vacuum *unreservedly* as a material, one has also the obligation to *think of it* as of a material of the daily life, as of a 'chunk' in the language of Katz. It seems to us that the electromagnetic theory has more to show than it appears at the first sight: not only it imposed the relativity, for instance, that turned the modern physics upside down, as it were, but also tells us how to turn our very intuition into concept, and that in a right way. As far as we are concerned, the message of the previous excerpt is quite clear: the existence or non-existence of the singular magnetic poles is pending on the necessity of *describing the electromagnetic field by Maxwell stresses.* However, this requires, indeed, more than the wave-mechanical idea of interpretation, and we shall return to it later on, with a special occasion.

Continuing on with our metrization procedure of the physical characteristics of Hertz's particles in the matter *per se*, the same rules apply for calculating the absolute metric in the case of three physical properties of material particles. To wit, instead of (1.12), we have, with (1.13), the absolute metric

$$(ds)^2 = \left(\frac{mdq_E - q_E dm}{m^2 - q_E^2 - q_M^2}\right)^2 + \left(\frac{mdq_M - q_M dm}{m^2 - q_E^2 - q_M^2}\right)^2 - \left(\frac{q_M dq_E - q_E dq_M}{m^2 - q_E^2 - q_M^2}\right)^2, \tag{1.14}$$

whence, by the transformation

$$m = q\cosh\psi, \quad e = q\sinh\psi, \quad q_E = q\sinh\psi\cos\theta$$
$$q_M = q\sinh\psi\sin\theta \tag{1.15}$$

we have a well-known form of the metric:

$$(ds)^2 = (d\psi)^2 + \sinh^2\psi(d\theta)^2. \tag{1.16}$$

This metric is formally identical with the metric found in the case of the Kepler problem (Mazilu *et al.*, 2019, equation (4.17)) or the metric of the relativistic velocity space (Fock, 1959). Obviously, the quantity q representing the values of quadratic form Q can be calculated from (1.15) and amounts to

$$q^2 = m^2 - q_E^2 - q_M^2. \tag{1.17}$$

From the concept of interpretation point of view, the condition $q \neq 0$ is a nonequilibrium condition, sending our thoughts to an ensemble which serves to interpretation. There are such ensembles of Hertz material particles for the interior of the absolute, as well as for the exterior of the absolute. As *per our convention*, the interior of the absolute describes the cosmic scale of matter, and in this case we have a well-known case indicating the nature of q. One notices, indeed, that, towards the center of the absolute, q approaches m, and this fact entitles us in considering it as a measure of the *inertial mass*, which in this case is not exactly equal with the gravitational mass. Rather, equation (1.17) can be rewritten as

$$\left(\frac{q}{m}\right)^2 = 1 - \left(\frac{q_E}{m}\right)^2 - \left(\frac{q_M}{m}\right)^2$$

showing *positively* that Einstein was, indeed, entitled to 'dismiss' the difference between the gravitational and inertial mass, because, as Weyl expressed it, we are always entitled to 'dismiss the electricity in the economy of the universe'.

The metric (1.14), by the way, is the typical absolute metric of a unit disk:

$$(ds)^2 = \frac{1}{(1-x^2-y^2)^2}[(1-y^2)(dx)^2 + 2xydxdy + (1-x^2)(dy)^2], \quad (1.18)$$

where the notations $x \equiv q_E/m$ and $y \equiv q_M/m$ are used. This is, again, a typical Beltrami–Poincaré metric of the unit disk, inasmuch as in the construction of the complex variable $h \equiv u + iv$, necessary for the description of the complex upper half plane:

$$u = \frac{y}{1-x}; \quad v = \frac{\sqrt{1-x^2-y^2}}{1-x} \quad (1.19)$$

with v obviously real, the position (x, y) has to be inside the unit disk. For details of the physical significance of the absolute metric of unit complex disk, see Lavenda (2012). In fact, it is now the moment to pinpoint the connection of this important 'odyssey' of Professor Lavenda, with the streak of ideas of our current presentation.

1.3 Lorentz Matter: Poincaré Interpretation of Electromagnetic Ether

As the reader may have noticed thus far, in this work and the work that precedes it (Mazilu *et al.*, 2019), we consider the name of Henri Poincaré for many other reasons besides the analysis of the Lorentz matter, to be contemplated immediately. Essential among these reasons is, of course, the non-Euclidean geometry, which, as we have seen before, in a Cayleyan — or absolute — approach, can be considered as a mathematical expression of the idea of confinement. This was the case of the center of force in the classical Kepler problem (Mazilu *et al.*, 2019, pp. 60–75), and in fact we shall describe here quite a few other cases. The analysis of Lorentz matter done by Poincaré turns out to be related to the class of problems which lead necessarily to admitting the non-Euclidean geometry as a natural geometry of the matter.

It is in this respect that we find noteworthy that, for quite a few years now, the worldwide physics community has benefitted from the exquisite production of Professor Bernard Lavenda of the University of Camerino, Italy, already cited above (Lavenda, 2012). This work is, indeed, quite an 'Odyssey in Non-Euclidean Geometries', as the subtitle describes it. And not just for the pure benefit of physics, moments of this odyssey are occasionally presented by its author as a kind of history of the internecine struggle of human knowledge along the painstaking path of gaining its modern status. On occasion, this struggle takes indeed social proportions, which always bring in a host of non-academic attitudes, predominantly among the bystander crowds, always bent into taking a non-rational approach to a purely rational stand. Yet, we are inclined to consider the production of Professor Lavenda as a call for rational judgment of the achievements of human genius at large, addressed especially to physicists. A call to take these achievements as lessons to be properly learned, and leave aside any social overload — be it actual or assumed — of a personality, for that is just incidental. In short, a call to avoid making any 'saints', as Einstein once used to say, and trying to judge impartially! A difficult task in most cases, which is why Professor Lavenda extends the presentation of issues over some seven hundred pages, worth considering for anyone interested in the very play of physics, and not just in kibitzing, even indifferently, to say nothing of imparting unwanted advise.

Here and now, this exquisite production retains our attention to a somewhat limited extent. Being mainly dedicated to the insufficiently recognized presence of non-Euclidean geometries in physics, in its development Professor Lavenda devotes a great deal of discussion to Henri Poincaré. As known, Poincaré was one of the main advocates of these geometries, with fundamental specific discoveries — no question about that. However, from the point of view of physics, he is occasionally even presented as having 'missed great opportunities', not only to bring this kind of discipline into the core of physics, but even to properly associate his name with the idea of relativity, for instance. Now, as we do not believe in saints, we are not here to establish, and so much less to sustain, any kind of 'seniorities', at least not of a social kind. Therefore, we use Professor Lavenda's production only in order to proclaim our stand more perceptively. Quoting:

> By the modern historical account of electromagnetism and relativity, *there were winners and losers.* Maxwell is said to have triumphed over Weber and Gauss, in formulating *a field theory of electromagnetism*, and over Lorenz and Riemann *in the formulation of his displacement current*, Einstein's absolute speed of light prevailed over Ritz's ballistic theory of emission, Lorentz's supremacy over Abraham and Bucherer in devising *a model of the electron* whose expressions for the variation of mass, momentum, and energy with velocity were later to be adopted in toto by relativity *as a model for all matter*, whether charged or not, and *Einstein's seniority in stating the principles of relativity* though they were previously enunciated by Poincaré. [(Lavenda, 2012); *our Italics*]

Thus, for once, if anything along the idea of 'winners and losers', we feel the urge of establishing 'seniority' to each and everyone of the great names cited in this excerpt, and many others in fact, by showing that every one of them has his own place in accomplishing the 'model of all matter'. Of course, when this matter is properly conceived, which for us here means: *conceived from the point of view of a scale relativity.*

Consequently, we need here too, and expressly at that, as we said, the presence of Henri Poincaré for a few fundamental achievements in theoretical physics, besides those acknowledged thus far, specifically related to non-Euclidean geometries as the one right above, but not recognized as such. And, paraphrasing Hendrik Antoon Lorentz who once expressed it (Lorentz, 1921), as much as we might wish to be thorough in the presentation, we will not be able to confer "... an idea anywhere near complete of how much the theoretical physics owes to Poincaré". Thus, as it is even impossible to construct the present work as an 'odyssey' through the various specific contributions of the great scholar, we shall dwell for a while around his personality only in order to *built a fluid interpretation for the concept of matter related to Lorentz's name.* For, as perhaps it became already obvious even for the casual reader by now, the matter in the ideal condition qualified by us as *per se*, does not admit a physical structure in the current acceptance of this concept, and we uphold the opinion that Poincaré was well aware of this fact. Aware and alert even to the point that he has given an authentic *interpretation* to the very light, in its capacity as electromagnetic field, an interpretation that can be associated with the great classical interpretation of ether, previously expressed by us through the apt words of Samuel Earnshaw.

Again, we think we even found the reason of this need for interpretation of a continuum like the electromagnetic field, in the fundamental fact that any *physical structure* is a mix of matter and space, so that the matter *per se* must be the 'place' where the space does not have access, as we defined it before. Such is the nucleus of a planetary model at any space scale, which, as we have shown (Mazilu *et al.*, 2019), should also be described by that non-Euclidean geometry of the unit disk, indisputably related to the name of Henri Poincaré. In this respect, as Professor Lavenda puts it...

> ...What *is incomprehensible* was Poincaré's need to 'adjust' the laws of physics *so as to preserve Euclidean geometry*, and *Einstein's later concurrence with him.* Was Euclidean geometry superior to non-Euclidean geometries to which Poincaré made so many outstanding contributions? Why *couldn't Poincaré connect with his fractional linear transformations* which preserve certain geometric properties and define a new concept of length in hyperbolic geometry with Lorentz transformations which he did so much work on? [(Lavenda, 2012); *our Italics*]

Well, as we have just mentioned above, the hyperbolic geometry is the absolute geometry of the unit disk after all, and it is known indeed — especially after Poincaré's own mathematical researches into analyticity — to be closely related to those 'fractional linear transformations'. Therefore, if as absolute geometry the hyperbolic geometry is the geometry of the center of force in the classical Kepler problem, no doubt that the answer to the questions raised here by Lavenda is somewhere in our sight. We only hope that the reasons for these episodes of theoretical physics will become 'comprehensible' as we go along with our presentation. Thus, delegating, whenever the case may occur, the presentation of coming into being of the non-Euclidean concepts within the realm of physics to the exquisite presentation of Professor Lavenda, we proceed to a narrower task here, namely that of revealing some inedit aspects of the fundamental physics professed by Poincaré. The way we see them, these aspects are necessary in order to come to a deeper understanding of those fractional linear transformations, thereby revealing why, specifically, Poincaré 'couldn't make the connection' between the linear fractional transformations and Lorentz transformations, envisioned by Lavenda. And, based on what we presented thus far, an overall reason can be uttered right away: while the universe as a whole can be thought of as a physical structure, the matter, in its *per se* condition, does not accept such a structure, but only an *interpretation.* The interpretation is done in terms of ensemble of Hertz material particles, endowed not only with one — as Samuel Earnshaw would say — but with *three forces of Newtonian type*, as we have shown above.

Regarding the specific case of interpretation that we have now in mind, a confession seems to be in order from our part, and we think this is the right place to pinpoint its idea, again, as related to a few words of Professor Lavenda himself. Quoting:

> Poincaré was infatuated with the break-down of Newton's third law, the equality between action and reaction, in his new mechanics. In a followup paper entitled, "The theory of Lorentz and the principle of reaction", Poincaré considers electromagnetic energy as a 'fictitious fluid' (*fluide fictif*) with a mass E/c^2 . The corresponding momentum is the mass of this fluid times c. *Since the mass of this fictitious fluid was 'destructible'*, for it could reappear in other guises, *it prevented him from identifying the fictitious fluid with a real fluid.* What *Poincaré could not rationalize* became 'fictitious' to him. [(Lavenda, 2012, p. 39); *Italics ours*]

To reiterate our opinion on the position of a genius as we stated it before, we maintain that the ideas have their own 'objective' dynamics, independent of personalities. However, only a genius can reach an idea as ultimate truth, even to the point where he succeeds to casting it into proper words. Beyond those words no one can 'rationalize', not even the genius himself. This is why we specifically think that the term 'fictitious' of Poincaré, should remain indeed as it is, for no one ever since, not even geniuses themselves, could further rationalize beyond Poincaré's reasoning. So, while openly agreeing with Bernard Lavenda's conclusion from the above excerpt, let us try to summarize a little bit different point of view, extracted from the very same works and words cited in his exquisite production.

The central production we are presently taking into consideration is the celebrated 1906 article from the Rendiconti del Circolo Matematico di Palermo (Poincaré, 1906). In fact, as we have already alluded to it, this well known and largely cited work of Henri Poincaré will not even be taken here in its entirety. More to the point, we consider it only insofar as it is related to the problem of gravitation, and the closely correlated problem of inertia, which were apparently a constant concern of the great scholar (*loc. cit.* §9). For, as we see it, this issue brings in — for the first time ever! — the *interpretation* of the electromagnetic field, given by Poincaré himself. That interpretation was used some five years earlier, in the work cited by Lavenda in the excerpt above (Poincaré, 1900), in order to describe *a possible structure of matter*, based exclusively on the phenomenology of electricity, as conceived by Lorentz.

This last work of the illustrious erudite is, indeed, all about a Lorentz model of the electromagnetic matter, which, having not included *the inertia*, may be suspected of departing somehow from the classical standards. And, sure enough, Poincaré discovered that this model *does not respect the third principle of Newtonian dynamics*, which is the reason he became 'infatuated' with it, according to Lavenda's expression. This deviation from that classical principle is a consequence of *space extension*, and to set things in order here, Poincaré had to admit that, *in the infrafinite range* of time — *i.e.* from the very old point of view of collisions, which to Newton helped *physically* endorse the forces responsible for planets' motion — the *energy* of the electromagnetic field behaves according to the laws of conservation in the case of fluids (Mazilu *et al.*, 2019). The quantum was not yet invented at the time: that 1900 article we are talking about here is, in fact, contemporary with the Planck's fundamental works, and the Einstein's article on heuristics in the problem of light would still have some five years to its publication. Thus, Poincaré had to issue an *interpretation* of his own for the electromagnetic field — which thus appears, again, *avant la lettre* as it were — in the exact sense to be defined by Charles Galton Darwin some three decades later (Mazilu *et al.*, 2019). This interpretation is referring to the genuine Maxwellian *electromagnetic field regarded as a continuum.* It also precedes, by a good three decades, the corresponding interpretation of Louis de Broglie (see Mazilu *et al.*, 2019, Chapter 3) wherein the light is, however, described not electromagnetically but *purely geometrically*, with the help of the classical theory of surfaces. The de Broglie's description, unlike Poincaré's, takes complete advantage of a certain notion of quantum, in full swing in theoretical physics of that time. This means, first and foremost, that Poincaré could not think the idea of fluid in the same terms as de Broglie did three decades later. Quoting:

> We can consider the *electromagnetic energy* as a *fictitious fluid*... that drifts in space according to POYNTING law. Only, we should admit that *this fluid is not indestructible*... which is what *precludes the possibility* of immediately *assimilating, in our arguments, our fictitious fluid with a real one.* [(Poincaré, 1900), *our translation and Italics*]

As we said, this is the *first interpretation ever of a physically manifested* — as opposed to classical ether, which is only hypothetically manifested — *continuum field*, specifically the light, albeit *in the electromagnetic representation*, whereby the idea of particle is only implicitly contained in the concept of a fluid declared by Poincaré as 'fictitious'. It is important, though, to always keep in mind on what account he confers this attribute to a fluid. Specifically, *it is not indestructible*! However, unlike Lavenda — who, in our opinion, goes a little too far with the modern Einsteinian connection between mass and energy, and takes them as freely substitutable for one another in natural philosophical reasonings — we are of the opinion that the Poincaré's 'fictitious' should remain attached to the fluid as in the original statement. Moreover, the same should happen with that 'destructible': it is referring to fluid *per se, not to mass.* If one calls for a deeper reason in order to understand this portrayal, it can be roughly presented as follows.

From a classical point of view, a fluid is conceived as an ensemble of material points, each one of them retaining its 'integrity' along any *imaginable* flow of fluid. Why? The answer rests on the fact that the classical material point — the epitome of fluid particle — *is by definition indestructible*, and a fluid of such material points should somehow preserve this essential classical feature. After all, this very basic feature of the classical material point, is transferred as such by Heinrich Hertz over to his material particle. As we see it, the indestructibility of a Hertz material particle is the only feature allowing for a proper Newtonian expression of the third principle of dynamics, hence for the possibility of defining static ensembles by a condition of equilibrium between the forces acting upon particles, as we have shown in the section just before the present one. Previously, we have characterized this kind of integrity in a fluid ensemble by saying that a classical material point maintains its identity in the form of a label, all along the fluid flow (Mazilu *et al.*, 2019). The mathematics involved in such a theory of fluids can be therefore regarded as describing the *invariance to a transformation of the labels* of a classical material point, and this is, indeed, the case occasionally (see, for instance, Salmon, 1982). Thus, at the epoch of physics we are in now with Poincaré, a real fluid could not be conceived but in terms of this kind of integrity, whereby *the label of a classical material point*, taken as a fundamental physical component of the fluid — the 'body', in a classical Newtonian stand — could be appropriated as the *position on a trajectory* of the point at a certain instant. Of course, this is a modern-day characterization (see Salmon, 1988, Jackiw, Nair, Pi and Polychronakos, 2004), involving the so-called Lagrangian description of the fluid.

In this respect, it is worth noticing that Poincaré closes one of his important works inspired by Joseph Larmor (Poincaré, 1895), with a phase space presentation (*loc. cit.* §§13ff) aimed at explaining the difference between an *ensemble characterizing a fluid per se* and *an ensemble characterizing an electromagnetic field.* It is, indeed, upon this old classical image of the flow of a fluid that Clerk Maxwell's theory of the *lines of force* (Maxwell, 1861, 1862) fell, with a sudden suggestion that a 'label' of a fluid particle can be very well appropriated from the electromagnetic field, not just *imagined* as a trajectory, like one usually does in the mechanics of fluids. The work from 1893 of Joseph Larmor — the one that inspired Poincaré in his production from year 1895 — and especially its later expansion over a good 250 published pages (Larmor, 1893, 1894, 1895, 1897), is all about the struggle of our knowledge to accommodate this state of the case, for both the theory of fluids and electromagnetic fields. And so is the very work of Henri Poincaré inspired by Larmor's approach, as its very title shows it, indeed.

Now, in the spirit of his source of inspiration, Poincaré critically examined the contribution of every major star of this drama of our knowledge, genuinely highlighting the essential *pros* and *cons* in recommending these contributions as *theories of matter*. Among the stars, Hendrik Antoon Lorentz has a special part (Poincaré, 1895, §7) with his recent theory on *electrodynamics of moving matter* (Lorentz, 1892), which brings us to the present issue. For this special topic of electrodynamics, the analysis of Poincaré discerned three conditions to be satisfied by any 'sound' physical theory of matter. Quoting again:

> 1° It must account for the experiments of Mr. Fizeau, *i.e.* for the *partial* (*original Italics here, a/n*) dragging of the light waves, or, *what comes to the same thing*, of the *transversal electromagnetic waves*;
> 2° It must conform the principle of *conservation of electricity and of magnetism*;
> 3° It must be compatible with the principle of *equality of action and reaction.* [(Poincaré, 1895, §9); *our translation and Italics, except as mentioned*]

Every major theory to date, analyzed by Poincaré in that year 1895, fails on at least one account of these three and, of course, Lorentz's theory takes no exception. So Poincaré's attitude is one of 'recognizing' and 'accepting the reality', we should say, but then trying to adapt a certain theory in the best possible way. More to the point, his proposal is to give up the hope of finding the perfect theory of electromagnetic matter, and try to improve on an existing one, on the choice of which he is, in fact, very specific:

> One must renounce to *elaborating a perfectly satisfactory theory*, and hold provisionally on the *least flawed of all*, which *seemed to be that of Lorentz*. This suffices for my task, which is *to deepen the discussion of the Larmor's ideas*. [(Poincaré, 1895, §11); *our translation and Italics*]

As he suggested later on (Poincaré, 1900), the criterion of his choice is semi-subjective, if we may say so: ≪*Les bonnes théories sont souples*≫. (Good theories are flexible; *original Italics, a/n*). However, we intend to intimate here the idea that Poincaré's choice is in fact *purely objective*, imposed by the very nature of the physical problem at hand. In other words, we aim to make the Poincaré's own criterion a little more precise: a good physical theory is not 'souple' *just by chance*!

First of all, let us see what is the content of the theory of Lorentz, and by what is it 'flawed'. It is an interpretation indeed, to start with, in the very sense of the later Darwin's definition of this concept, however without the necessary continuum to be interpreted. More precisely, it starts directly with the idea of an ensemble of particles, without assuming that these would be somehow 'detached', if it is to maintain the expression of Samuel Earnshaw. It also contains a suggestion of confinement of these ensemble of particles, with no hint whatsoever of what can happen in the hollow internal space of their confinement. Quoting, the structure of electromagnetic *matter* is conceived by Lorentz as a physical structure, based on the idea that…

> A *very large number* of little particles *carrying electric charges invariably attached to them*, are dispersed in the *volume of conductors and dielectrics*. They *cross the conductors in all directions* with *very high speeds*. In the *dielectrics*, on the contrary, they cannot experience *but small displacements*, and the *actions of the neighboring particles* tend to reduce them to their *equilibrium positions* from which they are swerving. [(Poincaré, 1895, §7); *our translation and Italics*]

One can see that, in this description, the dielectrics and conductors are, indeed, contemplated as *physical structures*, insofar as they have an internal space, in the form of a volume: electric charges are 'dispersed in the volume', where they can 'cross in all directions', *i.e.* they can move. Therefore, one can even imagine experiments in this volume, capable to account for such a physical structure, which thus can be characterized by chaotic flows of charges, whence we conclude that this should be a *fluid structure*. From this point of view, the flaw of theory can be revealed right away, even in simple words at that: it *does not account for the principle of action and reaction* — the third requirement from the list above. In the words of Poincaré himself:

> ...Consider a small conductor A, charged positively and *surrounded by ether*. Assume that *the ether is swept by an electromagnetic wave*, and that at a certain moment of time this wave touches A; the *electric force* due to perturbation acting upon the charge of A *will produce a ponderomotive force acting on the body* A. This ponderomotive force *will not be counterbalanced* from the point of view of the principle of action and reaction, by another force acting upon the ponderable matter. For, all of the ponderable bodies may be supposed as very remote and outside of the perturbed range of ether.
>
> One can get away with this by saying that *there is a reaction of the body* A *upon ether*; it is not less true (though) that we could, if not realize, at least conceive an experiment in which the reaction principle would seem in default, for *the experimenter cannot operate but on the ponderable bodies and does not know how to do it on the ether*. This conclusion would seem therefore hard to accept. [(Poincaré, 1895); *our translation and Italics*]

We need to stress again the conclusion here, in order to consider it closer later on: we can *assume an action* of the body upon ether, but experimentally *we do not know* how to accomplish it! This is quite a temperate expression of the general observation that the ponderable bodies *do not act upon ether*: the simple phenomenon of the motion of bodies is enough proof for this statement. In the excerpt above, we have however to deal with an electromagnetic wave, and this partially explains the phrase above: "does not know how to do it". For, if the ether is conceived as a support for the electromagnetic waves, this is the least we can state, leaving in reserve the fact that, someday, we may come to realize an electromagnetic action upon ether after all.

In the original work from 1892, Lorentz uses d'Alembert principle, then — as today for that matter! — a routine in the mechanics of continua (see Mazilu *et al.*, 2019, for a partial presentation of this issue). However, according to the experience of the final part of 19th century, he is forced to take special precautions, carried out by an equally special assumption, in order to cover some necessities obviously correlated with the *transport theory*. Thus, it is clear that the classical fluid should be somehow a model for matter. To wit, one has to assume that

> If, after *arbitrary movements*, the matter is reduced to its *primitive configuration*, and if, during these movements, *every element of surface* which is *steadfastly attached to the matter* was traversed by *equal quantities of electricity in opposite directions*, all of the points of system will be found in their *primitive positions* [(Lorentz, 1892, §57); *our translation and Italics*]

Notice that if one takes the 'element of surface steadfastly attached to matter' as referring to an infinitesimal portion of a 'wave surface', the situation indicated by Lorentz in this excerpt is the one to which Louis de Broglie has dedicated his 1927 analysis that introduced the idea of capillary tube as a model for the classical light ray. In a historical perspective, this may be taken as showing that there is no difference between optics and mechanics from the point of view of the diffraction phenomenology (Mazilu *et al.*, 2019). In view of this, we venture to assume that 'configuration' means here an ensemble of classical material points, so that when Lorentz says that an 'element of surface is attached to matter', we have to understand that this element of surface is determined by the positions of some material particles of Hertz's type. Lorentz himself finds that this assumption *is not always satisfied*, and by now we can even tell why, according to his own findings: there is a discrepancy between the time derivative and substantial derivative involved in the transport of energy (see *loc. cit.*, p. 424). However, Lorentz does not see in this a reason not to go any further with the model, and this shows to what extent was he going with the fluid as a model: whatever cannot be conceived for a fluid, cannot be applied to ether either. Quoting:

> If this hypothesis *cannot be admitted in the case of an ordinary fluid*, it *could not be applied to the electric fluid either*. However, this fact does not prevent our equations of motion from being accurate. Indeed, *the mass of this last fluid was supposed to be negligible*, and in calculating the variation δT (*kinetic energy, n/a*) only that kinetic energy was considered *which is specific to the electromagnetic movements*; it will suffice therefore that the material points liable of these motions, and *which are not to be confused with the electricity itself*, enjoy the property of returning to the same positions *if for each surface element the algebraic sum of the quantities of electricity* by which it has been crossed, is 0.
>
> Now, one is entirely free to try on the mechanism that produces the electromagnetic phenomena *any convenient assumption*, and while recognizing the difficulty of *imagining a mechanism that possesses the desired property*, it seems to me that we do not have the right to deny its possibility. [(Lorentz, 1892, §67); *our translation and Italics*]

Notice, additionally, the careful observation that the material points — the classical 'bodies' of dynamics — 'liable of motion' are 'not to be confused with electricity itself', a distinction which, we may say, brings forward the very observation of Poincaré about the impossibility of action upon ether. Also notice that the Lorentz matter thus defined, is the counterpart of the physical universe at large, as presented by Hermann Weyl in the excerpt above: 'the mass of *electric* fluid is supposed to be negligible' here, not the charge. This speaks of a universe where the charge is dominant, and therefore the mass is negligible in its 'economy' — like Weyl would say — not the charge, but the fact is that we cannot 'dismiss the mass', at least not the way we do it in the case of physical universe. As for the rest of the excerpt above, the most important thing, namely 'that mechanism... possessing the desired property' from the last sentence, was not to be 'assumed' anymore for, just about the period of time we are talking here, it was *physically accomplished in the form of the field generated via a periodic charge motion* by Heinrich Hertz [see the English translations collected in Hertz (1893), for the fundamental works which instituted the modern theory of electromagnetic field].

It thus turns out that the matter can carry charges, just the way it carries mass, however in such a case the image of ether, as described by Samuel Earnshaw seems impossible: in the ether we manifestly have (electromagnetic) waves not particles. It is in this instance, that one can conclude, borrowing the later words of C. G. Darwin, that *the ether is a continuum that needs to be interpreted.* One can therefore say that Poincaré actually just 'has not lost this opportunity', as Professor Lavenda would say: as a charge in motion, like those involved in the Lorentz view of the matter, *generates electromagnetic field*, and as this one can be made responsible for the mechanical imbalance, and consequently for an overall physical balance of the Lorentz matter, it is only necessary *to interpret the electromagnetic field itself as a fluid*, according to the old view of ether as matter. Which is what he tried indeed to accomplish, by the definition of that *fictitious Poynting fluid* (Poincaré, 1900). However, there is a drawback, contained in the old idea of *static equilibrium* left behind by this interpretation!

In order to understand the situation we need to swing back in history, to another interpretation of the material continuum, that of Augustin Louis Cauchy. In this interpretation, a static material continuum is viewed as an ensemble of 'molecules', this time 'detached' therefore, in the sense of Earnshaw, with static forces between them, whose *fluxes through any imagined plane* from this continuum are responsible for stresses. These stresses assumed the name of Cauchy ever since (see Moigno, 1868, pp. 616–721), for a clear account of the *theory of Cauchy*; see also Segev, 2017, and the works indicated there, for a modern theory of the *Cauchy fluxes*). Fact is that the static case cannot be rationally understood within framework of the classical mechanics, since an understanding, in this case, would need an interpretation of forces according to *Wigner's dynamical principle* (Mazilu *et al.*, 2019). As we have already signaled (*ibid.*), in the static case *inertia is conspicuously missing* according to our experience. The same conclusion may go, however, for the 'fictitious' Poynting fluid: from the point of view of the concept of interpretation, *the light has the same speed in any direction and everywhere* and, therefore, some classical material points representing a continuum that would be liable to interpret the light, are at 'rest' with respect to each other. And yet, the 'electromagnetic' fluid must account for the *creation* and *destruction* of *energy*, which exclusively brings in the *concept of inertia.* It is this fact that was instituted in physics by Einstein in 1905, in a phenomenological argument that brought to light the celebrated relation between the energy and mass, implied by Lavenda's excerpt above. For the moment, however, one had to insist on structural properties related to what came to be known later as *creation and annihilation processes*, related explicitly to energy. Quoting Poincaré again:

> In order to *define the inertia of the fictitious fluid*, one must agree that the *fluid which is created* in a certain point by it the transformation of energy, arises first *without speed* and that *it borrows its speed from the already existing fluid*; therefore, if the quantity of fluid increases but the speed remains constant, it will be nevertheless *a certain inertia to overcome*, because the new fluid borrowing speed from the old fluid, the speed of ensemble will decrease if a certain cause does not intervene in order to maintain it constant. By the same token, when *destruction of electromagnetic energy occurs*, it is necessary that the fluid, before it is destroyed, *looses its speed by imparting it to the subsisting fluid.* [(Poincaré, 1900), *our translation and Italics*]

Therefore, according to Poincaré, the inertia should be, at least partially, allocated to some *transport properties*, for which the Poynting's theorem (Poynting, 1884) offers the basis as an equation of continuity! It is this fact that makes the theory of Lorentz 'souple', and for this very reason it is held by Poincaré as the most flexible of the theories of the electromagnetic matter: *the possibility of interpretation*, in spite of the 'fictitious' character of the fluid used in accomplishing this interpretation. In fact, every fluid involved in a process of interpretation should be 'fictitious'! In the words of Lorentz himself, Poincaré constantly took his (Lorentz's) studies with a "kind interest" in which he (Lorentz) could always find a "valuable encouragement" (Lorentz, 1921). In turn, we think we could properly locate the inception point of this 'kind interest' of Poincaré in that work from 1895, inspired by the very same theme iterated for quite a long time period by Joseph Larmor, beginning with year 1893. In Poincaré's own words, his 1895 work...

> ...will merely include the summary of reflections *suggested* by the lecture of this important communication [*i.e.* (Larmor, 1893), *a/n*], which sometimes *took me quite far away* from the theory of Mr. Larmor. This is what justifies the title I just chose. [(Poincaré, 1895); *our translation and Italics*]

As it is at the point where Poincaré brought it, and completed with Einstein's ulterior findings that limit the interpretation proper by constraints due to the thermodynamics of the blackbody radiation (Einstein, 1965), the theory does not miss but a *static reference ensemble* and some *general thermodynamic equilibrium considerations* on the continuous electromagnetic field (destined to replace the Einstein's merely 'heuristic' considerations by some general and sound physical ones), in order to be an exquisite *theory of interpretation* as characterized before in the present work, according to the definition of Charles Galton Darwin. Which is what, by and large, we will try to provide in the remainder of this work.

It is along this historical path, heralded by grand ideas, that we finally stop now for a while at the last section of the Poincaré's work from 1906 volume of Rendiconti di Palermo. (Poincaré, 1906, §9). At that time, the issue at hand was the well known impossibility of exhibiting the absolute motion of the matter, considered as motion with respect to ether. The important idea here, not altogether realized by the modern theoretical physics, is that *the ether is a reference frame.* Reserving an acceptable elaboration on this issue, along the lines already suggested in Mazilu *et al.* (2019) a little later along the way of progressing of the present work, let us for now only try to appreciate the Poincaré's characterization of the Lorentz theory. In his own words...

> ...the theory of LORENTZ would completely explain the impossibility of revealing the absolute motion, *if all the forces were of an electromagnetic origin.* (*Italics ours here, a/n*)

> But there are forces to which one cannot assign an electromagnetic origin, like for instance the gravitation. It may happen, indeed, that two systems of bodies would produce equivalent electromagnetic fields, *i.e.* they would exercise the same action upon electrified bodies and currents, but nevertheless these two systems do not exert the same gravitational action on the Newtonian masses. The gravitational field is therefore distinct from the electromagnetic field. So, LORENTZ was compelled to complete his hypothesis, by assuming that *the forces of any origin, gravitation in particular, are changed by a translation* (or, if we like better, by a LORENTZ transformation) *in the same way as the electromagnetic forces.* [(Poincaré, 1906, §9); *our translation, original capitals and Italics, except as indicated*]

It is around this last hypothesis that Poincaré tried to build his theory of forces of non-electromagnetic nature. The cornerstone of this construction is the very same as that used in the *construction of special relativity*, but with a fundamental replacement: there it was the propagation law of light, here he needed *the propagation law of gravitation.* We should keep in mind, even from here, this important idea, to be later found as related only to Gödel's works on cosmology: *the gravitation is to be considered somehow as a wave phenomenon*, once it should be described 'in the same way as the electromagnetic forces'.

Therefore, Poincaré tries first to build a propagation law for the gravitational action, as is this usually done in the case of light in special relativity. It should be, says he, a function written in the implicit form as

$$\mathbf{\Phi}(t, \mathbf{x}, \mathbf{v_0}, \mathbf{v}) = 0 \tag{1.20}$$

to be used in order to define the time t, which, again as Poincaré states it, does not necessarily satisfy "the condition that the propagation should be done with the same speed in all directions". A little explanation of the symbols in this equation is in order. First, it is written in modern notations. For such a specific rendition of the Poincaré's 1906 paper — and for many other reasons in fact — we kindly advise at least a brief perusal of the articles of H. M. Schwartz, dedicated to this very work (Schwartz, 1971, 1972).

Now, equation (1.20) contains our special notations, whereby $\mathbf{x}$ is the current position at the time t of a material point acted upon by the gravitational force. This point is in a state of motion described by the initial velocity $\mathbf{v}_0$ and the current velocity $\mathbf{v}$. Insofar as the motion is accomplished as a consequence of the action of gravitation, the only concern should then be the expression of the local action of this gravitation. Poincaré takes it classically, as a force, F say, whose components at a certain moment of time should be functions of the variables of implicit function $\mathbf{\Phi}$, representing the condition of propagation of gravitation. Then he tries to build the mathematical expression of the force satisfying the Lorentz's assumption that it should behave under the transformation of time and coordinates *in the same manner as the electromagnetic force.* More specifically, Poincaré assumes the following premises:

1° The condition (1.20) should not be modified by the Lorentz group transformations.
2° The components of force F are affected by the transformations of this group *the same way* the corresponding components of electromagnetic forces are.
3° When two material points correlated by the propagating gravitation are at rest with respect to each other, the force F comes down to the "usual attraction law", *viz. Newtonian expression of the force*

At this point Poincaré interrupts the streak of hypotheses with an important observation, which we even find as essential:

> It is important to observe that, in this last case, the relation (1.20) becomes *pointless*, insofar as the time *t does not play any further part if the two bodies are at rest.*

We need to go a little deeper here, for the third hypothesis of Poincaré from the list above, contains the implicit assumption — or, in fact, only the suggestion — that *if two bodies are at rest with respect to each other, the gravitation does not propagate between them.* If true, this would mean that gravitation does not behave like light after all. A simple logic would then show that, insofar as the gravitation does not behave like light, it does not even make any sense to admit that the force of gravitation behaves like the force induced by the electromagnetic waves, to begin with. Assuming, of course, that the light is electromagnetic in its nature. But the issue is by far more intricate, for it depends on what kind of force of gravitation are we assuming here. Obviously Poincaré, like Einstein later, is referring to *a field of Newtonian forces satisfying Poisson equation.* And historically speaking the implications are much more profound.

First of all, Newton discovered the law of force carrying his name, by being enticed into explaining the Kepler motion. From the point of view of this very discovery, the "usual law" is far from describing a static situation, inasmuch as the material point upon which the gravitation acts *is forever moving*: the motion is just as permanent as the force. Strange as it might seem, this permanence is usually translated in physics as an action of the gravitational force only *in the infrafinite scale of time*. To wit, the very existence of force is taken as an instantaneous action of gravitation. One should not forget this important condition, as it could not be forgotten by Newton himself, who took the pain of validating the idea of gravitational force — which is a permanent force — with the help of collision forces — which are forces whose action is accomplished in an infrafinite time scale — acting toward a common center (Brackenridge, 1995, p. 25). The force itself, however, should have as magnitude one of the expressions given by us previously (Mazilu *et al.*, 2019) whereby the parameters involve the initial conditions of the motion (Mazilu *et al.*, 2019). Obviously, the "usual law" is effective only in the cases where the Kepler motion starts from rest, a fact which, by the way, would be able to explain physically that proviso of Poincaré according to which the interpretative fluid, the one supportive of inertia, "arises first without speed" after which "*it borrows its speed from the already existing fluid*". On the other hand, however, this imposes definite conditions on the very foundations of relativity, one of which directly concerns the subject matter of the present work.

The idea of an 'already existing fluid' asks for a characterization of a moving cloud of Hertz material particles in the structure of revolving body of Kepler motion. In order to unveil this description, assume that between the *times of motion* of different particles under the action of gravitation there is a homographic transformation of the nature of those given previously (Mazilu *et al.*, 2019) which we transcribe here as

$$t' = \frac{\alpha t + \beta}{\gamma t + \delta}, \quad x' = \frac{x}{\gamma t + \delta}, \tag{1.21}$$

where t' is the time of motion and x' is the coordinate of motion. This equation incorporates the following philosophy, à la Poincaré as it were: the propagation of gravitation, just as the propagation of light, *can be described by a phase.* If this phase is to be locally measurable, its value should be revealed, according to Berry–Klein scaling theory, by a local ensemble of harmonic oscillators, in which case the phase must be a homographic function in the time of oscillator (Mazilu *et al.*, 2019). The succession of local times of the recorded signal is then well defined. Now, if the propagation is accompanied by motions of some bodies, each one of these moves in different directions, while 'vibrating' at the same time, like the light. From an algebraic point of view, the first equality (1.21) can be taken as a consequence of the fact that a homographic transformation between the times of motions representing *the same* propagation, appears as only natural, if the phase of the propagating field is to be physically characterized. Then the second equality from (1.21) represents nothing more than the space coordinate, along an arbitrary direction, of the motion connected with the propagation in question.

According to this philosophy, we can replace the purely classical static condition, as requested by Poincaré, with the condition that *the two bodies participating in propagation have the same time of motion*, which is expressed by differential equation $dt' = 0$, in (1.21). This then happens if and only if a differential condition [equation (2.47), Mazilu *et al.*, 2019] is satisfied, which brings out the idea that the sequence of times revealed by the propagation of gravitation represents an ensemble of time moments — as it is just natural for a propagation — but described statistically by a distribution having quadratic variance function. If, moreover, the second equation (1.21) is accepted as expressing the transition between the coordinate of a general motion and the coordinate of propagation of gravitation, then the only particles giving an interpretation of this situation should be free particles whose motion along the directions of the two bodies is given by an invariant function (Mazilu *et al.*, 2019, equation 2.45) describing the sl(2,R) action (1.21). No doubt, the freedom here can be taken both in its classical connotation — the equation of variance can be understood as equation of motion of a free particle in *any* Euclidean reference frame — as well as in a wave mechanical connotation — the free particle Schrödinger equation is invariant to the action given by equation (1.21).

It is at this point that the name of Albert Einstein rightfully enters the theoretical stage of relativity. For, if this line of ideas is correct, then Einstein's characterization of relativity by the concept of simultaneity (Einstein, 1905) should satisfy it, insofar as the light in his acceptance — *i.e.* as an electromagnetic field — satisfies it. It is, therefore, in this sense that *the special relativity is... special.* Whence we can infer that the above line of ideas should contain its natural generalization needed by Laurent Nottale's scale relativity theory. In a word, if the gravity is propagating, we need to find *a general description of the physical concept of propagation*, a description valid for *both gravitation and light, as similar physical phenomena in the sense of Poincaré.*

Let us stop here for the time being, in a note of recognition of the lesson served to us by Poincaré, through the model of matter constructed by Lorentz. We shall strive to complete the theory of Lorentz in the spirit of *idea of interpretation.* The concrete way is by considering the concept of surface as mandatory, as Lorentz has shown by his assumption already quoted above:

> ...it will suffice therefore that the material points liable of these motions, and *which are not to be confused with the electricity itself,* enjoy the property of returning to the same positions *if for each surface element the algebraic sum of the quantities of electricity* by which it has been crossed, is 0.

The gravitation has to be described by propagation, like the light itself, no question about that. This fact needs the general concept of wave, and with it the *concept of surface*: this is the rigorous follow up of Fresnel's theory of wave surface, as well as of Louis de Broglie's ray theory which, as we have seen, is a necessary step in completing it conceptually. Poincaré has, in fact, just shown positively, among other things to be disclosed in due time, how 'for each surface element' the quantity of electricity is zero. We therefore, have to assume that our duty, within the present work, is only to explain it rationally, and for this we need a proper insight into a suitable theory of gravitation. Only after that we can elaborate on a general concept of physical surface.

1.4 Einstein's Matter: Authentic Matter Interpretation

In order to avoid any possible misunderstanding, we need to acknowledge from the very beginning that the term 'authentic' here, refers exclusively to 'matter', having nothing to do with 'interpretation': until further outcomes will emerge — like, for instance, the idea of reverse interpretation we mentioned before a few times — this is the one defined for the necessities of wave mechanics (Darwin, 1927). For instance, the ether was always thought to be matter, but it is not authentic matter, *i.e.* is not that kind of matter that could be revealed directly to our senses; the electromagnetic field is also thought to be matter, but it is not authentic matter: this is why it needs an interpretation in the first place. As well known, the interpretation existed even from the old times, and in fact it entered the natural philosophy by the idea that any piece of matter revealed to our senses, can be contemplated as an ensemble of constitutive parts, somehow connected together. The ether was hard to characterize accordingly, for no possibility exists to 'break' it in parts, as it were; the same goes for the electromagnetic field, a fact duly recognized by Henri Poincaré, as shown above. However, they were considered as matter, albeit described as such only *by interpretation*, as we have seen, for instance, in the excerpt from Earnshaw above. But the 'parts' which served for such an interpretation remained, by and large, vaguely characterized from a physical point of view. This is why, in history, the ether needed so much theoretical elaboration. One can even say that Poincaré added a 'real' ether — the *electromagnetic* one — to the categories of matter described by interpretation, but still, this was ether, not 'authentic' matter, *i.e.* matter of the primordial kind, revealed to our senses. Only a complete Lorentz model would have been able to account for such matter. Speaking from the point of view of the theoretical physics it generated, the ether is, undoubtedly, the prototype of classical idea of matter, from which we can even borrow some of the principles of description of matter in general. However, we need to stress it again, this is not the authentic, primordial matter, from which the idea of interpretation sprung, and limiting our review to ether, would only mean that we will not be able to recognize that physics had its fair share in the interpretation of the matter as revealed by our senses. The facts we have in mind for illustrating this issue — and thus learning the due lessons! — are exclusively related to the Einsteinian doctrine, largely unrecognized as demanding anything, even remotely close to an interpretation. It produced, however, a few exquisite cases of interpretation, which properly considered as such will set us along a sound way to a *scale* relativity theory.

We have already quoted before many of Einstein's ideas in our context here. They were all directed toward improving the presentation of the general relativity *as he conceived it, viz.* according to the thesis that *the matter controls the metric of spacetime.* However, in the mathematical rendering of this thesis Einstein, and, after him, anybody else ever since, describes the gravity *in an entirely classical way.* More to the point, the metric tensor is a solution of some system of partial differential equations — the Einstein's field equations, replacing the classical Poisson equation — which, naturally, *ask for boundary conditions in spacetime.* It is these conditions then, that are the essential issue of Einstein, and he always relates to them in a way or another, in almost any of his discussions on gravitation. The Newtonian problem of matter *per se*, in its most striking aspect that 'the matter does not equally fill the space at its disposal' (Newton, 1974), is circumvented by the theory of general relativity, and this in quite a natural way we should say, but only mathematically, nevertheless. Quoting:

> ρ is the *mean density of matter*, calculated for a region which is *large* as compared with the distance between neighboring fixed stars, but *small* in comparison with the dimension of *the whole stellar system*. [(Einstein, 1917), *footnote on the second page of the article; our Italics here, n/a*]

Now, from the very same mathematical point of view, in calculating a mean density this way, one has to ask, first and foremost, for the knowledge of the metric tensor of the host space and/or for some measure of 'the whole stellar system', capable of describing accurately enough the spacetime extension conditions of that system. But this is not all of it: we have quoted Feynman before, on the impossibility of calculating the very mean density this way (Feynman, 1995; Mazilu *et al.*, 2019). This impossibility, however, has also a positive connotation. For, it revealed to knowledge that the physics can uphold here this classical concept of density, only by adding yet another differentia to it: *the counting of constitutive matter formations*, according to the idea that the whole matter, *viz.* the matter as a universe, is a physical structure. Still, this very concept becomes uncertain, inasmuch as it obviously depends on the space scale where we perform the counting, through a specific fundamental physical structure of the universe, presented directly or mediately to our senses at that space scale. For instance, Einstein was considering the stars as fundamental constituents of the universe to be contemplated in counting, but today one considers galaxies, and even metagalaxies, as such fundamental matter formations at a cosmological scale.

Related to this issue, we need to recognize that the constitutive physical components — necessary to accomplish the counting process in establishing the mean density — remain forever undecided due to their obvious scale dependence. Consequently, it seems that an *a priori* geometry, that would render the variation of the metric tensor in such a way as to eliminate the matter represented by the density of a physical structure from the general-relativistic scenario, should be made available, and in fact is not quite out of the ordinary. That geometry, though, should not be the usual Euclidean geometry. We have illustrated this subject here, within the very Newtonian natural philosophy, by the mathematical idea of confinement, as a normal outcome of a Cayleyan geometry. This, we have shown, led quite naturally to a well-known non-Euclidean geometry — specifically a Lobachevsky one — of the space containing the kind of matter assumed to generate the central field of the very classical dynamical Kepler model. However, the fact that the matter changes the geometry of space was first ostensibly 'professed', as it were, by general relativity, and was actually presented in great explanatory detail even from the beginnings of the Einstein's theory (Flamm, 2015), as one of its natural achievements. Worth noticing, in order to reveal the state of the case from the point of view of general relativity, is that the details of explanation *are referring to motion*, and that they are principally incomplete. That perception — *viz.* that the matter changes the geometry — occurred in connection with one of the first general relativistic models of what we call here authentic matter, that generates a gravitational field in general relativity: *a sphere of ideal incompressible fluid* (Schwarzschild, 1916).

A small digression is in order here: considerations like these, motivate us to take as fundamental for knowledge the idea of a *general natural philosophy* — a natural philosophy that makes no distinction between Newtonian and Einsteinian viewpoints. From such a perspective, while the Newtonian mathematical principles allow us to naturally introduce the non-Euclidean geometry in the description of matter (Mazilu *et al.*, 2019), the Einsteinian mathematical principles allow only an *interpretation* of that matter, via a concept of fluid, as required by a scale relativity theory in Nottale's stand. So, it is the work of Schwarzschild just cited that we shall analyze first from this very point of view, for it contains the concept we need — that of incompressible fluid — and with it we see the implicit content of the modern concept of *interpretation of a continuum filling the space* in the Newtonian sense. This will take us indeed, along a path where the Einstein theory proves to be all about the *interpretation of the very spacetime continuum* in each and every one of its instances.

Now, along such a path, the story can, in fact, begin anywhere. However, we think that in order to make it more comprehensible — by raising, for instance, even the interest of a casual reader, we should say — we start with a rather intriguing note of the illustrious Felix Klein, excerpted from a letter addressed to Albert Einstein, where Karl Schwarzschild's work mentioned above appears cited in a context related to our present discussion. The intriguing part of the note, is that the work of Schwarzschild cannot be associated to the subject on which Klein was reporting but only in concept, specifically through the concept of interpretation, as defined for the necessities of the wave mechanics, and that association has deep consequences for the relativity itself. Among these, the chief one is that the concept of interpretation itself should be amended with yet another differentia: the already mentioned *reversibility*, to be properly explained in due time. Quoting, therefore:

> In order to give a physical turn to my letter after all, I note that de Sitter's ds^2 appears implicitly already in Schwarzschild's paper of *24 February 1916*. One just has to set $\chi_a = \pi/2, c = 2$, $R = \sqrt{(k\rho_0/3)}$ in formula (35) there, in order to have de Sitter's ds^2. Formula (35) relates, of course, to the *interior* (*original Italics, n/a*) of the *sphere at rest* considered by Schwarzschild *of gravitating liquid of constant density*. Formula (30) is thus applicable, which yields $p = \rho_0$, hence *a steady pull*. (Klein to Einstein, in *The Collected Papers of Albert Einstein*, Volume 8, Princeton University Press, Document 566; *our Italics, except as specified*]

Now, the intriguing part of this excerpt seems to be quite obvious: as well known, and widely discussed in the specialty literature and, in fact, not only there, Willem de Sitter's solution is referring to a special Einsteinian construction of the metric of spacetime, whereby the matter *may not even exist*. Klein's observation aims at giving a physical meaning to that spacetime, which is why he brings in the Schwarzschild solution here. However, that solution is originally referring to *a sphere of ideal fluid of constant density* representing *authentic matter* — not even ether of any kind — according to precepts of Newtonian physics, and *so much the more* according to those of general relativity. And if Klein uses such a metric in order to give physical reasons to an empty spacetime, this fact cannot be taken but only as *an interpretation of such a continuum*, the way this concept was defined by Charles Galton Darwin. In fact, that 'steady pull', which is taken by Klein in the excerpt above *as defining the density of this empty spacetime* based on pressure, emphasizes, in our opinion, this very circumstance. Let us show, in details, what this issue is all about, and explain what we see in this moment of human knowledge. For, we have here, indeed, a moment of human knowledge, and a great one at that!

Everything started in this problem with Einstein's *Cosmological Considerations* (Einstein, 1917). This work was dedicated, as it seems just natural according to its title, to the... cosmological problem, of course, but obviously related to the boundary conditions for the metric tensor of the spacetime at *the edge of the universe* — the Einstein's constant concern, as we mentioned before. The problem should inevitably occur within Einsteinian natural philosophy of approach of the gravitation and inertia, just as it inevitably occurred in the classical Newtonian approach of the topic, represented by the Poisson's equation, from which the Einsteinian point of view sprang. Only, this time, a fundamental fact harshly jumped our intellect, for the humanity *became aware of infinity of the universe.* The endlessness thus took a positive turn in physics *as a precise object of study*: one cannot define the right cosmological boundary conditions but on the edge of the universe as a physical object, and, as much as one needs it, this edge cannot be properly defined by any means! In the words of Einstein himself:

> In *my treatment of the planetary problem* I chose these limiting conditions in the form of the following assumption: it is possible to select a *system of reference* so that at spatial infinity all the gravitational potentials $g_{\mu\nu}$ *become constant.* But it is by no means evident *a priori* that we may lay down the same limiting conditions *when we wish to take larger portions of the physical universe into consideration.* In the following pages the reflexions will be given which, up to the present, I have made on this fundamentally important question. [(Einstein, 1917); *our Italics*]

In other words, one might say that Einstein felt as having to accommodate the 'ongoing' dynamics of astronomical discoveries, enlarging 'daily', as it were, our knowledge and, with it, the 'size' of the perceived universe. And just for anticipating things to a certain extent, we may add even by now that the Cosmological Considerations represent an admission of the fact that one cannot construct limiting conditions appropriate for the requests of Einsteinian theory regarding the metric tensor. To wit, the overall conclusion of that work is this: in order to consider it as a viable cosmology, the general relativity should refer to a universe conceived as a finite space filled with matter, in the genuine Newtonian meaning of this statement.

Indeed, against all odds, and in fact quite understandably we should say, Einstein never appeared inclined to give up the mathematical way of approaching the problem, *i.e.* by a system of partial differential equations. Accordingly, he found a method to avoid the problem of boundary conditions for the metric tensor, simply *by making them unnecessary*. Specifically, he has split up the spacetime continuum into space and time, and considered that *the space resulting from this splitting must be finite*, because, actually, even from the point of view of our experience there is no other possibility. What, then, can be the relation between the spacetime and space proper? Mathematically speaking, for Einstein that relation boils down to an *embedding of space into the spacetime*. However, as it turns out, the procedure leaves the corresponding *problem of time* in suspension. As a matter of fact this issue was to be naturally expected for the procedure: if the space of the 'disunion' thus accomplished is not the space we usually conceive, *i.e.* a Euclidean space, neither should be its time. However, the time here should still preserve its *relativistic* differentia as a concept. Specifically, it has to be, at least to a certain extent that will be explained here later on, the 'cosmic and phenomenal time' in Hermann Weyl's phrasing excerpted by us in the beginning of the present chapter. Aside for this fact, Einstein makes a choice of the embedding procedure that is entirely in the spirit of our subject matter here, *viz.* the scale relativity theory.

Referring the deeply interested reader to some original works for detailed mathematical considerations [for the geometric justification of the method, see Weyl (1923), and also Cartan (2001)], it is sufficient to say here that Einstein sees the accomplishment of such an embedding of space as *a restriction of a four-dimensional manifold* of quadratic type — whereby the coordinates of events are denoted ξ_μ — specifically a hypersphere of constant radius R:

$$\xi_1^2 + \xi_2^2 + \xi_3^2 + \xi_4^2 = R^2 \tag{1.22}$$

to the hyperplane $\xi_4 = 0$, which can be taken as reference. In broad strokes, the embedding is accomplished as in the known classical case of a surface in the regular space: the metric of the space thus obtained is constructed by restricting the four-dimensional metric to that of the hyperplane $\xi_4 =$ const. Then, *if the four-dimensional metric is also Euclidean:*

$$(ds)^2 = (d\xi_1)^2 + (d\xi_2)^2 + (d\xi_3)^2 + (d\xi_4)^2 \tag{1.23}$$

using equation (1.22), the metric of the three-dimensional space becomes

$$(ds)^2 = \gamma^{ij} dx_i dx_j, \quad \gamma^{ij} = \delta^{ij} + \frac{x^i x^j}{R^2 - \sum_k x_k^2}, \tag{1.24}$$

where $\xi_k \equiv x_k$ are the coordinates of a position in the chosen hyperplane of the hyperspace. Here our 'contravariant' writing of the metric tensor is not merely imposed by the dummy indices rule of summation, but has a physical reason which shall be explained a little later. For the moment though, let us take notice that the quantities accessible to measurement are here the *contravariant coordinates* with respect to the metric (1.24), viz.

$$x^i \equiv \gamma^{ij} x_j = R^2 \frac{x^i}{R^2 - \sum_k x_k^2}$$

and these are essential, for instance, in a physics implemented within the framework of Carlton Frederick's stochastic setting of the natural philosophy (Frederick, 1976). Perhaps for now it is too much to say 'accessible to measurement' — in view of the fact that we have not defined yet, according to the well established axiomatic custom in physics, what is the 'measurement' itself — but we have at least a prescription of evaluating such coordinates, which specifies them as 'gauges' in a precise physical sense, everywhere in the universe, in the world of the small, as well as in that of the large. In other words, *the prescription is universal*, and comes with that conclusion of the Newtonian natural philosophy, regarding the confinement, to which we alluded before. Let us detail our argument further.

1.5 Non-Euclidean Geometry as a Gauge Geometry

The prescription we are talking about is referring, of course, to the very same 'treatment of the planetary system' from the perspective of which Einstein once chose his reference frame in deciding the 'limiting conditions'. Our observation is that an Einstein-like choice is implicit in the mathematical treatment of the Kepler problem by classical dynamics. We just have to uncover it, with no need of any further hypotheses. Therefore, from this point of view, one can say that Einstein just followed, up to a certain moment, we have to admit, a natural course of knowledge, provided we consider this course as an objective dynamics of ideas. This fact can make Einstein's procedure a right one, in the first place. However, there is more to it, and in quite a mathematical way for that matter. Indeed, in the section $x_3 = 0$ of the metric space thus obtained by Einstein, let us take the coordinates as defined, up to a constant scale factor, by:

$$x_\alpha \equiv v_\alpha, \quad R^2 \equiv (k/\dot{a})^2,$$

where the index α runs through values 1 and 2. Then, it is pretty obvious that the metric tensor (1.24), describing this section of the space is, again up to a factor, the inverse of the matrix of quadratic form representing a Keplerian orbit, given by us before (Mazilu *et al.*, 2019). The eigenvalues of this inverse are, therefore, *the magnitudes of the semiaxes of the orbit*, so that they properly represent *length gauges*, just as they should. This observation has far-reaching consequences, from mathematical, as well as physical points of view.

First of all, from a physical point of view, the mathematical setting thus conceived is entirely analogous to that instituted in the Fresnel's theory on the physics of light, by his ellipsoid of elasticities

(Fresnel, 1827). In the case of a Kepler problem, however, unlike the case of light, the coordinates are decided by the *initial conditions of the motion* describing the orbit, conditions that define quantitatively, through some constant integrals of the motion — specifically, the so-called *Laplace–Runge–Lenz vector* — the space extension of the source of forces sustaining the motion in the corresponding dynamical problem. This dynamics characterizes exclusively the field of Newtonian forces, *i.e.* central forces having a magnitude inversely proportional with the square of distance. Therefore, by *Mariwalla's theorem*, the argument is valid no matter of the space and time scale (Mariwalla, 1982): it is valid for the planetary atom in microcosmos, just as much as it is valid for the planetary system proper at the cosmic scale. This is the physical 'prescription' we were talking about before, for the covariant coordinates of Einstein, and implicitly that of the metric tensor of space, given in equation (1.24).

At this point we find appropriate — and even necessary, we should say — a digression on the Fresnel's physical theory of light, which brings us to the crux of physical connotation for this approach to Kepler problem. That theory — *viz.* Fresnel's — needed gauging, and this gauging was initially based, as we said, on the concept of *ellipsoid of elasticities.* The physical motivation rests upon the fact that the dynamics in such a theory of light — which, at the time of Fresnel, would make a physical theory out of it — is only incidental. Let us elaborate a little more on this statement, since it is important to understand its meaning for our knowledge in general. Recall that Fresnel incorporated *the local phenomenon of diffraction* into the phenomenology of light, in order to complete the former classical phenomenology, based on just *reflection* and *refraction* phenomena. The newly added diffraction phenomenon revealed the importance of periodicities in the local manifestation of light, which in turn brought out the *idea of phase* and thus the trigonometric functions with it. With trigonometric functions, the second-order differential equation became, naturally, part and parcel of the mathematical rendition of the theory. Along this mathematical rendition, the second principle of a dynamics involving elastic forces then came just as normal from a natural philosophical point of view, for the second time derivative of a coordinate is usually associated with an acceleration. This way, the old idea of Robert Hooke about the transversality of the light phenomenon (see Hooke, 1665, pp. 55–67), has gotten an apparently sound theoretical proof based on the very phenomenology of light, without 'inventing hypotheses', as Newton himself would say (Mazilu & Porumbreanu, 2018). We shall return to the details of this specific issue later on, as it is essential in introducing a proper fractal view of the world.

Now, in order to make this dynamics 'lawful', so to speak, in the case of light, one last proviso was needed: *the phase, involved as independent variable in the trigonometric functions, had to be linear in time.* A condition which brought *the frequency* in, and with it *the wavelength*, thus generating right away a whole new experimental technology, but leaving nevertheless behind what the dynamical principle really needed for a sound physical theory: first, the elastic properties of the medium supporting the light and, secondly, the interpretation of light, which obviously requires the idea of particle, and therefore some inertial properties. The Fresnel's ellipsoid of elasticities pretty much fills in for the first aspect of this problem, while the second one was left in suspension ever since, being replaced with *ad hoc* creations every now and then, and so is it, actually, even today to a large extent. A proper dynamical use of the second principle of dynamics in the matters of light came in handy only later on, with the advent of the electromagnetic theory of light.

Now, along the same line of thinking through a phenomenology, in the microscopic space — where the planetary atom replaces the planetary system proper, from the cosmic scale — the electromagnetic field *must be included in the habitual phenomenology*, as it were, insofar as it is present among the 'routine' phenomena at this space scale. Or so it seems anyway, as long as Kepler dynamics is taken as the regular dynamics of the microscopic space range. This means that, inasmuch as the electromagnetic field is a gauge field proper, it should be the *actual appearance* of a certain past in a Kepler problem describing the atom, just as the Newtonian force is the actual appearance of the past conditions for the planetary system proper. Only *as such* is the electromagnetic field manifest as a *transition field* in the sense of the philosophy of Niels Bohr. We shall take up these issues a little later along this presentation, from the very same point of view of the concept of interpretation.

Coming back to our main concern here, it is useless to say that Einstein did not follow the problem of motion the way we just presented it, and as one might be, in fact, led to understand it from the previous excerpt, for this is by now almost a common knowledge. Instead, he followed the problem of correlation between field and matter, as he saw it earlier, on the occasion of building the general relativity, to start with. Classically, this correlation is given by the equation of Poisson, which makes the potential a fundamental characteristic of the field, provided the density of matter should not be a problem. And from this point of view the Newtonian universe is doomed to non-existence. For, such a universe is spatially finite, 'although it may have an infinite mass'. In it, the radiation of stars may well travel radially outwards, with no possibility of return. And insofar as, going beyond the Poincaré interpretation, the radiation may be thought of as a physical structure, so may, in fact, behave the very fundamental authentic matter structures: they may very well be expelled outwards themselves, with no possibility of return, by a sort of statistical process, for instance of the kind we know today as a *Penrose process* of extraction the energy, but from the black holes (Penrose, 2002). Quoting:

> We might try to avoid this peculiar difficulty by assuming *a very high value for the limiting potential at infinity.* That would be a possible way, if the value of the gravitational potential *were not itself necessarily conditioned by the heavenly bodies.* The truth is that we are compelled to regard the occurrence of any great differences of potential of the gravitational field as contradicting the facts. These differences must really be of so low an order of magnitude that *the stellar velocities generated by them do not exceed the velocities actually observed.*
>
> If we apply *Boltzmann's law of distribution* for gas molecules *to the stars,* by *comparing the stellar system with a gas in thermal equilibrium,* we find that the Newtonian stellar system cannot exist at all. For there is a finite ratio of densities corresponding to the finite difference of potential between the centre and spatial infinity.

A vanishing of the density at infinity thus implies *a vanishing of the density at the centre.* [(Einstein, 1917); *our Italics*]

This excerpt suggests that Einstein may have felt that *the potential in matter asks for a separate quantitative definition.* Perhaps not in the way we recently presented this issue, *i.e.* a potential defined by the amplitude of a wave via a stationary Schrödinger equation (see Mazilu *et al.*, 2019, equation (2.35)), but by something like an Emden–Fowler equation for instance, as in the Thomas–Fermi method, subsequent to Einstein's own theory, if it is to maintain the guise of a physics based on partial differential equations in this picture. Such a situation was however unbearable for Einstein, as the last sentence of the above excerpt shows.

1.6 De Sitter's Spacetime

It is at this point where we give the floor to the critical work of Willem de Sitter, because it seems most appropriate in unveiling the true nature of the whole involvement of some great minds in this equally great moment of human knowledge (de Sitter, 1917). For details one can follow also de Sitter (1916), especially the *Third Paper* of that series. It is time, indeed, to take notice of the fact that, from the point of view of a scale relativity, the Einstein's procedure of embedding the *manifold of positions* — the space — in the *manifold of events* — the spacetime — just like Fresnel's procedure of construction the wave surface from infinitesimal pieces [see Hamilton (1841), for a geometrically unitary theory of such construction], carries the special meaning of the *mathematical transition from infrafinite to finite scales*, or vice versa, if it is to use the suggestion of Nicholas Georgescu-Roegen about the quantitative hierarchy of scales. In this connection, it is important to notice that de Sitter's line element of the universe can be produced as a Cayleyan metric of the spacetime, just like the metric involved in the Kepler problem (Castelnuovo, 1931). For an independent discussion of the physics of de Sitter's universe, one can consult Tolman (1929).

In the embedding procedure, however, there is an important notice to be taken: the geometrical structure of the space of events at the infrafinite level [equation (1.23)] is, from a metric point of view, the same as that at finite and transfinite level [equation (1.22)], specifically, Euclidean. True, the identity is just a particular case of invariance to the transition of scales, but it is an invariance nevertheless. Also true, the space positions are just particular events, to wit, some *simultaneous* ones, but events nevertheless. However, while thus warranted — from the point of view of a general natural philosophy, we should say — such a procedure breaks the symmetry of the theory of relativity, whereby the time coordinate and the location coordinates play a similar part in the theory, reflected in the covariance and the symmetry of expressions of the field equations. Specifically, the *time sequence* defining the common time of the positions in the space thus defined, is arbitrary, and this might give our intellect that unwarranted freedom that Hermann Weyl was afraid of. And Weyl's fear was not in vain: that freedom was, indeed, taken later on, with the occasion of the well-known case of the Gödel's universe (Gödel, 1949, 1952); until further elaboration on the Gödel moment's significance for the human knowledge, (see Mazilu *et al.*, 2019).

This is why Willem de Sitter applies Einstein's procedure *to a five-dimensional finite quadratic manifold* — an abstract manifold to a large extent, taking into consideration the fact that its algebraical definition does not mean too much from a physical point of view — just in order to preserve the symmetry of the manifold of events in the exact form Einstein did it for the symmetry of the manifold of positions. Specifically, de Sitter uses the Euclidean quadratic five-dimensional manifold of equation

$$\xi_1^2 + \xi_2^2 + \xi_3^2 + \xi_4^2 + \xi_5^2 = R^2 \tag{1.25}$$

and the corresponding Euclidean metric, in order to find the metric of the spacetime in the Einstein-like form

$$(ds)^2 = g^{\mu\nu} dx_\mu dx_\nu; \quad -g^{\mu\nu} = \delta^{\mu\nu} + \frac{x^\mu x^\nu}{R^2 - \sum_\alpha x_\alpha^2}, \tag{1.26}$$

where, this time, the Greek indices run through four values. The observation of Felix Klein from the excerpt that started our presentation of this issue, is referring to such a metric, for which Klein found two important things. In the first place, he found that it is not non-Euclidean as it appears at the first sight from the form of the metric tensor (1.26). A lawful change of coordinates makes it Euclidean, and according to the Einsteinian natural philosophy such a universe is devoid of matter. This fact somehow convinced Einstein that de Sitter's solution might be a proper solution of a relativistic cosmology. However, to Einstein's own disappointment, it firmly installed in physics the so-called *cosmological constant*, an apparent artifact. In the second place, Klein found that a price needs to be paid for this Euclidean character of the spacetime, namely *a special description of the time sequences for such a universe.* In other words, one can say that the de Sitter's universe is not Euclidean quite 'gratuitously': the concept of time, even as 'phenomenal', to use Weyl's expression, needs to have a special differentia. This implies its definition as a coordinate by a certain function of the coordinates, exactly as expected from an Einsteinian point of view.

This is a very subtle point of difference between Newtonian natural philosophy and Einsteinian natural philosophy, on which we need to insist a little, especially in order to better understand the position of Newtonian time. That time was defined, as a concept, independent of space. The concept even missed one of its differentiae, for it was not even necessary. Einsteinian natural philosophy shows why: the Newtonian space was, in fact, empty of matter. Now, taking into consideration the Klein's perspective on the de Sitter's universe, which brings in a special time sequence related to the empty space, one can say that the Newtonian spacetime is just a particular one, whereby the time sequence is completely arbitrary.

Let us consider a few more details here in order to better evaluate the situation [see (Weyl, 1923), *loc. cit.*]. The first move in the choice of some cosmological boundary conditions would be the *a priori* metric tensor of the spacetime at infinity, specifically the empty spacetime metric of the special relativity:

$$\begin{matrix} -1 & 0 & 0 & 0 \\ 0 & -1 & 0 & 0 \\ 0 & 0 & -1 & 0 \\ 0 & 0 & 0 & 1 \end{matrix} \tag{1.27}$$

Regarding this limiting case, both Einstein and de Sitter agree entirely: it is not a proper choice because it requires a special reference frame, of the kind mentioned by Einstein in the excerpt above. Consequently, according to Einsteinian doctrine, this tensor cannot be taken as either invariant integral of the field equations, or at least as boundary conditions for the metric tensor. Giving, finally, the floor to Willem de Sitter himself, in an excerpt referring to this critical situation, we read:

> ...The values (1.27) do not satisfy this condition. The most desirable and the simplest value for the $g_{\mu\nu}$ at infinity is evidently *zero* (*original Italics, a/n*). EINSTEIN has not succeeded in finding such a set of boundary values, and therefore *makes the hypothesis* that *the universe is not infinite*, but spherical: *then no boundary conditions are needed*, and the difficulty disappears. From the point of view of the theory of relativity it appears at first sight to be incorrect to say: *the world is spherical*, for it can by a transformation analogous to a *stereographic projection* be represented in a *Euclidean space*. This is a perfectly legitimate transformation, which leaves the different invariants ds, G etc. unaltered. But even this invariability shows that also in the Euclidean system of coordinates *the world, in natural measure, remains finite and spherical.* If this transformation is applied to the $g_{\mu\nu}$ which EINSTEIN finds for his spherical world, they are transformed to a set of values which at infinity degenerate to
>
> $$\begin{matrix} 0 & 0 & 0 & 0 \\ 0 & 0 & 0 & 0 \\ 0 & 0 & 0 & 0 \\ 0 & 0 & 0 & 1 \end{matrix} \tag{1.28}$$
>
> It appears, however, that the $g_{\mu\nu}$ of EINSTEIN's spherical world (and therefore also their transformed values in the Euclidean

system of reference) do not satisfy the differential equations originally adopted by EINSTEIN, viz:

$$G_{\mu\nu} = -k\left(T_{\mu\nu} - \frac{1}{2}g_{\mu\nu}T\right).$$

EINSTEIN thus finds it necessary to add another term to his equations, which become

$$G_{\mu\nu} - \lambda g_{\mu\nu} = -k\left(T_{\mu\nu} - \frac{1}{2}g_{\mu\nu}T\right). \quad (1.29)$$

Moreover, it is found necessary to suppose *the whole three-dimensional space to be filled with matter*, of which the total mass is so enormously great, that compared with it all matter known to us is utterly negligible. This *hypothetical matter I will call the "world matter"*.

EINSTEIN *only assumes* three-dimensional space to be finite. It is in consequence of this assumption that in (1.28) g_{44} remains 1, instead of becoming zero with the others $g_{\mu\nu}$. This has suggested the idea *to extend EINSTEIN's hypothesis to the four-dimensional time-space*. We then find a set of $g_{\mu\nu}$ which at infinity degenerate to the values

$$\begin{matrix} 0 & 0 & 0 & 0 \\ 0 & 0 & 0 & 0 \\ 0 & 0 & 0 & 0 \\ 0 & 0 & 0 & 0 \end{matrix} \quad (1.30)$$

Moreover we find the remarkable result, that now *no "world matter" is required* [(de Sitter, 1917); *our Italics, except as indicated*].

In other words, Willem de Sitter 'felt' that the Einstein's procedure of embedding *is somehow correct* as a 'physical' philosophy, but it *is not applied correctly*: for a relativist it is quite important that the spacetime, not the space, should be properly described by the equivalence of coordinates, as understood in their simultaneous linear transformations. Even in spite of the fact that the five-dimensional quadratic manifold (1.25) remains in suspension now, regarding its origin! And indeed, using the Einstein's modified equations (1.29), de Sitter finds that, in the metric (1.26), there is a non-trivial solution of the field equations, satisfying (1.30) for

$$\rho = 0, \quad \lambda = 12/R^2, \tag{1.31}$$

where ρ is the density of matter, and λ is the cosmological constant. Therefore, while in the classical theory of gravitational field the cosmological constant is pendant on a certain aspect of the presence of matter — the uniformity in its distribution, in order to allow for the infinity of the world — in the general relativity this constant should be there even regardless of the existence of matter! Moreover, it is this constant which is connected with the curvature, and not the density of matter, at least for the de Sitter's particular model of the world. Let us provide some details of the mathematical procedure of that moment of knowledge, toward this conclusion.

Hereafter we assume the spherical symmetry, with a proper system of coordinates mapping the space, as used by Einstein himself. De Sitter assures this by setting the metric from equation (1.24) in a manifestly spherical form:

$$(ds)^2 = (cdt)^2 - R^2\{(d\xi)^2 + (\sin^2\xi)\cdot(d\Omega)^2\}. \tag{1.32}$$

Here we used the definitions of the Cartesian coordinates with respect to the spherical angles of colatitude and longitude (θ, ϕ), with $(d\Omega)^2$ conveniently denoting the line element measure of the usual unit sphere, and $r \equiv R\cdot\sin(\xi)$. This spacetime metric is a solution of the modified cosmological field equations (1.29) for the conditions $\rho \sim \lambda$ and $\lambda = R^{-2}$, quite reasonable from the Einstein's point of view. The metric of empty spacetime is also a solution of these equations with $\rho = \lambda = 0$, which, again, seems quite normal according to Einsteinian natural philosophy. However, there is also a possible solution of the cosmological equations (1.29), starting from the metric (1.26) of the spacetime with one more 'spherical' variable, ζ say. This metric can be written in the manifest spherical form (de Sitter, 1917):

$$(ds)^2 = -R^2\{d\zeta^2 + (\sin^2\zeta)\cdot[(d\zeta)^2 + (\sin^2\zeta)\cdot(d\Omega)^2]\}. \tag{1.33}$$

But then, by the transformation [(de Sitter, 1918), p. 229]

$$\tanh(ct/R) = R\tan\zeta\cos\xi, \quad \tan(r/R) = R\tan\zeta\sin\xi \tag{1.34}$$

the metric (1.33) can be written in the notable form,

$$(ds)^2 = [\cos^2(r/R)](cdt)^2 - \{(dr)^2 + R^2 \sin^2(r/R)(d\Omega)^2\}, \quad (1.35)$$

which is, again, *solution of the same cosmological field equations.* The result (1.31) of Willem de Sitter goes with this form of the metric. Obviously, this last line element does not represent an empty spacetime according to Einsteinian doctrine. And to add insult to injury, as it were, the density of matter is zero, with a non-trivial cosmological constant correlated with the curvature. This would mean, first, that the matter does not control the curvature, at least not in a reproducible way, as the Einstein's natural philosophy requires, and, secondly, that the cosmological constant is not just an *ad hoc* remedy in the theory, as it was intended in the first place, but has to be there: it is part and parcel of the theory. These incidents motivated the conclusion of a certain *lack of physical meaning* of the metric tensor. Quoting:

> We can also *abandon the postulate of Mach*, and replace it by the postulate that at infinity the $g_{\mu\nu}$ or only the g_{ij} of three-dimensional space, shall be zero, or at least invariant for all transformations. This postulate can also be enounced by saying that it must be possible for the whole universe *to perform arbitrary motions, which can never be detected by any observation.* The three-dimensional world must, in order to be able to perform "motions", *i.e.* in order that *its position can be a variable function of the time*, be thought *movable in an "absolute" space of three or more dimensions* (*not* the time-space x, y, z, ct; *original Italics here, n/a*). The four-dimensional world requires for its "motion" *a four-(or more-) dimensional absolute space*, and moreover an extra-mundane "time" which serves *as independent variable for this motion.* All this shows that the postulate of the invariance of the $g_{\mu\nu}$ at infinity *has no real physical meaning. It is purely mathematical*! [(De Sitter, 1917); *our Italics, except as indicated*]

A harsh conclusion for the Einsteinian natural philosophy, if we keep in mind that this philosophy places the stakes precisely on the fact that the physical meaning of the metric tensor of spacetime is a key point in the very physics of matter. In an Einsteinian context, the Mach's postulate ensures the fact that inertia is caused by the matter located, spatially speaking, *beyond the local experimental accessibilities.* The de Sitter's conclusion shows that this cannot be the case, if it is to judge — still within the framework of the Einsteinian philosophy! — from the point of view of a metric which seems relativistically more appropriate, at least when the boundary conditions enter the stage. And when we say 'relativistically more appropriate', we have in mind a relativistic mentality considering the universe as a *manifold of events*, without including the characteristic of relative *simultaneity of events* in order to describe a state of space *per se*. In other words we have a universe which is unique, whereby the simultaneity of events is somehow 'implicitly' involved, for instance as a differentia within the concept of a phase, like in the propagation of light, or in that of a wave function of the Schrödinger type for describing the fundamental constitutive structure of that universe.

The first reaction of Einstein, after analysis of de Sitter's conclusions, was to notice that those conclusions may, in fact, be somehow artificial, insofar as the metric tensor of the spacetime does not satisfy *in the whole universe* some natural requirements of the theory (Einstein, 1918). Indeed, the theory requires that the matrix of the metric tensor of spacetime should be non-singular, in order to calculate the contravariant metric tensor, routinely necessary, as known, for the construction of the Einstein's field equations. Therefore the determinant of this tensor should be, everywhere and at any time, nonzero. From (1.33), this determinant comes down to

$$g \propto \sin^4(r/R) \cdot \cos^2(r/R) \cdot \sin^2 \theta \tag{1.36}$$

with obvious singularities for $r = 0$, $\theta = 0$ and $r/R = \pi/2$. Therefore, in the points of this spacetime with de Sitter's coordinates having such values, the metric tensor is not an invertible matrix in order to be properly used in constructing the field equations, so that the gravitational equations are not valid. Einstein notices that the first two of those singularity points are removable by a proper choice of the space coordinates; however, the third one persists even after this operation is done, and one cannot see any possibility to remove it within de Sitter's embedding procedure, so that he concludes:

> If the de Sitter solution were valid everywhere, it would show that the introduction of the " λ-term" *does not fulfill the purpose I intended.* Because, in my opinion the general theory of relativity is a satisfying system only if it shows that the physical qualities of space are *completely* (original Italics here, *n/a*) *determined by matter alone.* Therefore, no $g_{\mu\nu}$-field must exist (that is, no space-time continuum is possible) without matter that generates it.
>
> In reality, the de Sitter system (2) [the metric (1.36) here, *n/a*] solves equations (1) (*these are the vacuum Einstein equations having a cosmological term:* $G_{\mu\nu} - \lambda g_{\mu\nu} = 0$, *n/a*) everywhere, except on the surface $r = (\pi/2)R$. There — as in the immediate neighborhood of gravitating mass points — the component g_{44} of the gravitational potential turns to zero. The de Sitter system does not look at all like a world free of matter, but rather like a world whose matter is concentrated entirely on the surface $r = (\pi/2)R$. This could possibly be demonstrated by means of a limiting process from a 3-dimensional to a surfacelike distribution of matter. [(Einstein, 1918); *our Italics, except as indicated*]

As one can notice right away, Einstein got a new problem here, namely that of dealing *apart* with the cosmological constant. This constant was introduced simply, in the spirit of Einstein's initial procedure of constructing the general relativity [see, for this, Einstein (1916, §16): specifically, by extending the field equations to include also the 'classical' Seeliger amendment of the Poisson equation, in order to save the fundamental philosophy, which states that 'matter prevails over geometry'. As it turns out, however, this new constant tends by itself to ruin the philosophy it is called to save, 'from the inside' as it were, rather than saving it. Anyway, his overall conclusion is that, until contrary proven — which, according to Einstein, and not only with him, would mean to prove that the singular (hyper) surface is an illusion due to the particular 'mapping procedure' represented by the coordinates used — the de Sitter universe *is a universe still containing matter.*

That proof came, indeed, but not altogether 'contrary', as it were. Concerned with this issue, Einstein inquired here and there, approaching the German mathematicians of the time (see *The Collected Papers of Albert Einstein*, Volume 8, Princeton University Press; especially the English rendering of the volume, pp. 351–357, and the correspondence indicated there). Among them, Felix Klein — long-ago preoccupied with the non-Euclidean geometry, and especially with its cosmological connection — has given an answer related to the *a priori* choice from equation (1.25), left behind by Willem de Sitter in applying the Einstein's procedure (Einstein, 1916). From an SRT perspective, Klein's answer covers two points of interest, the mathematical details of which we delegate to some extensive specific works (Klein, 1918, 1919). For now, we just need the essence of these points.

First of all, he upholds the idea that equation (1.25) of the manifold chosen to represent the finite scale is, at least occasionally, *a problem of non-Euclidean geometry* in a precise stance: all it needs is to be defined just by an *algebraical* sum of exact squares. This would mean, first and foremost, that the five-dimensional quadratic manifold is not quite so 'abstract' from a physical point of view. He even indicates a 'proof' of this fact in the Schwarzschild solution, as shown in the excerpt above from the letter addressed to Einstein. True, there is a danger here, arising with the idea that this may be thought of as a kind of 'backfiring' argument, insofar as that solution is obtained from Einstein's theory *with no cosmological constant*, while de Sitter's solution is specifically related to the very existence of the cosmological constant. Nevertheless, in the first place, we know today that such a non-Euclidean geometry can be built *a priori*, therefore independently of any physics, to represent mathematically the *condition of confinement* (Mazilu *et al.*, 2019). Moreover, the classical Kepler problem 'endorses', as it were, the idea in a genuinely physical way, involving the matter indeed, in one of its most essential features, involving the concept of interpretation, as we have shown above. Secondly, as a consequence of this non-Euclidean geometry, Klein has the merit of representing the *time sequence* necessary in rendering the spacetime as a manifold of events, *by a continuous functions* on submanifolds of the manifold of events (1.25) representing the finite scale of the world. Again, some elaboration on these issues seems necessary.

Notice that the metrics (1.24) and (1.26) can be obtained from (1.32) and respectively (1.33) by a usual 'mapping', as de Sitter himself shows it: projection of a 'sphere' to a 'plane' (de Sitter, 1917). Consequently, the metric in this dilemma, namely (1.33), is indeed intimately related to such a 'mapping' procedure. From this perspective, Klein took notice that if one chooses the *real coordinates*

$$\begin{aligned} &\xi_1 = R\sin\omega\sin\theta\cos\phi, \quad \xi_2 = R\sin\omega\sin\theta\sin\phi, \\ &\xi_3 = R\sin\omega\cos\theta, \quad \xi_4 = R\cos\omega\cosh(ct/R), \\ &\xi_5 = R\cos\omega\sinh(ct/R) \end{aligned} \tag{1.37}$$

naturally satisfying the quadratic constraint:

$$\xi_1^2 + \xi_2^2 + \xi_3^2 + \xi_4^2 - \xi_5^2 = R^2 \tag{1.38}$$

the very de Sitter metric from equation (1.33) becomes the metric form with constant coefficients

$$-(ds)^2 = (d\xi_1)^2 + (d\xi_2)^2 + (d\xi_3)^2 + (d\xi_4)^2 - (d\xi_5)^2. \tag{1.39}$$

In a word, the finite manifold of 'hyper-events' and the infrafinite one do have the same algebraical structure, but 'quasi-Euclidean', not genuinely Euclidean, as in the seminal case of Einstein. And so the problem of this five-dimensional manifold introduced by Einstein, gets an outstanding genuine solution, based on the very Einstein's theory of general relativity, once we accept that it is quasi-Euclidean with some natural limiting conditions on the metric tensor. However, there is still a catch, but even this turns out to be very fortunate, leading to a generalization of the idea of propagation, possibly even as required by Poincaré's conclusions based on Lorentz's theory of matter. Namely, the *time sequence* in this universe will then be given by a time coordinate to be calculated through formula:

$$t = \frac{R}{2c}\ln\frac{\xi_4 + \xi_5}{\xi_4 - \xi_5}. \tag{1.40}$$

This induces Felix Klein into further noticing:

> It is amusing to picture how two observers living on the quasi-sphere and *equipped with differing de Sitter clocks* would squabble with each other. Each of them would assign finite time ordinates to some of the events that for the other would be lying within infinity

> or that would even show imaginary time values. [*The Collected Papers of Albert Einstein*, Volume 8, Princeton University Press, Document 566; *our Italics*]

Perhaps concerned, nevertheless, with the fact that this construction could be easily construed as just a mathematical tweaking and, as such, it could be placed itself into the category from which it tried to pull the Willem de Sitter's singularity, Klein suggests its physical explanation based on Schwarzschild solution, which turns out to be an *interpretation proper*, as signaled beforehand. Thus, he is right in choosing this interpretation but, in order to prove this, we need to elaborate a little upon the very Schwarzschild's solution in question.

1.7 Schwarzschild Matter: Sphere of Ideal Fluid

The Schwarzschild solution chosen by Klein, is basically not the one nowadays well-known for staying at the foundation of the modern concept of black hole. In this instance, it represents the "gravitational field of a homogeneous sphere of finite radius, consisting of *an incompressible fluid*", as Karl Schwarzschild himself states. An excerpt from his original work could, again, be useful in clarifying not only the necessity of this solution within the framework of natural philosophy, but even the general attitude adopted by physics ever since, and *made possible only by the advent of the general relativity*, with all its problems that specifically motivated the human mind. Here it is:

> As another example concerning Einstein's theory of gravitation, I calculated the gravitational field of *a homogeneous sphere of a finite radius, consisting of incompressible fluid.* The specification "consisting of incompressible fluid", is necessary to be added, due to the fact that *in the framework of the relativistic theory, gravitation depends not only on the quantity of the matter, but also on its energy* and, for instance, *a solid body having a specific state of internal stress would produce a gravitation different from that of a liquid.* [(Schwarzschild, 1916); *our Italics*]

The italicized words of this excerpt contain, essentially, the difference between the new Einsteinian natural philosophy, brought about by general relativity, and the old Newtonian one: *it is the internal state of the matter that determines the external gravitational field.* This is, in fact, quite a positive addition of the Einsteinian natural philosophy, which made possible the subsequent idea of gravitational collapse, usually accompanying the different presentations of the concept of black hole. Moreover, this elucidates why the Schwarzschild solution — which refers to matter as a homogeneous incompressible fluid — can be, indeed, taken as physically justifying the de Sitter's system. Namely, the fluid here plays only the explanatory part in interpretation: the density, necessary for accomplishing Einsteinian natural philosophy in the de Sitter's case, is just mathematically *defined* by a parameter describing the internal state of the fluid, namely the pressure. This system is obviously referring to a universe without matter, once the density of its quantity, reckoned according to Einstein's theory, turns out to be zero, so that a physical structure does not make any sense for it. Consequently, the homogeneous Schwarzschild fluid can be considered here as only *giving an interpretation to a valid solution of Einstein field equations in general*, entirely analogous to the interpretation of its classical epitome, the ether. It represents what can be called the *ether of matter*, by contrast with the generally accepted ether, which is actually the *ether in space.* This is how the ether was classically explained and understood anyway (Earnshaw, 1842), and this is what Henri Poincaré himself felt as necessary in order to complete the theory of Lorentz, as we have shown previously. But there is more to it than the concept of interpretation, as defined starting from wave-mechanical necessities, would necessarily ask!

There is, indeed, the idea of a *reverse interpretation*, as it were, which we kept mentioning thus far. This is first seen by the fact that equations (1.38) and (1.39) are obtained by Klein *from an already existing valid solution of Einstein's cosmological equations*, the de Sitter solution, not by an axiomatic definition like the one used by Einstein. Thereby, the concept of fluid as characterized, for instance, through the property of indestructibility mentioned by Poincaré as instrumental in defining the classical concept, *needs to be interpreted as a continuum*, a medium *for which the pressure parameter makes sense* as a parameter of state. Thus, it is the density of Schwarzschild fluid that needs to be defined as a state variable by the pressure, not the other way around. The quintessential example of the kind is the Lorentz matter, defined directly as a molecular fluid, but missing the quality of continuity of a medium supporting waves. It is our opinion that the cosmological constant answers to this objective need of our knowledge. For instance, the Poincaré's interpretation of the electromagnetic continuum offers a necessary fluid. In fact, the case is exactly reciprocal: the whole Lorentz fluid needs a reverse interpretation by a continuum, which turns out to be the real electromagnetic field, in its capacity of a scaling field. The reality of this approach is, in fact, a matter of experiment, and was actually accomplished by Heinrich Hertz in 1893, who, according to this view, actually proved that the continuum is real. Another notorious case of such an interpretation can still be cited: it is the Schrödinger wave function, on which we shall elaborate further, as we go on with this work.

Coming back to our current subject, we can say that, in short, the Karl Schwarzschild's result referring to this physics is the following. First, the equation of state of the sphere of incompressible fluid of density ρ and pressure p — equation (30) mentioned by Felix Klein in his letter to Einstein — is calculated by Schwarzschild from the Einstein's field equations — without any cosmological constant — as a benefit of introduction of an 'angular' coordinate χ, and given by an 'equation of state', as it were, of the form:

$$\rho_0 + p = \rho_0 \frac{2\cos\chi_a}{3\cos\chi_a - \cos\chi}. \tag{1.41}$$

The field equations are referring to a gravitational field in the interior of a *matter sphere* having a metric of Einstein type (1.32) but with variable coefficient of $(dt)^2$, that we designate here as the 'interior Schwarzschild metric', which is not what is currently designated as such in general relativity, but:

$$(ds)^2 = \left(\frac{3\cos\chi_a - \cos\chi}{2}\right)^2 (dt)^2$$
$$-\frac{3}{k\rho_0}[(d\chi)^2 + \sin^2\chi(d\Omega)^2]. \qquad (1.42)$$

This is that 'equation (35)' mentioned by Felix Klein in the excerpt that started our story. For the gravitational field in the exterior of the sphere, Schwarzschild gets the exterior solution, obviously. This is actually the solution usually taken as the prototype metric in the discussion of the modern concept of the black hole:

$$(ds)^2 = (1 - \alpha/R)(dt)^2 - \frac{(dR)^2}{1 - \alpha/R} - R^2(d\Omega)^2,$$
$$R^3 = r^3 + \rho. \qquad (1.43)$$

Here r — the radial coordinate proper — and α are calculated from the expressions:

$$r^3 = \left(\frac{k\rho_0}{3}\right)^{-3/2}\left[\frac{9}{4}\cos\chi_a\left(\chi - \frac{1}{2}\sin 2\chi\right) - \frac{1}{2}\sin^3\chi\right],$$

$$\alpha = \left(\frac{k\rho_0}{3}\sin^2\chi_a\right)^{-1/2}$$

with the constant χ_a corresponding to the radius r_a of the fluid sphere. The solution (1.43) is formally the same with the one previously obtained by Schwarzschild, in the work on the gravitational field of the *classical material point*. The only difference with respect to that case is that the expression of ρ from equation (1.43): in the case of fluid sphere it is

$$\rho = \left(\frac{k\rho_0}{3}\right)^{-3/2}\left[\frac{3}{2}\sin^3\chi_a - \frac{9}{4}\cos\chi_a\left(\chi_a - \frac{1}{2}\sin 2\chi_a\right)\right]$$

while for the classical material point it is $\rho = \alpha^3$.

Schwarzschild closes his work with a series of remarkable observations. However, what startles most is a conclusion that warrants entirely, from the point of view of a general natural philosophy — making no difference between the Newtonian and Einsteinian points of view — the embedding procedure of Klein, as described above. Namely, if we take $dt = 0$ in the metric (1.42) of the material sphere, we get the metric of the space in which the matter resides, *but described from the physical point of view* — *i.e.* a homogeneous incompressible fluid of pressure p etc. — on which Schwarzschild concludes:

> This is the line-element of the so-called non-Euclidean geometry of a spherical space. The spherical space geometry *holds also in the internal region of our sphere.* The curvature radius of such a spherical space is $\sqrt{(3/k\rho_0)}$. *Our sphere has formed not all of the spherical space, but only a region in it; this is because* χ *cannot grow up to* $\pi/2$, *but grows up only to the boundary limit* χ_a. Concerning the Sun, the curvature radius of the spherical space, which determine the geometry of the interior of the Sun, would be equal to about *500 radii* of the Sun...
>
> It is an interesting result of Einstein's theory that it calls for the *reality within gravitating spheres* of the geometry of *spherical space*, which *hitherto had to be regarded as a mere possibility.* [(Schwarzschild, 1916); *our Italics*]

In the case of transition to the de Sitter's metric, the coordinate χ covers the entire *a priori* range at its disposal, *i.e.* 'grows up to $\pi/2$', and the density is physically decided by the equation of state (1.41). Again, in our opinion, this bestows the status of a tool of interpretation upon this 'Schwarzschild fluid', necessary in accomplishing any interpretation of the matter *per se*, provided *the density of matter thus interpreted is taken as equivalent to a stress state.* According to this principle, on which we shall return in due time here, the de Sitter's continuum can, indeed, be properly interpreted — *i.e.* according to the very concept of interpretation as necessary to wave mechanics — as a Schwarzschild fluid in 'steady pull', as Klein states it in his letter 566 from Einstein collection already cited above. Taken as such, the Schwarzschild's solution can rightfully offer an interpretation even to *the ether in matter*, for instance, and therefore so much the more to *a vacuum*, which, according to todays' physical standards is matter. However, it also has the virtue of a 'reverse interpretation': the fluid, having 'detached constitutive parts', as Earnshaw would say, *is reversely interpreted as a physical continuum* of pressure p characterizing its 'internal state'. All the better, it can offer interpretation to an *electromagnetic ether* in fact, therefore, in general, to a de Sitter world, as Felix Klein suggests. The factual conclusion would then be that the reverse interpretation offers the possibility of constructing an affine reference frame (Mazilu *et al.*, 2019) based on the internal state of the matter, as opposed to its quantity. Before further elaboration on this statement, a few more words on Schwarzschild solution.

The physics called upon by Klein's interpretation of de Sitter's world has still another connotation related to the very Schwarzschild solution allowing that interpretation. Indeed, we have here the first modern reference to the idea of *space with matter* in a certain state, which can be appreciated as a theoretical physical structure, once it has a space extension 'measured' by the planets' orbits. Indeed, that '500 radii of the Sun' represents a spherical space around the Sun, enclosing the internal planets of the solar system up to just about the mid-distance between Mars and Jupiter. This is manifestly an essential example of a physical structure, insofar as it represents matter penetrated by space. Now, if the non-Euclidean geometry is the one that dominates the space in which the solar matter exists, this geometry should be of a special type: it should be *a hyperbolic geometry of the second kind*, in a modern nomenclature apparently due to Klein himself, *i.e.* a Lorentz geometry in the jargon of the modern theoretical physics, but for a *one-sheet hyperboloid* [see de Sitter (1916), the *Third Paper*, footnote on page 10; for the modern connotation see Duval and Guieu (2000)]. From the point of view of the absolute geometry, however, this is not to say that it should be any different from the spherical geometry mentioned by Karl Schwarzschild and used by Felix Klein in his calculations, for they are formally identical [see Pierpont (1928) for the analytic geometry of the ruled quadrics].

We cannot close this issue of interpretation without marking the Einstein's closing position in the debate around de Sitter's findings. This position is usually qualified as ambiguous, to say the list: it is claimed that Einstein never publicly acknowledged the fact that he was not right, nor he published his corrected position in the critique of de Sitter's work. Quoting from the letter of Einstein, written as response to Klein's observations from Document 566:

> De Sitter's world is, *in and of itself*, free of singularities and its spacetime points are all equivalent. *A singularity comes about only through the substitution providing the transition to the static form of the line element.* This substitution changes the *analysis–situs (emphasis in the original, a/n)* relations. Two hypersurfaces
>
> $$t = t_1 \quad \text{and} \quad t = t_2$$
>
> intersect each other in the original representation, *whereas they do not intersect in the static one.* This is related to the fact that, *for the physical interpretation, masses are necessary in the static conception, but not in the former one.* My critical remark about de Sitter's solution needs correction; *a singularity-free solution for the gravitation equations without matter does in fact exist.* However, *under no condition could this world come into consideration as a physical possibility.* For in this world, *time t cannot be defined in such a way that the three- dimensional slices t = const. do not intersect* one another and so that these slices are equal to one another (metrically). [*The Collected Papers of Albert Einstein*, Volume 8, Princeton University Press, Document 567; *our Italics, except as mentioned*]

According to this brief account, Klein has shown, in fact, that the interpretation touches the de Sitter's conception: 'the former one', in the excerpt above. The 'static world' may not even need masses for interpretation, for it is actually referring to the very masses as simple quantities of matter [see equation (1.14)]. Fact is that Einstein never abandoned his position in this argument, which he may even have considered of no consequence for the Einsteinian natural philosophy.

Whatever was essential has already been published: *we have to do here with an interpretation*, and as such the issue must be turned to an entirely another forum of discussion. Fact is that the *cosmological definition of matter needs a static stance*, and that stance has to be geometrically a quadratic manifold at any scale: infrafinite as well as finite or transfinite.

Chapter 2

Recovering Some Forgotten Roots

The idea of hyperbolic geometry of the second kind, mentioned above, brings about another important moment of human knowledge centered around the personality of Albert Einstein, which further clarifies the position and the profound implications of his results from *Cosmological Considerations.* In hindsight, this even adds to the elucidation of position and implications of the very general relativity theory with respect to classical physics, from the point of view of a general natural philosophy, as we already signaled before. Due to its tremendous importance for our knowledge, we feel like having to dedicate to this moment a substantial portion of this presentation, involving this and the next chapters, for some fairly detailed explanations and proper conclusions. On its historical side, this moment can be safely related to the name of Kurt Gödel, and was, in fact, conceived on the basis of classical natural philosophy (Gödel, 1949, 1952). It had — and actually still has! — a broad impact on the human intellect at large, reflected in a vast literature that covers the many aspects of the problem, and ranges from highly specialized theoretical works down to science fiction subjects. In our opinion though, this moment has a basic natural philosophical significance for human knowledge, originating in the fact mentioned in passing by us before, namely that the space needs to be defined *as a state* by a *manifold of positions.* This will be succinctly explained in what follows, using that concept of reverse interpretation briefly touched by us here, just a few pages before. On this note, it should appear as no surprise that it triggered a fundamental revision of the very classical Newtonian mechanics and cosmology alike (Heckmann and Schücking, 1955, 1956). This is actually, by and large, the major idea we are following in our presentation.

Concerning the concept of interpretation *per se*, this moment could also have a fundamental significance on many other levels. To wit, start for instance by noticing a quite suggestive fact, explaining our approach in this presentation: at the time when he began being involved in the subject matter of general relativity, Gödel was, socially speaking, already a notable logician, known mainly for his *incompleteness theorem*, and its impact into the *theory of ensembles* [see the *Collected Works*, as indicated under (Gödel, 1949, 1952)]. Basically, this theorem states that there are truths whose decision is out of the language expression in any axiomatic system. This is a fact which — being almost explicit in a mathematical theory — is only implicit in a physical theory as, hopefully, our very presentation here will eventually make it clear. However, more important than this recognizance of a human spiritual 'weakness', as it were, the incompleteness theorem places the human spirit in an appropriate critical position, in order to habilitate it for looking upon any axiomatic system it creates from *the outside*, as it were. Such an attitude almost necessarily implies either an interpretation *per se* — whence, in our opinion, the occurrence of the problem of 'completeness' in connection with the theory of ensembles — or a reverse interpretation, as the case may happen to be, and this is actually what happened in 1949, on the occasion of 70th birthday of Albert Einstein, when Gödel was invited for a contribution (Einstein, 1949). On that festive occasion Gödel has shown that, seen from outside, the general relativistic system *misses the concept of space as an ensemble properly defined according to a specific concept of completeness.* The idea of interpretation suggested by Klein's observations described by us before, was apparently not taken face value, and therefore was not properly developed into a 'full-fledged' theory. His contribution to the *Einstein Festschrift* (Gödel, 1949c) leaves no doubt about this issue [see, however, Malament (1995) for an adequate expression of the fact in all its aspects, philosophical and technical]. Specifically, the Gödel moment revealed that, as it was at that time, *the theory of relativity*

(a) was not *properly concerned* with the second of the differentiae of the time concept — that of *time as a sequence*, or the 'phenomenal time' as Weyl once said — and what Einstein actually showed in his 1917 *Cosmological Considerations* was how a space can be defined as an ensemble by the *condition of simultaneity* [see Rindler (2009) for further details];
(b) should consider the imbedding procedure as universal. On this occasion Gödel adds the hyperbolic geometry of the second kind to the possible field of quadratic metric space–time manifolds [see Gödel (1949a); also Malament (1995), for a modern connotation, and Ozsváth and Schücking (2001) for a 'gauge expression' we shall also use in the present work];
(c) like all the other cosmological theories, including the modern Newtonian one (McCrea and Milne, 1934), did not consider properly its starting point, *i.e.* the very Newtonian theory.

In a word, the general relativity was *incomplete*, and based on the fact that, like any human system of knowledge it is an axiomatic system in need of interpretation, we dare to attach this incompleteness to the general logical concept, connected to the name of Kurt Gödel. While the point (a) cannot be elucidated but only after a 'full-fledged SRT' is developed, and (b) turns out to be a purely technical problem to be undertaken properly in the due course of our present work, we settle in this introduction into subject only for the point (c) of this list, as being, physically speaking, of essence. Two exquisite works have been chosen to assist us on this short route, both of which have explicitly the purpose to construct, or at least describe, rotational cosmologies the Einsteinian way: *i.e.* starting from Poisson equation [see (Ozsváth & Schücking, 2001); (Rindler, 2009)]. An intuitive imagery may also be helpful here, and we chose another two just as exquisite works, possibly very helpful in suggestions (see Németi, Madarász, Andréka, and Andai, 2008; Buser, Kajari, and Schleich, 2013).

2.1 Introducing the Subject: Time, Space and the Idea of Compass

There can be no question about the fact that Einstein used the Newtonian point of view as the inception point of his theory of general relativity (Einstein, 1916), and also in correcting this very theory in order to be properly applied to cosmology (Einstein, 1917); Gödel also used this point of view in constructing his cosmology [see especially, Gödel (1949a)]. However, rarely, if ever, is it noticed that both Einstein and Gödel — and anyone yet occupied with general relativity for that matter — did not take the Newtonian point of view *in Newton's stand*, but in Poisson's, *i.e.* via the known second-order partial differential equation involving the density of matter (Poisson, 1813). Distinctively though, Gödel's approach to cosmology, while plainly using the idea of a metric universe as, in fact, Einstein did, and even takes the very same 'Poisson stand', as it were, has an almost explicit reference to a 'Newton stand', by insisting in the construction of a cosmology for a universe with rotating matter, closer than ever to a Keplerian image of the planetary motion.

It is on this issue that we want to concentrate here. In order to better understand it, one needs to understand the difference between what we call 'Newton stand' and the 'Newtonian point of view' in cosmology. We recommend, as a first step, a small article of Don Lemons referring to a cosmology that 'Newton himself would understand' (Lemons, 1988). In fact, the Newtonian cosmology *per se*, as understood nowadays, was created under the incentive of the very Einsteinian cosmology (McCrea and Milne, 1934), and from the very same 'Poisson stand'. We really doubt though, that Newton would have been able to understand Don Lemons' proposal, which is made from the very same position: the cosmological term, necessary in order to put things in order with the idea of infinity of the universe, is due to a mean density, while the gravitation *per se* should be connected to the fluctuations of density. But this is besides the point here. Fact is that Lemons' work seems to us the clearest, and especially the shortest presentation of the issues of the modern cosmology. For further details on modern cosmology we advise the study of detailed works Heckmann (1942), and also Harrison (2000).

In fact, speaking of the 'Newton stand', we take the one of Lemons' (and, in fact, of many others!) face value, and will go a little deeper in history, in order to bring again, hopefully to its proper light now, a forgotten issue. This concerns the very concept one usually claims to have been removed by the general relativity from classical natural philosophy: *the force.* Indeed, as well known, and quite frequently and incautiously uttered in fact, mostly in the literature of popularization, the general relativity 'replaced the force with the curvature'. In so doing it started — as we regularly expressed it here — from the byproduct of the Newtonian mechanics, involving the idea of continuum in a tractable mathematical way, *viz.* Poisson equation. This equation describes the forces *inside matter*, where the mass density is taken as a *continuous* function of position, and therefore the matter is considered as continuously distributed in space. Needless to say, the Poisson equation has not much to do with inertia, so that in order to construct a dynamics, the general relativity had to appeal to external geometrical principles. And, as it turns out, these geometrical principles are much simpler than those based on which the concept of force had been invented in the first place. One of the side issues of this simplicity is, indeed, that apparent elimination of the force from a relativistic scenario.

Specifically, according to Newtonian dynamics, in the absence of forces the trajectory of a moving body is a straight line. Since in the classical space which contains the bodies — *i.e.* the Euclidean space — the straight line is a geodesic, one can easily figure out a natural generalization. Namely, if one assumes that the existence of forces is equivalent to a change in the nature of space, then the motion of a particle in a field of forces should be described by a geodesic in the space equivalent to the Euclidean one, but defined by the existence of those forces. This is what *almost* happened in historical reality, for there was a little twist on proceedings, insofar as nobody could tell precisely what is that nature of a space equivalent to the Euclidean one, in cases where forces are present, and how can it be discovered. Instead, however, everybody could tell that, considering *the forces as consequences*, with the potential as a primary concept, there would be no need of changing the geometry of space, because from the knowledge of potential one would be able to operationally supply the forces *within their very classical environment*, the Euclidean space. Provided, of course, a special description of the matter in that classical environment is granted, via the *boundary conditions* naturally required by the equation which defines the potential. A prerequisite which, again, seemed to be in line with the classical natural philosophy. And thus, the emphasis has been shifted on potential, all the more so as it did not demand any formal change in description of the arena of presence of the matter: that arena remained the very same classical Euclidean space. Such was, for instance the whole attitude of Henri Poincaré, as shown earlier, in the Chapter 1 of this work.

However, even the concept of potential did demand something gnoseologically significant: *an addition to the concept of time, that arose mainly with the electrodynamic phenomenology* in Maxwell's take. As we repeatedly stated here, the time concept has *two differentiae.* One of them is the *continuity*, which was almost exclusively used in the classical dynamics and, in fact, was used as such even by Einstein himself in creating the mathematical basis of general relativity; the other one is the sequencing — the 'phenomenal' aspect of time, in Weyl's phrasing — which, while classically was still related to continuity, became a critical feature of the time concept only with the advent of electrodynamical phenomenology. The fundamental expression of this version of the concept of time was the physical idea that *the simultaneity* of events is strongly dependent on the place in space occupied by matter, and this led to the theory of special relativity (Einstein, 1905). Starting from this, with the supplementary identification of the *length* — which is a material characteristic — with a *distance* — which is a spatial characteristic — allowed Einstein to draw the conclusion that the space–time, in general, is a non-Euclidean continuum, whose geometrical description *can be based on a metric.* Therefore, he further concluded that the space–time should be a Riemannian continuum (Einstein, 2004). The essential condition of the possibility of such a conclusion — the identification of the length with a distance — was totally forgotten, and, in fact even today it does not even count for our knowledge: *everyone speaks carelessly of the distance instead of length or vice versa, as of a natural fact.*

The conclusion of a Riemannian space–time continuum gained significant reinforcement along the classical idea just mentioned by us before: the space containing forces should not be a Euclidean space. Only, now, the case was not about the space, but about the space–time, and one was not talking directly of force, but of the potential. Fact is that, if we assume that in the explanation of a universe the motion is essential, and therefore the matter is actually characterized by *ensembles of events* rather than by *ensembles of positions*, the manifold of residence of the matter should be actually a space–time, not a space. And the geodesics of this space–time can inherently incorporate the classical potential via the concept of curvature. One more step, incarnated, as it were, in the identification of the *gravitational acceleration* with an *inertial acceleration*, was then necessary to Albert Einstein, in order to construct the general relativity as a theory of gravitation. And thus we have today a theory of general relativity, whereby the metric tensor of the space–time metric is to be determined by the distribution of the matter in space–time. Only later on, in cosmological applications, the price to be necessarily paid for this idea started showing up, and with the works of Gödel it became, in fact, even *critically apparent*, in spite of the optimistic observations of Hermann Weyl, due to an 'incompleteness' issue: in cosmology *the compass of inertia may point differently from the compass of gravitation*, a fact which the general relativity ignores at the axiomatic level. One might say that, even though the inertial and gravitational mass may be identical as values, they are not *physically* identical!

An orthodox physicist can surely raise against this last statement the argument that, from a quantitative point of view, the two kinds of mass are only *locally* revealed by measurement, no matter what we think is their origin. This means that, even if the Mach's principle is to be taken in the sense that the *inertial mass* is controlled by the whole universe, the quantitative identity surely can still be read out as a physical identity, as Einstein did indeed, for there is no other possibility of measuring than locally. This is why the Gödel's model universe was considered first and foremost as directed against the Mach's principle (Ozsváth and Schücking, 1969). However, we think that it is not necessary to go as far as inventing relativistic models in order to prove that the Mach's principle should be taken *cum grano salis*, so to speak. First of all, what is today known as Mach's principle, is actually the Einstein's creation, making the grounds of theory of general relativity. And, secondly, the things were never different, even in the very classical natural philosophy of Newton himself: this is why the classical mechanics itself needed a 'Mach principle', in the first place. It was, again, only the concept of potential and the classical idea of body without space extension, that led to the original suggestion of Ernst Mach, which Einstein transformed into a principle, and followed it, we should say, faithfully. But in the beginning, when the concept of potential was missing, and that of force was reigning free over the natural philosophy, the things were entirely different. In fact, this is how we perceive the situation here and now, with the benefit of a century of modern theoretical physics' achievements. For, at the moment when the idea was really presented as such to our spirit, people were not even able to think this way.

2.2 Necessity of a Cosmology "Newton Himself would Understand"

The expression of Don Lemons, referring to Newton's understanding (Lemons, 1988), should not be taken as just a parable intended to educate us. Such a cosmology is utterly necessary indeed, and we are set here to show the reasons. For once, the overall reason is that only with reference to forces would anyone be able to properly understand the difference between the *compass of gravity* and the *compass of inertia*. But then, with the addition of Gödel's ideas, one would be able to understand even more: between the two compasses, it is always necessary a specific *compass of light*, in order to settle the difference between their indications according to a certain space scale.

Start by noticing what were the Newton's reasons for refusing to the fixed stars the role that Ernst Mach assigned to them later.

For, it is hard to imagine that, inventing the force as he did, Newton was not aware of the fact that their action has to be universal: in fact, he acknowledged that, himself, repeatedly. In order to make the point, a significant excerpt would do the best job:

> COR. I. The fixed stars are immovable, seeing they keep the same position to the aphelions and nodes of the planets.
>
> COR. II. And since these stars are liable to no sensible parallax from the annual motion of the earth, they can have no force, *because of their immense distance*, to produce any sensible effect in our system. *Not to mention that the fixed stars, everywhere promiscuously dispersed in the heavens, by their contrary attractions destroy their mutual actions*, by Prop. LXX, Book I. [(Newton, 1974), Volume II, Book III: The System of the World, Prop XIV, Theorem XIV; *our Italics*]

This clearly shows that Newton would not accept that the fixed stars act in the classical universe, which was clearly the Copernican universe (Harrison, 2000). A little digression is, nevertheless, in order here, regarding the main reasons of this denial: it is a purely quantitative reason, formulated in the exact spirit of Newton's original definition of the forces, *i.e. by measurement* (*Principia*, Book I, Proposition VII, Corollary 3).

Notice the first move of Newton regarding the local action of fixed stars: it can be seen in that "because of their immense distance". This seems, indeed, just normal for his natural philosophy: the force of universal influence among celestial bodies *fades away* with an inverse power of the distance between them. But then he might have realized that the fixed stars, being distributed randomly not only on the surface of the accepted celestial sphere, but also 'in depth', so to speak, this quantitative aspect needs to be somehow amended. So he added the possibility that the fixed stars are dispersed in the sky "everywhere promiscuously" and that, even if they could raise influence, either upon each other or upon bodies of the universe, "their contrary attractions destroy their mutual actions". Thereby — the way we understand the notion today — as 'fixed', *the stars should form a static system.* These two aspects of the classical cosmology — fading of the forces with the distance, and the 'promiscuity' of distribution of the 'fixed stars' no matter of distance — had to be dealt with explicitly along the time, from apparently two different angles, aroused by the general relativistic point of view.

First comes the idea that the inertia might not fade away as steeply as the gravitation, *i.e.* it may be a comparatively 'long distance' force [see Sciama (1953); see also Sciama, (1969, Chapter 3)]. As Dennis Sciama presents it, this theory of inertial forces is inspired by the only *perceived action* at distance realized via a field: the field of a charge in motion, *i.e.* the electromagnetic field. Basically, Sciama uses the known modern argument that, in the field of a charge in motion, there are two components to be differentiated from the point of view of 'fading': one fading with the square of the distance, the other fading with the distance itself (Jackson, 1998), Chapter 14, Eqs. (14.13–14.14). However, we find it even more significant to introduce this idea from some alternative points of view, more complete we should say. First, it can be presented in connection with an older point of view, settled originally along the same lines [(Langevin, 1905); see also Poincaré (1906)], but introduced by the *language of waves.* We extracted here the final remarks of Paul Langevin from the 1905 work just cited, insofar as they are particularly significant with respect to the subject matters of force and field, in spite of the fact that they are concerned with the electromagnetic field. In fact, with a proper conjunction, they can be taken as an indicating that a genuine field — a Yang-Mills field, as it were — must have the properties of the electromagnetic field. Quoting, therefore:

> — The points upon which I think useful to insist, are the following:
>
> 1° The electromagnetic perturbation produced in the medium by an electrified particle in motion is composed of two parts that propagate with the speed of light, starting from the emission center.
>
> The first part, or the *velocity wave*, which exists only in the case of rectilinear and uniform motion, depends only on the velocity of mobile; it contributes in developing around this one a wake whose energy varies with the velocity, which therefore contains the kinetic energy related to the electrified center, and which accompanies this one in its displacement, modifying itself if the motion is accelerated.

2° This modification is produced through the intermediary of second part of the perturbation, the *acceleration wave*, having at *any distance* from the emission point the properties of transversality and equality of the electric and magnetic energies, which correspond to the free radiation.

This *acceleration wave* transports at a great distance, where the velocity wave becomes negligible, a finite energy proportional to the square of acceleration and increasing indefinitely with the velocity when this one approaches that of light. The polarization properties of this wave are particularly simple when the velocity is small.

The velocity wave does not transport any energy at great distances; the *energy of the corresponding wake* only follows the center in its displacement.

3° The relative energy of the two waves of velocity and of acceleration, *i.e.* the *energy of change*, represents precisely the provision of energy necessary to reorganize the wake and make it correspond to the new velocity. In all the cases accessible to experience it is enormous in comparison to the *radiated energy* representing the intrinsic energy of the acceleration wave, *i.e.* the defect necessary to reorganize the wake.

4° The energy of change, delivered by the external field producing the acceleration is, in its expression, akin to the work of the external forces, the only energy exchange that the usual equations of Mechanics bring on stage.

The *radiated energy*, which the external field must equally provide, and which is of a different form, is not contained in the laws of dynamics and this would enact their modification, if the smallness of the radiated energy with respect to the change energy does not make this correction insignificant in all the experimental cases.

The preceding considerations seem to cast some light on the intimate mechanism of the phenomena of inertia and radiation. [(Langevin, 1905); *our translation, original Italics*]

The last statement in this excerpt, mentions only *en passant* the problem of inertia due to electromagnetic radiation *per se*. It was not quite so clear, at that moment in history, to what extent the theory of electromagnetic radiation should be taken as a standard in the analogy of action at a distance. This is why the work of Langevin is dedicated only to the 'inertia of electromagnetic nature'. And again, as it is not clear at this very moment of our elaboration how specifically the waves of Langevin may enter in the economy of the concept of inertia, we keep the episode momentarily in reserve, in order to come back to it later. For, we want to present the Sciama's idea from yet another perspective, having a closer relationship with the cosmological theories.

2.3 A Thermodynamical Argument: Classical Ideal Gas as a Model

Indeed, Sciama's idea of distance behavior of the inertia, can also be presented in connection with the very *Cosmological Considerations* of Einstein. There is a phrase we have already extracted before from this work of Einstein's, to wit: "If we apply *Boltzmann's law of distribution* for gas molecules *to the stars*, by *comparing the stellar system with a gas in thermal equilibrium*, we find that the Newtonian stellar system cannot exist at all". This approach of the matter in universe is typical for the classical cosmology, be it relativistic or of any other guise: in theoretical physics approach, the stars' system is taken as *analogous to an ideal gas*. The point of view has been extended as such, to include images having the galaxies and metagalaxies as analogs of classical molecules. Obviously, such heavenly appearances can hardly qualify as classical material points, if it is to judge by their direct observations, but they can be used as material points in a physical theory at least *at a certain space scale* (Popescu, 1982, 1988).

However, this issue apart for a while, we need to notice for now that Einstein himself seems to have been always concerned with another, correlated question, which appears to us as quite reasonable: to what extent the very *classical ideal gas is entitled to be a standard concept in an analogy* like the one necessary in developing a cosmological theory? He was astutely aware of the importance of a solution to this problem, and from time to time he has undertaken himself tasks toward its solution. The work from 1905 that set the ground for the modern concept of photon [see the modern English rendition (Einstein, 1965)] is the most notorious among his works dedicated to such a task. It can be considered, in fact, as a perceptive critique of the very classical concept of ideal gas, in order to make it applicable to the problem of light. Let us, therefore, see what differentia the classical concept of ideal gas has critically missed, and how was this problem dealt with in such a critical occurrence, both in the quantum theory, as well as in the classical thermodynamics. Then we shall come back to cosmology with an improved model of ideal gas, in order to settle the problem of space and time scale from yet another perspective, unitary from the point of view of the physical theory.

In the case of ideal gases, our physical experience was long time ago incorporated into the well-known equation of state of the gas:

$$pV = \nu RT. \tag{2.1}$$

Here p denotes the pressure; V is the volume of the enclosure containing the gas; ν is the number of gas mols; T is the absolute temperature; and R is the characteristic constant of these gases. In order to understand the symbols in this equations, we need to explain them in terms borrowed from geometry and mechanics. It seems quite obvious that the volume V does not need an explanation, and momentarily we leave this issue aside for a later tackling, going for the explanations of the pressure and temperature, which happen to be purely mechanical. Start, therefore, with the Newton's second law, involving inertia in the description of motion:

$$\mathbf{f} = m\ddot{\mathbf{r}}. \tag{2.2}$$

Here $\mathbf{f}$ is the force acting upon a molecule of *inertial* mass m, considered as a classical material point, and $\mathbf{r}$ is its position vector, in terms of a Euclidean reference frame, arbitrary with respect to its origin on the ensemble representing the gas, and also concerning its orientation.

Now, the forces acting upon an arbitrary molecule of the ensemble representing the gas are, phenomenologically speaking, of three types: forces of *direct collision between molecules*, forces of *collision between molecules and the walls* of the enclosure, and forces of *action at distance between molecules.* It is worth noticing again that, according to Georgescu-Roegen's quantitative hierarchy of scales, as applied to the time scales of the action of forces, we have here two types of forces acting in the *infrafinite range* of time, and one type of force acting in the *transfinite range*. This last force is therefore a permanent force. It turns out that we can deal with these forces in a mathematical way, with significant results. To wit, from equation (2.2) we can get the following relation, *valid on suitably chosen time intervals*, therefore considering the action of forces on some suitable *finite* ranges of time (Jeans, 1954)

$$\sum(1/2)mv^2 = -(1/2)\sum \mathbf{r}\cdot\mathbf{f}. \tag{2.3}$$

The right-hand side of this equation is called by Rudolf Clausius the *virial of forces* (Clausius, 1870). This is why equation (2.3) is usually called the *virial theorem*. The sum in the right-hand side of this equation is performed over the whole ensemble of molecules surrounding one of them, the one considered momentarily as the origin of the reference frame used to perform the calculations. So, the virial theorem shows that the mean kinetic energy of the gas molecules equals the virial of forces acting upon them. This last quantity is easier to calculate in order to describe some reasonable experimental conditions, using just as reasonable theoretical motives. Let us therefore calculate it.

From the perspective of virial theorem, the two parts of virial involving the collisions, represent transitions between two time scales: the infrafinite time scale of the collisions, and a finite time scale of averaging, which is characteristic to the states of the gas within enclosure. If we are to extend our considerations over to cosmology, this would mean a rational extension of the time averaging to one further time scale, the transfinite scale. The transition needs to be 'rational', even as smooth as it gets, because the *averaging procedure* itself — which is, naturally, only defined on finite time intervals — is highly theoretical to any finite scale. Therefore, the process of transition must be rationally done, in order to insure its correctness, which is obviously not the case in cosmology up to this point in history, except for isolated cases. Now, again, following the historical routine, the part of the virial due to collisions between molecules, can be eliminated from the point of view of the forces, by reasons of molecular chaos: the mean virial of all such forces is zero. Notice that this elimination of the forces acting in the *infrafinite range of time*, is a consequence of that condition of 'promiscuity' used by Newton himself in order to eliminate permanent forces of stars, *i.e.* those forces acting in the *transfinite range of time.*

On the other hand, the part of virial due to the collisions with the walls of enclosure can be estimated over an appropriate *finite* time scale, as an area rate of forces representing the pressure. This is, indeed, quite a practical explanation of the pressure: the elementary force exerted by the flux of molecules acting *simultaneously* on the wall over an elementary area dA. It is given by

$$-p(\hat{\mathbf{a}}\,dA),$$

where $\hat{\mathbf{a}}$ is the internal unit vector orienting the elementary area. Assuming that we are allowed to replace the sum in (2.3) by an integral over the whole area of the enclosure, we have:

$$\left(\sum \mathbf{r}\cdot\mathbf{f}\right)_{\text{wall}} = -\oiint p(dA)(\hat{\mathbf{a}}\cdot\mathbf{r}).$$

This integral can be estimated by Gauss theorem — further assuming appropriate conditions, of course — and thus estimated amounts to $(-3pV)$ so that the contribution of the walls to virial is

$$-(3/2)pV. \tag{2.4}$$

Let us estimate now the part of the virial of forces due to the distant interaction among the molecules. Assume that the distant action is due to permanent forces that are central, having magnitude that depends on the coordinates of the point of their application only through the distance between this point and the acting molecules. Incidentally, such forces are necessarily conservative (Burns, 1966). Any pair of molecules should have a virial contribution given by

$$r\,f(r), \tag{2.5}$$

where $f(r)$ is the magnitude of force acting between them. Collecting (2.4) and (2.5) in (2.3), we get

$$pV = (1/3)\sum mv^2 + (1/3)\sum\sum r\,f(r), \tag{2.6}$$

where the double sum extends over all the different pairs of gas molecules.

This last result occasioned to James Clerk Maxwell an interesting speculation, which eliminated totally, and apparently forever, the action at distance between molecules of ideal gases (see Maxwell, 1965, Volume II, p. 422). Namely, equation (2.6) shows that the pressure is given both by the kinetic energy of the individual molecules and by their permanent distant interaction. At a given temperature one can compare (2.6) with the experimental relation (2.1), in order to verify it 'experimentally' so to speak. If the pressure is given by interaction between molecules, these should act repulsively. Newton gave already such an explanation in his *Principia* (see Newton, 1974, pp. 290–302). Unfortunately, it seems that his conclusion cannot be rationally supported: if at a given temperature, the product of pressure into volume is constant, as equation (2.1) requires, the double sum in (2.6) must be constant. This condition cannot be satisfied *but by forces with magnitude inversely proportional with the distance between molecules.* These are those long distance forces which, according to Dennis Sciama, should be forces of inertia or, if electromagnetic, they should be due to acceleration waves according to Langevin's phrasing. But then, they would cause observable motions among molecules which would not be rectilinear, contrary to the assumptions of the classical ideal gas model. So, there are a few spots here, where our speculative knowledge has to be 'patched', so to speak. The first among these is that there is the problem of action at distance: is it accomplished 'wave-wise', in the manner described by Langevin for the electromagnetic waves? If so, then how a classical material point moving along a straight line with constant velocity — as the classical model of ideal gas requires — can create 'waves', which obviously have the makeup of a continuum?! We will come later on, with some other similar occasions, to these topics, but for now let us turn to Maxwell's decision.

The reason of this decision can be basically summarized to following: as here we have to do with long distance forces, these forces would allow to a material point to act preponderantly upon the more distant material points, than on the close points. Such an action will be able to create *observable disturbances* in the gas volume, which thus will be *eternally far from any mechanical equilibrium.* This seems, however, not to be the real case in a thermal equilibrium defining the absolute temperature, so that the long distance forces, as well as any other forces in fact, have to be eliminated from among the molecules of an ideal gas. Thus the only contribution to virial is that provided by the kinetic energy:

$$pV = (1/3) \sum mv^2. \tag{2.7}$$

Starting from this moment, the things went on as well known. From (2.1) and (2.7), we get an estimate for the mean kinetic energy of a molecule

$$\overline{\varepsilon}_c = (3/2)kT \tag{2.8}$$

where k is here the Boltzmann's constant. Thus we have *an explanation* of the temperature, concomitantly with an explanation of pressure. It is thereby explained *as a statistic* related to the ensemble of rectilinear motions of the gas molecules. To wit, *the temperature is the statistic representing the mean kinetic energy over the ensemble of molecules of gas.*

We need to insist a bit longer upon this 'Maxwell moment', as it were, with some concluding remarks, for it is quite significant, inasmuch as, in our opinion, it contains some guiding demands for the whole knowledge, not only for physics. The general observation is that Maxwell performed a 'clean-up' among the forces between molecules supporting the classical model of an ideal gas: it only accepts forces at an infrafinite time scale, *viz.* forces of impact in collisions. True, the consequences of these forces are eliminated by the idea of molecular chaos, just like the chaos envisioned by Newton for the forces at transfinite time scale, but the fact is that they *were indeed considered.* The forces at transfinite time scale — the permanent forces — are eliminated, unlike those from the Newtonian case, only by 'decree' so to speak, based on the law of ideal gases. The forces with action at finite time scale were not even considered in the explanations of either pressure or temperature: they implicitly appear in the time average called 'virial', therefore via a potential.

In the realm of cosmology, the idea was thereby encouraged that we have to do with only *one universe*, and the forces at transfinite scale act only on bodies at that scale: *there cannot be a scale relativity of forces*! Further on, there are no waves in *the reality* of an ideal gas, for that would mean a nonequilibrium of some sort, involving the forces. However, this does not mean that there are no waves in the model: if we are to apply it at the transfinite scales, as Einstein noticed in his *Cosmological Considerations*, the model has to contain, somehow, waves, according to Lorentz and Poincaré: the gravitation, for example, should act like light in its electromagnetic stance, according to Poincaré. So much the more, for the concept of interpretation, the ideal gas has to contain waves. But how? We shall return soon to this important issue, raised by the idea of continuum, and answered by the Gödel's model of universe. Meanwhile, let us see where the theory, as it was left by Clerk Maxwell, leads. In fact, we all know from history where it headed, we just need to highlight some of the most important critical points. And these are all related to the long distance forces from the virial of Rudolf Clausius!

2.4 Sufficiency, as a Physical Concept

It so happened that in applying the concept of temperature defined by equation (2.8) for the ideal gas model, a few limitations have been discovered at the beginning of the last century, which proved to be of essence. One of these is referring, for instance, to the well-known application of the absolute temperature to the explanation of the behavior of specific heats at very low temperatures, close to absolute zero on the Kelvin scale. On this occasion, Arthur Holly Compton made a few observations that may be construed — and we actually do it here — as indicating that equation (2.7) — and therefore (2.8) — must, in fact, also contain a constant nonvanishing term (Compton, 1915). This would mean that in equation (2.6) the second term — the term depending on permanent forces — should have a contribution, and therefore the long range forces of Langevin and Sciama type should be present even in the very ideal gas model! Let us elaborate on the essentials of this issue.

Start with the observation that, currently, one usually considers that in order to explain the behavior of thermal capacities of solids at low temperatures we need the concept of *zero point energy*, heralded even within the formalism of old quantum mechanics (Einstein and Stern, 1913). Worth mentioning at this juncture, is the fact that by the middle of the last century, the zero point energy came to be physically explained by the property of the electromagnetic radiation of having a part of its spectrum *independent of temperature* (Marshall, 1963, 1965). This is a thermodynamic property and, fact extremely significant, it seems to be exclusively specific to electromagnetic radiation — a continuum — and cannot be found to the ideal gas, the interpretative ensemble that inspired and, in fact, generated the rational thermodynamics (Boyer, 1969, 1975). In our opinion, based on these works of Timothy Boyer, the message of Compton's work just cited above is that the electromagnetic continuum and the ideal gas should be tied up together somehow, in a kind of interpretation that Poincaré once searched for. Let us, therefore, closely elaborate on this opinion of ours, so much the more as this elaboration takes us along a path where some important concepts are occurring quite naturally!

The essence of Arthur Compton's work can be described quite simply in our view: it ascribes to a classical molecular ensemble the character of the ensemble describing the thermal equilibrium radiation. To be more precise, such a molecular system is being described in terms of a *probabilistic ensemble having an exponential probability density with the variance function depending quadratically on the ensemble mean.* Indeed, in the light of Poincaré's attempt to an interpretation of the electromagnetic field, Planck's quantization gains an outstanding meaning: it shows that the electromagnetic field cannot be interpreted but in terms of a density which must be taken, at some point along the theoretical streak of reasons, *as a probability density.* This probability density belongs to a class of densities characteristic to physics — the class of so-called *exponentials* (see Lavenda, 1990, 1992) — but is quite particular in fact, insofar as it has a quadratic variance, when this statistic is considered as a function of another statistic: the ensemble mean. This kind of exponential probabilities came to be systematically recognized in theoretical statistics only late in the last century [see Morris, 1982; for the kind of statistical theory involved in the Planck's quantization see Mazilu (2010); also Mazilu and Agop (2012)].

From this point of view, one can say that Compton's work even added something new to our knowledge. That something is the necessity of properly understanding the *idea of sufficiency* in building a statistic, even to the point where this concept can be appropriately defined in words. Indeed, Compton uses the concept — largely vehiculated in the theoretical physics of the epoch — of *agglomeration of molecules* in order to rewrite the thermodynamics of solids into a form to be used for the calculations, then quite impercipient, of the specific heats at low temperatures. This fact brings on stage the *temperature fluctuations* forgotten, and even banished we should say, by the classical definition of thermodynamical equilibrium ensemble in Maxwell's acceptance, as described above. However, for a proper edification on this issue, let us follow the very walk of reasoning of Arthur Compton himself, in constructing the necessary statistic for describing the specific heats at low temperature.

Remarkably, this trek of our knowledge follows, quite faithfully we should say, the path initiated by Max Planck in constructing the necessary statistics of the first quantization ever. And even if discovered specifically for the thermal radiation, the Planck's statistics carries the burden of universality of its ancestor, the Maxwell's statistic for the ideal gas. Specifically, it can be applied to any physical ensemble which can be portrayed in two parameters — frequency and temperature — not only in temperature, like that ancestor. Obviously, the atomic solids can be considered as such ensembles of harmonic oscillators. This fact was realized even from the introduction of the idea of light quantum in 1905, and carefully elaborated by Einstein himself in 1907, being fruitfully used by him in order to explain the behavior of the specific heats of solids at low temperatures (Einstein, 1907). Yet, for a certain category of physicists — among which we can certainly place Compton at that time — the relative success of the quantum theory needed to be taken *cum grano salis*, and that from quite a rational point of view in fact: *the statistical explanation of things cannot be a deterministic explanation*, in the classical spirit of the word 'deterministic', *i.e.* with a meaning close to physically 'lawful', as it were. However, such an explanation of the behavior of solids was beyond any hope. Especially, after the recent works of Einstein and Hopf, which proved that, even if used in the manner of Josiah Willard Gibbs, the classical statistics (Gibbs, 1883) could not bring any progress regarding the explanation of quantization (Einstein and Hopf, 1910). [See Milonni and Shih (1991) for a pertinent history of the problem.]

The Compton's work we are referring to here (Compton, 1915) is aligned to the idea that something is missing from the very classical statistical theory, so that in order to make it proper for procedures in thermodynamical problems, that something should be found at any rate. What was specifically missing became quite clear from the natural philosophical point of view. It was indeed obvious in the phenomenological difference between a solid and an ideal gas, whose kinetical theory led to settling the definition of absolute temperature in the manner shown before. Obviously, from the point of view of the *molecular kinetic theory*, a solid cannot exhibit the same number of degrees of freedom for motion as a gas: its molecules are not free to move unconditionally. The idea of *agglomeration of the degrees of freedom* was aroused, according to which in a solid, part of the degrees of freedom of the component material particles become 'agglomerated' in the precise sense that they *cannot participate to the definition of the kinetic energy*. In a suggestive word, they cannot represent a motion significant in participating to the statistics necessary in defining the absolute temperature, because in fact they have nothing to do with the idea of 'freedom' in the sense it is understood in classical mechanics.

This is the moment when the *idea of sufficiency* becomes necessary to statistical physics. The concept can be understood, the easiest way, if we call in the *Maxwell demon* (Szilard, 1929): if the definition of the temperature by the ideal gas would be indeed universal — and therefore the temperature itself would be as 'absolute', as it is usually labeled — then such a demon would be able to construct a *perpetuum mobile* by selecting the molecules moving much faster than the majority of them. Fact is that such a device cannot practically exist, so that we should assume that even for the ideal gases, and even in Maxwell's acceptance, the definition of temperature is not quite as absolute! The reason stands upon the idea of sufficiency just mentioned, which was appropriately defined, in the theoretical statistics first, a little later than the Compton's work (Fisher, 1922), and in physics even a decade later (Szilard, 1925) [see also Mandelbrot, 1956). The issue is that for an ideal gas *in the classical connotation*, the kinetic energy of the interpretative molecular ensemble represents a *sufficient statistics* for the physical parameter temperature. In the spirit of Ronald Fisher's definition for sufficiency (*loc. cit. ante*), this would mean that, once the temperature of a gas volume is quantitatively defined, it is the same no matter how much from that gas we take into consideration. We can even talk of the temperature of such a volume containing a single molecule, which would warrant the idea of a *perpetuum mobile*, to be operated by an appropriate demon, as in the fairy tales (see Maxwell, 1904, pp. 338ff). Lucky or unlucky for us, Leo Szilard 'exorcised' such a demon!

Along with the development of energetical method, the concept of degree of freedom started standing out in relief more and more clearly, including also the idea that some kinds of motion, other than pure translation may exist, capable of participating to the definition of temperature. The notion of agglomeration of degrees of freedom picks up a more precise sense: the agglomeration does not involve just the density of atoms or molecules. Compton, for instance, takes up the idea that the agglomeration is classical, related to the density of atoms, but he regards it as "any state of *association of the atoms* on account of which *degrees of freedom for thermal motion disappear*" (*loc. cit.*, p. 377, footnote 2; *our Italics, a/n.*). Therefore, he revives the classical *spatial vicinity*, suppressed in the statistical ensemble that defines the absolute temperature in the case of ideal gas — suppression perpetuated, quite naturally we should say, in the quantal oscillator ensemble that defines the physics of a solid — by the definition of the agglomerate. Simply put, with the words of Arthur Compton himself:

> If the relative energy between *two neighboring atoms* in a solid falls below a certain critical value, *the two atoms become agglomerated* so that *the degree of freedom between them vanishes*; but as soon as the energy increases again above the critical value, *the degree of freedom reappears.* [(Compton, 1915); *our Italics*]

Therefore, in order to become 'agglomerated', the atoms must have a relative energy small with respect to a threshold value, under which the energy itself does not contribute anymore in defining the temperature. Of course, in view of the statistical definition of the absolute temperature, the energy participating in that definition qualifies as kinetic energy. Once it goes under the threshold value, it is just a potential energy: from a statistical point of view, the two kinds of energy cannot be interchanged freely. The very interchange 'phenomenon is nevertheless reversible: *at threshold*, the potential energy can go very well into kinetic energy, reestablishing the thermal degree of freedom. It is not too hard to see that the procedure based on this phenomenon allows for a statistics of energy, whereby the space vicinity, brought for a moment to the fore by the concept of relative potential energy, is again suppressed by the very same. Whence the necessity of the concept of degree of freedom.

Here we have to deal, in fact, with what came to be known today in the theoretical physics as a *superstatistics*, a concept having its origin in the very statistical method that validated theoretically the idea of energetic quantum (Mazilu and Porumbreanu, 2011). Indeed, in deducing the limit distributions of the spectral energy of thermal radiation — the Rayleigh-Jeans and Wien radiation laws — we have two levels where the statistics should be applied. The first level is the description of the structural element of the ensemble representing the thermal radiation — the harmonic oscillator — while the second level describes the distribution of this structural element within the physical system of which it is constitutive part. For any system having the harmonic oscillator as constitutive element, the statistic of this last level is represented by frequency, and the representative distribution of this parameter is *a priori* taken as uniform, replicating the case of energy from the molecular ensembles. In the work of Compton referred to here, the structural element is an ensemble of agglomerates representing the solid, energetically described by equation

$$U \equiv U_n + U_a = nRT + (3N - n)\gamma.$$

Here U_n stands for the 'non-agglomerated' energy contained in n degrees of freedom contributing to the definition of the temperature T, and γ is "the potential energy of an agglomerated degree of freedom", so that U_a is the 'agglomerated' energy. One considers that the system contains N atoms which, taken as classical material points in the classical sense, give a total of $3N$ degrees of freedom, out of which $(3N - n)$ are agglomerated.

At this point, Compton takes notice of the fact that the ratio $\pi \equiv n/(3N)$ is a measure of the probability of existence of a *thermal degree of freedom*, while $(1 - \pi)$ is the probability of existence of an *agglomerated degree of freedom*. This probability can be calculated if we use the standard procedure for the case of distribution of velocities — according to which the kinetic energy is distributed with zero mean and variance proportional with the temperature — and, moreover, we assume that the values of potential energy are distributed exactly as those of the kinetic energy. We have here the germ of the modern method of the 'effective potential', constructed from the partition function at thermal equilibrium. Anyway, the final result of the Compton's work is that the probability π is an *exponential* defined by

$$\pi = \exp(-\tau/T), \quad \tau \equiv \varepsilon/(2R), \tag{2.9}$$

where ε is the threshold energy of an atom, just mentioned, and R is the gas constant. Using this result, Compton calculates the specific heat at constant volume in the form

$$C_v = C_\infty e^{-\tau/T}(1 + \tau/T), \tag{2.10}$$

where C_∞ is the value of this quantity at a high equilibrium temperature. This formula is to be compared with other ones based on the theory of quanta, and the comparison seems to favor it, even by comparison with Einstein's 1907 formula. Einstein himself was, of course, aware of the shortcomings of his formula, otherwise one cannot explain his dedicated statistical studies in the problem of radiation, and especially the work of 1913 of Einstein and Stern, where the zero point energy was explicitly introduced in calculations. Compton, however, does not mention this last article issued two years before his own work, earlier enough to be noticed by him. We cannot believe, therefore, that he would not know about the production, so much the more as he appears as very familiar with the use of quantum hypothesis in the theory of solids (see Compton, 1916): it is rather plausible that he ignored it. The zero point energy was indeed an invention of our spirit that proved physically viable only much later in time, through the advent of experimental proof of the existence of *Casimir effect* (Casimir, 1948; Sparnaay, 1958).

2.5 A Statistical Interpretation of the Classical Action

Speaking of the threshold transition between kinetic and potential energy, necessary in sustaining the Compton's argument, we feel like detailing an important issue of physics by the way of concluding this argument: *the constitutive element of the interpretative ensemble.* In our opinion, the first quantization solved a crisis related to this issue, and the concept of sufficiency is an apt illustration of the outcome of this solution. The constitutive element we are talking about here is quite simple: the harmonic oscillator. It is next in a line of simplicity, as it were, following the simplest case known, *viz.* the classical free material point, which is the constitutive element of the ensemble representing an ideal gas. Arthur Compton's work, raised an important issue though, that can be traced back to the very idea of a classical Lagrangian, and to the equation of motion correlated with it. In order to fruitfully follow this idea, the harmonic oscillator appears as the richest structure, when it comes to pertinent lessons. Let us illustrate this statement.

To this end, we shall take into consideration not quite a simple, undamped, harmonic oscillator, but a damped one, for reasons soon to become obvious. Also we shall limit those considerations to the one-dimensional case for now, a more complicated two- or three-dimensional case being touched later on our development, insofar as these cases are related to some fundamental mathematical issues. So, in the phase plane the equations of motion of a one-dimensional damped harmonic oscillator are as follows:

$$\dot{p} = -2\frac{R}{M}p - \frac{K}{M}; \quad \dot{q} = p. \tag{2.11}$$

The second equality here is obviously a definition of the 'momentum'. The system (2.11) is not yet a Hamiltonian system, as one is usually accustomed with, because its matrix does not define an evolution in the Hamiltonian sense, *i.e.* it does not represent an involution (in order to represent an involution the matrix should have a null trace). To make this fact even more obvious we write equation (2.11) in a matrix form

$$\begin{pmatrix} \dot{p} \\ \dot{q} \end{pmatrix} = \begin{pmatrix} -2R/M & -K/M \\ 1 & 0 \end{pmatrix} \begin{pmatrix} p \\ q \end{pmatrix}$$

and considering the physical parameters of the oscillator as constants, we can exhibit here the Hamiltonian as a quadratic form:

$$\frac{1}{2}M(p\dot{q} - q\dot{p}) = \frac{1}{2}(Mp^2 + 2Rpq + Kq^2). \tag{2.12}$$

Now, in order to make something out of the constancy of the physical parameters M, R and K, these need to be first physically explained. However, for now, we just take the statement as it is, with no explanation, postponing it for later. Thus, equation (2.12) shows that the energy — the quadratic form from the right-hand side of this equation — is basically the rate of variation of the physical action, represented as elementary area in the phase plane. This was actually the case all along the history of theoretical physics of the harmonic oscillator, so the conclusion has nothing of new or original. What we want to stress here is that the energy does not necessarily need to be conserved in order to be taken as the rate of variation of action, as historically seemed to have been the case. Actually, there is no conservation law of the energy here, if the frequency of the oscillator is not a constant itself. However, we can still ask ourselves if there is a conservation law and, in case there is, what would that be.

In order to reveal that law, notice that equation (2.12) can be cast in the form of a differential equation of Riccati type:

$$\dot{w} + w^2 + 2\lambda w + \omega_0^2 = 0, \quad w \equiv \frac{p}{q}, \quad \lambda \equiv \frac{R}{M}, \quad \omega_0^2 \equiv \frac{K}{M}. \tag{2.13}$$

The parameters λ and ω have to be explained, just like the originals from which they are calculated. Notice that equation (2.13) is always a consequence of the Hamiltonian system representing the motion of harmonic oscillator:

$$\begin{pmatrix} \dot{p} \\ \dot{q} \end{pmatrix} = \begin{pmatrix} -R/M & -K/M \\ 1 & R/M \end{pmatrix} \begin{pmatrix} p \\ q \end{pmatrix}. \tag{2.14}$$

This is quite a general aspect of the relationship between the Hamiltonian dynamics, as described in the phase plane, and the Riccati equation (Zelikin, 2000). One can recover equation (2.12) from equation (2.14), simply by constructing from this last equation the differential 1-form representing the area rate in the phase plane of the harmonic oscillator. Therefore, equation (2.12), just as the equation (2.13), can be thought of as a consequence of Hamiltonian dynamics in the phase plane of the harmonic oscillator. As for equation (2.12), it can be directly integrated, showing explicitly that *the energy is not conserved in this case*, indeed. However, in this case, we still have a conservation law, discovered by H. H. Denman, but it is a lot more complicated, properly involving a phase (Denman, 1968):

$$\frac{1}{2}(Mp^2 + 2Rpq + Kq^2) \cdot \exp\left\{ \frac{2R}{\sqrt{MK - R^2}} \tan^{-1}\left(\frac{Mp + Rq}{q\sqrt{MK - R^2}} \right) \right\}$$
$$= const.$$

One can see here that the energy is conserved in the classical sense only if, among the physical parameters of the oscillator, *the damping coefficient* is zero. This was already known from the old, but in a form hiding the true identity of the energy. Specifically, this property is usually translated into the fact that the energy is conserved as long as the oscillator is *isolated*. This creates the impression that the mass, as well as the elastic rigidity, would have to be intrinsic properties, unexplainable by interactions, which is not the case altogether: at least occasionally we may need to add some provisos. Let us insist a little more on this issue.

Taken face value, however, the theory of harmonic oscillator reveals one of the most interesting properties of the energy, related to the conservation law, from which an observation developed especially by Louis de Broglie ensues. Namely, the equation of motion of the *undamped* harmonic oscillator is a direct consequence of the property of stationarity of the *time average of difference between the kinetic and potential energies* — the Lagrangian — over the whole period of the motion. Indeed, the time integral of the Lagrangian between two time moments gives the action between the two moments of time, whose stationarity leads to the equations of motion. Assuming, then, that the *physical time is a uniformly distributed statistical variable*, the action can be, indeed, construed as a time average of the difference between kinetic and potential energies, as de Broglie once noticed (de Broglie, 1961, 1962). Therefore, one can say that the undamped harmonic oscillator *is the system which distributes the two kinds of interaction energies* — kinetic and potential — in such a way that the average of their difference over any time sequence included in the period of motion is stationary. This means that the property necessary to Compton's argument *was always there in physics*, in passing from one energy to another, except for the idea of 'threshold'. This concept is characteristic to the close environment, so to speak, representing the awareness on agglomeration in general, which in the case of harmonic oscillator takes the form of elastic Hookean interaction. We just need to see how the statistic is hiding here, and what is its physical nature.

Before going any further, let us stop for a moment, in order to pinpoint an important idea already mentioned quite a few times in different junctures of our discussion in this work. Namely, the concept of time here deviates significantly from the regular concept of classical dynamics, by assuming a differentia which brings it closer to the time of special relativity. Let us emphasize once again that, if we have to describe a general concept of time, then we have to assume that this concept must have *two* differentiae: that revealed first in the classical case, related to the property of continuity of motion, and the one associated mainly with the special relativity in describing an electrodynamical universe, *viz.* the 'cosmic and phenomenal' time in Hermann Weyl's phrasing, whereby the time is a parameter of *global ordering of events.* It is in this last instance that the time is defined by the idea of sequence, which is a special case of an ensemble of time moments. As the Feynman's development of quantum electrodynamics shows, such an ensemble may not even be necessarily a causal sequence in the classical sense. All it needs is only to remain deterministic from the physical point of view, as shown by the space–time theory of positrons (Feynman, 1949).

Now, continuing on with our discussion of the time statistics related to Lagrangian here, not quite the same mathematical argument can be applied in obtaining the classical equation of motion of a *damped* harmonic oscillator [equation (2.52), Mazilu *et al.*, 2019, for instance]. This one does not involve in its physical structure *a direct transition*, as it were, between the long distance inertia forces and the elastic forces of a close environment reflected in the time average of the Lagrangian. However, the physics underlying the case can be saved by the *very same statistical argument*, for the statistics is essentially the same from a general theoretical point of view. Only its type changes, in quite a precise manner though: by the character of its basic distribution density. Indeed, the common observation is that the equation of motion of the damped harmonic oscillator can be obtained by making stationary the action related to the Lagrangian

$$L(q, \dot{q}, t) = \frac{1}{2}(M\dot{q}^2 - Kq^2) \cdot e^{2\frac{R}{M}t}. \tag{2.15}$$

According to its definition, the action corresponding to this Lagrangian is given by an integral like

$$A_R(t_0, t_1) = \int_{t_0}^{t_1} (M\dot{q}^2 - Kq^2) \cdot \exp\{2(R/M)t\}\, dt, \tag{2.16}$$

where the physical parameters have the same meaning as before. The associated variational problem of this action leads to a *Caldirola-Kanai Hamiltonian* from a purely physical point of view

(Caldirola, 1941, 1983; Kanai, 1948). This Hamiltonian turns out to be no more the sum between kinetic and potential energies, as they appear in the Lagrangian from equation (2.15). The participation of physical parameters to time variation is the main thing to be noticed here: depending on the sign of damping coefficient, in a phase plane description the inertial mass increases while the elastic stiffness decreases, or the other way around. Therefore, there is still an interdependence between the terms of the Lagrangian, but with *the notable participation of the physical parameters of the oscillator.* The Hamiltonian may not even be a conserved energy, as in the case of undamped harmonic oscillator. However, the physics embodied in equation (2.16) can be, as we said, saved by statistics, for the action integral can still be construed as a time average of the difference between the two well-defined energies. On this occasion though, we have to deal with a different probabilistic measure of the time domain.

Indeed the exponential factor from the integrand of (2.16) can be interpreted as an *exponential distribution density describing the ensemble of time sequences* inside the time interval between the moments t_0 and t_1. In the case of undamped harmonic oscillator, the action is $A_0(t_0, t_1)$ — the index 0 of the action is referring to the value of the damping coefficient — and the exponential factor is 1. This particular action can be, indeed, interpreted statistically as a mean over a *uniform distribution* of times in a sequence, as stated before. The difference between the two cases — zero and nonzero damping coefficient — rests only upon the exponential factor in the expression of action integral, which, from a statistical point of view, *is thus not an attribute of the oscillator per se, but of the time domain.* One can say that an evolution for the damped harmonic oscillator means an ensemble of events characterized by sequences of *equally probable times* in a certain time interval, whereby the 'equally probable' attribute in a sequence is defined not by a uniform probability distribution, as in the case of undamped oscillator, but by *an exponential distribution proper.* At least for the case of damped harmonic oscillator, the physical character of time is, first and foremost, plainly a statistical property. This property is the one that allowed Richard Feynman the construction of his sum over paths, to begin with. This explains why the great physicist has placed so much physical emphasis upon harmonic oscillator. In hindsight, this emphasis cannot be explained but only by taking into consideration the results of Berry and Klein regarding the forces of Newtonian type (see Mazilu *et al.*, 2019, Chapter 9). In order to better illustrate the issue at hand, we extract a couple of phrases from the Abstract, giving the customary short presentation of work, of the Feynman's 1942 famous dissertation. This excerpt contains an observation explaining the importance that the approach of the wave mechanics initiated in that work bestows upon harmonic oscillator:

> As a special problem, because of its application to electrodynamics, and because the results serve as a confirmation of the proposed generalization, *the interaction of two systems through the agency of an intermediate harmonic oscillator is discussed in detail.* It is shown that in quantum mechanics, just as in classical mechanics, *under certain circumstances the oscillator can be completely eliminated*, its place being taken by a *direct, but, in general, not instantaneous, interaction* between the two systems. [(Brown, 2005); *our Italics*]

There is not too much to say over these words, in order to see in them the future results of Berry and Klein, indeed: the oscillator is present in any conservative Hamiltonian approach whereby the time is specially defined under condition of invariance of Newtonian forces [see equation (9.42) and the discussion around it in Chapter 9 of Mazilu *et al.*, 2019]. Involving in its physical structure parameters from 'two worlds', as it were — the far away part of the universe and the closest of its part — the oscillator is the best suited physical structure for describing the interaction between two systems. More importantly though, as we shall see here, this is the property that allows us to turn the special relativity into a universal theory, which thus can lie at the foundations of that special mathematics associated with the scale relativity physics.

It is in order to make this statistical property of time into a physical property, that we need to exhibit the physical reasons for changing the time sequence statistics. This too, will help in a proper understanding of the explanation of the physical parameters of harmonic oscillator, inasmuch as the change of the time statistics seems to be, at least to a certain extent, intrinsic to the *physical properties* of the harmonic oscillator. Indeed, considering the oscillator only, the Caldirola–Kanai Hamiltonian corresponding to the Lagrangian from equation (2.15) indicates, as we mentioned above, the variability with time of the physical parameters reflecting interactions with the remote and, respectively, close environment of the particle representing the oscillator as a physical structure: the *inertial mass* and *the elastic stiffness*. The case from equation (2.16) is only a particular one among those which led to the *classical idea of gauging*. Let us elaborate for a moment on this very idea, even though in a well-known manner, just in order to open the doorway for some more profound and general arguments, possibly involving even the concepts of threshold energy and that of agglomeration, from which we started here.

The equation of motion (2.11) cannot be obtained quite directly from the variational principle applied to action (2.16): one still needs some definite conditions at the ends of the time interval. The first one, and the most important among these conditions is that the trajectories of motion all end in the same position at the time ends, *i.e.* all pass through the same endpoints:

$$\delta q|_{t_0} = \delta q|_{t_1} = 0. \tag{2.17}$$

Further on, the idea of cycle of the harmonic oscillator triggers the condition that the evolution starts and ends in the same point:

$$q(t_0) = q(t_1). \tag{2.18}$$

Moreover, if the situation is described in the phase plane of the harmonic oscillator, we need a condition like this for the velocities also. It is therefore a matter of problem setting, to decide which specific conditions we need to take at the ends of time interval, in order to apply them over the variational principle, in order to define it properly. However, conditions like (2.17) and (2.18), involving the ends of the time interval, or some variations on such conditions, are essential in any formulation of that principle. When we consider them, the Lagrangian proves not to be unique from the point of view of the variational principle: it is defined up to an additive function which represents an exact time derivative, and takes the same values at the ends of time interval. In order to show this, it is better to reason on a general Lagrangian, explicitly dependent on time, like in equation (2.15), but in a more general manner, and then, based on this treatment, to evaluate our specific case given by equation (2.15).

Let us therefore apply the variational principle in order to obtain the equation of motion for a Lagrangian of the functional form $L(q, dq/dt, t)$. The physical action is given as the definite integral:

$$A(t_0, t_1) = \int_{t_0}^{t_1} L(q, \dot{q}, t)\, dt.$$

The principle of stationary action — the Hamilton principle — shows that for the real motion, the variation of this action, taken into consideration the conditions (2.17) must vanish. Let us write this variation explicitly. Denoting as usual by δ the variation operator, the principle of stationary action can be written as:

$$\begin{aligned} \delta A(t_0, t_1) &= \delta \int_{t_0}^{t_1} L(q, \dot{q}, t)\, dt \\ &= L(q, \dot{q}, t)|_{t_0}^{t_1} + \int_{t_0}^{t_1} \delta L(q, \dot{q}, t)\, dt = 0. \end{aligned} \tag{2.19}$$

The variation of the Lagrangian due to the position in the phase plane is given by

$$\delta L(q, \dot{q}, t) = \partial_q L(q, \dot{q}, t)\, \delta q + \partial_{\dot{q}} L(q, \dot{q}, t)\, \delta \dot{q} \tag{2.20}$$

with an obvious notation for the partial derivatives. The second term here involves the velocities, and if we assume that the operator of variation commutes with the operation of differentiation, then we can write

$$\begin{aligned} \partial_{\dot{q}} L(q, \dot{q}, t)\, \delta \dot{q} &= \partial_{\dot{q}} L(q, \dot{q}, t) \frac{d}{dt} \delta q \\ &= \frac{d}{dt}(\partial_{\dot{q}} L(q, \dot{q}, t)\, \delta q) - \frac{d}{dt}(\partial_{\dot{q}} L(q, \dot{q}, t))\, \delta q. \end{aligned}$$

Therefore, the variation (2.20) becomes

$$\delta L(q,\dot{q},t) = \frac{d}{dt}\{\partial_{\dot{q}} L(q,\dot{q},t)\delta q\} + \left\{\partial_q L(q,\dot{q},t) - \frac{d}{dt}\partial_{\dot{q}} L(q,\dot{q},t)\right\}\delta q$$

so that the variation of the physical action from equation (2.19) can be written in the form

$$\delta A(t_0,t_1) = (L(q,\dot{q},t) + \partial_{\dot{q}} L(q,\dot{q},t)\delta q)\Big|_{t_0}^{t_1} + \int_{t_0}^{t_1} \left\{\partial_q L(q,\dot{q},t) - \frac{d}{dt}\partial_{\dot{q}} L(q,\dot{q},t)\right\}\delta q\, dt = 0. \tag{2.21}$$

Under the conditions from equation (2.17), and for equal values of the Lagrangian for the end times t_0 and t_1, the first term of this variation vanishes. Now, in some quite general conditions, the second term of the variation from (2.21) vanishes too, if the Lagrangian satisfies the well-known Euler–Lagrange equations:

$$\frac{\partial L(q,\dot{q},t)}{\partial q} - \frac{d}{dt}\left(\frac{\partial L(q,\dot{q},t)}{\partial \dot{q}}\right) = 0. \tag{2.22}$$

Applying this recipe to the Lagrangian from equation (2.15), which obviously depends on time explicitly, we get the equation of motion for the damped harmonic oscillator in the form (2.11).

Notice, however, that in order to get the Euler–Lagrange equation we need the assumption that the Lagrangian has equal values at the end times of the motion, otherwise a redundant term would remain in the variation (2.21), which would allow in no condition to extract those equations. But an ambiguity still persists: in the very same working conditions, we can add to the Lagrangian any function of time just as well, provided it has equal values at the ends of time interval, and our conclusions do not change. In other words, within our working conditions, the Lagrangian is defined up to an additive function of time, which is the time derivative of a function having equal values at the ends of the time interval, but otherwise arbitrary.

This is the basis of a well-known classical *gauging procedure*. However, we read it here a little bit differently, having in mind the

Riccati equation: *we can reduce the Lagrangian to a perfect square, by gauging it in the manner just described*, and this reduction has a significant meaning. The procedure is well known and largely exploited in the control theory (Zelikin, 2000), so that we can shorten the story. The cycling condition (2.18) now enters the play. All one needs is to add to the Lagrangian from equation (2.15) the term represented an exact derivative:

$$\frac{1}{2}\frac{d}{dt}(w \cdot e^{2\frac{R}{M}t} \cdot q^2) \tag{2.23}$$

where w is a continuous function of time, and then ask that the final Lagrangian should be a perfect square. In view of condition (2.18), the final equations of motion do not change. However, the new Lagrangian of the gauged harmonic oscillator proves to be a little more complicated as being defined by

$$L(q, \dot{q}, t) = \frac{1}{2}M \cdot e^{2\frac{R}{M}t}\left(\dot{q} + \frac{w}{M}q\right)^2 \tag{2.24}$$

provided w satisfies the following Riccati equation:

$$\dot{w} = \frac{1}{M}w^2 - 2\frac{R}{M}w + K. \tag{2.25}$$

Obviously, under this condition, the Lagrangian (2.24) leads to the same equation of motion as Lagrangian from equation (2.15), if we use the condition (2.25) in the results of the corresponding variational problem. However, the Lagrangian (2.24) has the property of the classical prototype of the Lagrangians — *the kinetic energy* — namely of being a perfect square which describes a particle with its mass exponentially variable with time and the velocity redefined appropriately.

What is the reason of this reading of ours? Again, the point here is to physically interpret — with the interpretation defined in the sense of Charles Galton Darwin — a *simple constitutive system* like the harmonic oscillator, which usually entails some allegedly fundamental interactions in its physical structure, in order to carry this interpretation over to an ensemble of oscillators. Fact is that in constructing such an ensemble we always need, as we said before, *the constitutive element of this interpretative ensemble*, which is the harmonic oscillator. However, this comes with strings attached: the interactions involved in the explanation of its parameters. The mass is here inertial, and physics assigns to it an interaction involving the *remote part of the universe*, specifically that part located at infinity, whatever this infinity may be. The elastic stiffness is of a *deformational nature*, and the physics associates with it the *close part of the universe* representing a static environment, like any deformation. The damping term would then represent a *transition* between the two parts of the universe. It is according to this view, that the problem of interpretation needs to be solved, and the Lagrangian from equation (2.24) provides such a solution: it allows us to identify the harmonic oscillator *with a free particle.* What the gauging procedure accomplishes, is that it creates a Lagrangian which is a *perfect square*, and such a Lagrangian carries the identity of a free particle, like its classical counterpart, the kinetic energy. To wit, the Lagrangian from equation (2.24) represents an ensemble of 'free' particles with mass exponentially variable with time, each one of them characterized by a 'transported' velocity $[dq/dt + (w/M)q]$, depending linearly on the solutions of Riccati equation (2.25). Then, it is this last equation that needs a sound interpretation, which turns out to be statistical: $\dot{w}$ is the variance function of an exponential family of distributions having quadratic *variance function*, for which w is the *mean* (Morris, 1982). The distribution of this ensemble varies itself in time, and thereby the time itself represents a parameter indexing the family of probability densities, in much the same manner in which the temperature marks an ensemble of molecules in a thermal equilibrium. Mention should be made that even an undamped oscillator can be non-trivially made this way into a free particle.

The previous elaboration reproduces the gist of Compton's statistics involving the transition between kinetic and potential energies at the threshold. As it turns out, this is an old problem involved in the very dynamics of the harmonic oscillator, which was the first to raise doubts on the issue of sufficiency related to the definition of the absolute temperature. Thus, it further turns out that the Planck's quantization is by far not the only lesson we need to learn from this moment of our knowledge. The most important of them all, we should say, is the fact that *we need to account for the very structure of the constitutive element of the ensemble serving for interpretation*, as defined for the necessities of the wave mechanics.

2.6 Average Lagrangian as a Statistic: The Transition Amplitude

No doubt, the theoretical physics of harmonic oscillator is quite a prolific workshop of natural philosophy, still awaiting for important conclusions to be drawn from it for the benefits of knowledge at large. As we have seen, one of the most interesting properties of such a fundamental physical structure, is that *property of transition* between the energies, that we have mentioned in the previous section quite a few times. Specifically, the Lagrangian is the difference between two mathematically well-specified energies, and the motion of oscillator itself can be considered as a stochastic time process, distributing those energies in such a way that the time average of this difference is constant. This is the content of the principle of stationary action anyway. Now, why would we need the interpretation of such a fundamental physical structure? The reason is quite simple: this should be a structure that can be taken into consideration in a process of counting; however, depending on the space scale of the world under consideration, it also admits interpretation. The notorious case is that of the first quantization, but there are more palatable cases at hand right away: the remote stars or galaxies — fundamental components of the large scale universe — can hardly be considered as simple structures from a physical point of view.

Now, the damped variant of harmonic oscillator clarifies the nature of the mechanism of transition between the two energies of the oscillator. It is a mechanism external with respect to the harmonic oscillator itself, referring to an ensemble of time sequences, and specifically concerns the statistic of the very time average of

Lagrangian, which is the classical Hamiltonian action:

$$\langle L(t_0, t_1)\rangle \propto A(t_0, t_1) = \int_{t_0}^{t_1} L(q, \dot{q}, t)\, dt. \tag{2.26}$$

As the ends of the time interval on which this average is calculated always belong to a time interval, they are themselves statistical variables which can be taken as members of a random sequence of time, so that the average, and therefore the corresponding action, is indeed a statistical variable. Now, if we are to construct a *Madelung transformation* [equation (2.4), Mazilu *et al.*, 2019] in order to build an interpretation for such an ensemble, we can start with the *complex exponential of the action*:

$$\exp\{i\lambda A(t_0, t_1)\}, \tag{2.27}$$

where λ is a real parameter. In view of the time statistics in which the classical action is involved by the reasons just presented above, the idea finds suggested that this exponential factor can be taken, according to human knowledge at large, as an auxiliary in building the *characteristic function of the average Lagrangian* over the time interval interval $[t_0, t_1]$. This fact has been *suggested* as early as the fourth decade of the last century by Dirac, and even inspired the Feynman's approach of wave mechanics (Dirac, 1933). In this capacity, it defines the characteristic function of the average Lagrangian according to the statistical definition:

$$\varphi_A(\lambda) \equiv E[\exp\{i\lambda A(t_0, t_1)\}] = \int e^{i\lambda x} dF_{A(t_0,t_1)}(x). \tag{2.28}$$

Here dF_A is the elementary probability of the values x of time average Lagrangian $A(t_0, t_1)$. The problem to be solved now is to find the elementary probability necessary in order to calculate this characteristic function. There is not too much to report on the solutions of this problem, as it has never been stated as such in physics. Our incentives for these observations come from an entirely different direction: *the theoretical statistics*. We have shown (Mazilu *et al.*, 2019) that in an Euclidean environment, the wave function is a typical imaginary Gaussian. Its classical epitome is the well-known *Fresnel integral*, which generated the idea of *Fresnel transform* (James and Agarwal, 1996). The equation defining this wave function according to non-stationary Schrödinger equation is precisely of the form (2.28). One can say then that the de Broglie 'application function' (Chapter 9) might be a reflection of the existence of that statistics necessary in calculating the characteristic function of the statistic we call now 'mean Lagrangian', therefore the wave function. When following, as we do actually, the solution of a de Broglie kind of problem (de Broglie, 1935), defined by the words of the great scholar as 'application of the charge upon field' or as 'application of the field upon charge', the uncovering of such a statistics is of major importance. As far as we are aware, there is but one case of the kind in the specialty literature (Albeverio and Hoegh-Krohn, 1974), analyzing the connection between the non-stationary Schrödinger equation and its classical counterpart, *the heat equation*. This last equation is connected, in the work just cited, with the stationary Schrödinger equation in the stochastic approach (Nelson, 1966). However, the heat equation *per se* is well known to provide a reliable way of construction of the probability densities, that has been generalized in quite a few significant ways. One such way of generalization interests us especially, being associated with a remarkable property of the solution of heat equation — therefore of the non-stationary Schrödinger equation — discovered by Paul Lévy by the middle of the last century (Lévy, 1943, 1944). This property is in turn related to the study of the Brownian motion, a subject matter on which Paul Lévy has equally remarkable fundamental results (Lévy, 1954, 1965).

Fact is that the solutions of the heat equation caught the attention of the scientific community especially after the Einstein's 1905 works on Brownian motion (Einstein, 1956). In a one-dimensional space this can be described statistically as a stochastic process characterized by the probability density:

$$p_x(x;t) = \frac{1}{\sqrt{4\pi t}} e^{-\frac{x^2}{4t}}, \quad \frac{\partial}{\partial t} p_x(x;t) = \frac{\partial^2}{\partial x^2} p_x(x;t). \tag{2.29}$$

This density is a Gaussian of variance t and mean zero. t here is not a time moment but a time interval. In order to make this clear, we need to adopt the stochastic reading (Lévy, 1943), according to which we have to do with a random function of time, $X(t)$, whose increment $X(t'') - X(t')$ for $t'' - t' > 0$ is a Gaussian variable having the variance $t'' - t'$. This fact can be expressed by the probability relation:

$$Pr[X(t'') - X(t') < x\sqrt{t'' - t'}] = \frac{1}{\sqrt{2\pi}} \int_{-\infty}^{x} e^{-\frac{\xi^2}{2}} \, d\xi. \qquad (2.30)$$

The Gaussian here is not a solution of the standard heat equation given in (2.29), but of a heat equation having a coefficient 2 in front of the time derivative. Fact is, that no matter of the constant coefficients in front of the derivatives in the general heat equation, it can be reduced to the standard form by a homothetic transformation effected simultaneously on the time and space variables, so that the final solution preserves the Gaussian form (Skinner, 2016). In what follows, we only reproduce the results of Paul Lévy from the work just cited above, based on the definition (2.30) of the Brownian process.

If in a time interval $[t_0, t_1]$, the general Gaussian variables $X_0 \equiv X(t_0)$ and $X_1 \equiv X(t_1)$ are known to belong to the same Brownian process, then the general variable

$$X(t) = \mu(t) + \sigma(t)\xi(t) \qquad (2.31)$$

can be calculated, for any value of t belonging to that interval. Here $\mu(t)$ and $\sigma(t)$ are the parameters of the general Gaussian process — playing the parts of mean and standard deviation — and $\xi(t)$ is what Lévy calls *reduced Gaussian variable*, characterized by a zero mean and unit standard deviation, as in equation (2.30). Indeed, knowing that $X_1 - X_0$ is a Gaussian variable of variance $t_1 - t_0$, the variables $\mu(t)$ and $\sigma(t)$ can be calculated by the formulas

$$\mu(t) = \frac{(t - t_0)X_1 + (t_1 - t)X_0}{t_1 - t_0}, \quad \sigma^2(t) = \frac{(t - t_0)(t_1 - t)}{t_1 - t_0}. \qquad (2.32)$$

The usual values describing a Gaussian process, solution of the heat equation, are obtained for $t_1 \to \infty$. Based on these values, one can calculate the correlation coefficient of two reduced Gaussian variables $\xi(t)$ and $\xi(\tau)$, with t and τ from the interval (t_0, t_1), such that $t < \tau$. The result is [Lévy, *loc. cit.*, equation (9)]:

$$R_\xi(t,\tau) = \frac{(t-t_0)(t_1-\tau)}{(t_1-t)(\tau-t_0)}. \tag{2.33}$$

The right-hand side of this equation is, up to a sign, one of the six cross ratios of the four numbers involved in its expression. This construction is known to be invariant with respect to any 'linear fractional transformation', using the expression of Bernard Lavenda (homographic transformation in the old expression of Paul Lévy), whence the following theorem:

> All the *stochastic properties* of the function $\xi(t)$ from equation (2.31), in its definition interval (t_0, t_1), are invariant with respect to any *linear fractional transformation* performed concomitantly on t_0, t_1 and any value $t \in (t_0, t_1)$. The only restriction is that, if t_0, t_1 and their correspondents through linear fractional transformation are finite, the *interior of the interval* (t_0, t_1) has to correspond to the *interior of the interval* delimited by their correspondents. [(Lévy, 1943); Théorème 1]

In other words, the stochastic properties of reduced Gaussian variable are dictated in relation to the linear fractional transformation of its variance, which in turn is connected to the measure of time. Now, let us use this result for a heuristic guidance in our own proceedings.

2.7 The Motivation for a Space–Time Transformation

This last result of Lévy compels us to a reevaluation of the SL(2,R) action in two variables given in equation (1.21). That equation is actually a hint benefited from the classical Galilean kinematics (Mazilu and Porumbreanu, 2018) that seems to have a universal connotation, for which that kinematics is just a particular case: the transformation represents the content of time and space coordinates on a direction, expressed in terms of two uniform motions. Taken as just a group action, with no physical reference to its content whatsoever, the transformation (1.21) possesses a general invariant function: any arbitrary continuous function of the algebraic expression $x^2/(at^2+2bt+c)$, where a, b, c are three constants (Mazilu *et al.*, 2019). Now, from this perspective, the classical Galilean suggestion further reveals another truth, primarily connected with theoretical statistics: the second-order polynomial from the denominator of the fundamental invariant expression of the action (1.21), can be taken as representing the infrafinite measure of the time range and, possibly, even a fractal measure of the time, if it is to remain in the finite ranges of time. This can be seen by differentiating the first of the equations of action (1.21) in order to find the ensemble of time moments t corresponding to $t' = constant$. The result gives the infrafinite measure of time dt as a quadratic polynomial in the finite measure of time t, with its differential coefficients representing an **sl**(2,R) coframe. Now, both the infrafinite measure of time dt, as well as the finite measure t, can have a *statistical meaning* facilitated by the fact that the **sl**(2,R) coframe is usually represented by conservation laws. This means that there is a Riemannian space represented by this algebra, whose geodesics are described by constant rates with respect to the arclength, and these rates are given by the three components of the coframe. The statistical meaning in question is the following: the ensemble of time moments t corresponding to $t' = constant$, is a statistical ensemble of mean t, with a variance given by the ratio between dt and the elementary arclength of the **sl**(2,R) algebra.

These ensembles are what we have called before sequences of equally probable times, to be used in the calculation of the characteristic functions of the Lagrangian. The probabilistic distribution functions describing them are of exponential type, having quadratic variance when considered as a function of the ensemble mean (Morris, 1982). The first striking example of such distribution in physics was the Planck's distribution, that led to the first quantization ever (Mazilu, 2010). We think that this classical case still contains many valuable suggestions, particularly the way of introducing the fractals into the mathematics of theoretical physics (*loc. cit. ante*, §4); the Planck's quantization can be considered, indeed, as the epitome of fractalization. But what we want to emphasize here is the fact that there are six types of quadratic variance distribution function in the one-variate statistical case, both discrete, like the Planck's distribution, and continuous, and these seem to cover most instances of physical interest that occurred in theoretical physics lately. These are the following distributions (Morris, 1982): Gaussian, Binomial, Negative Binomial, Poisson, Gamma, and Generalized Hyperbolic Secant. One can find these distribution functions well described in many statistical compendia, of which we can recommend (Johnson, Kotz, and Kemp, 1992) and (Johnson, Kotz, and Balakrishnan, 1994). Among these, the Gaussian, Gamma, and Generalized Hyperbolic Secant are of continuous type, while the Binomial, Negative Binomial, and Poisson are of discrete type. Planck's original distribution is a Negative Binomial. The Gaussian can be counted among these distributions — even though its variance is a constant and does not depend explicitly on the mean — on account of the general algebraical fact that a constant can always be considered a second-degree polynomial. However, it plays a more important part in the physical statistics than this participation to a classification.

The Lévy's theorem above, allows us to characterize the values of a random function of time like $X(t)$, as space ensembles prone to interpretation by free, of partially free, classical material points. That interpretation emerges if we notice that the Riccati equation expressing the condition $dt' = 0$, which describes the ensembles of times t corresponding to the same time t' is invariant through any linear fractional transformation. In view of the theorem of Paul Lévy, we are then entitled to write the Gaussian characterizing such ensembles as

$$p_x(x; dt) = \frac{1}{\sqrt{4\pi dt}} \exp\left(-\frac{x^2}{4(dt)}\right), \quad dt = \omega^1 t^2 + \omega^2 t + \omega^3, \tag{2.34}$$

where $(\omega^1, \omega^2, \omega^3)$ is the coframe of $\mathbf{sl}(2,\mathrm{R})$ algebra (Mazilu *et al.*, 2019). A few observations are now in order, for the better understanding of this stochastic theory.

Thus, in the infrafinite range in time, the exponent of Gaussian used in the theory established by Paul Lévy, is actually the *invariant function* of the SL(2,R) action (1.21). What is the meaning of this situation? In order to reveal it, we follow closely the work of Vittorio de Alfaro and his collaborators, already cited above in connection with the transformation (1.21) (de Alfaro, Fubini and Furlan, 1976). It is, indeed, the time to recall the Berry–Klein definition of the scale invariance of forces, leads, for the Newtonian forces, to a scale factor for coordinates, satisfying the equation:

$$\ddot{\ell}\,\ell^3 = const.,$$

where an overdot means time derivative as usual. In this equation, the 'scaling length' defines a transition between times in the infrafinite range of Mariwalla type given in equation (1.2):

$$d\tau = \frac{dt}{\ell^2(t)}.$$

In the new time τ, the motion described by a certain Hamiltonian appears as a Newtonian motion made under conservative forces. Now, assume that the gauge length is established by a statistical procedure. This is a typical procedure, for instance in the ideal gas case, where the mean free path of the molecules plays an important part in the description of the model. And, if we take the ideal gas as a model for the matter in a certain universe, then it is necessary to carry all of the characteristics of the model over into the physical description of that universe. This is exactly what happened historically, only with a significant twist on proceedings, forced upon our knowledge by the general relativity. Namely, in the universe at large we need to consider always the distance between the fundamental constituents of the universe — the galaxies, for instance — which has to be taken as a gauge length for the model. This seems to be the whole morale of the Hubble law that accounts quantitatively for the apparent radial recession of the galaxies with respect to us (Hubble, 1929). In this specific case, the distance between places in the universe is gauged by the speed of galaxies, and it appears in the redshift of spectra.

These facts indicate that in the description of a universe, a statistical estimate of the distance between the fundamental constituents of that universe can always serve as a gauge length for a dynamics under conservative forces in that universe. Provided, of course, that statistical estimate satisfies the second-order Berry–Klein differential equation given above. Let x be such a statistical estimate of distance, to be inserted in the Mariwalla transformation between time differentials. Then x can be interpreted as a standard deviation. Indeed, de Alfaro, Fubini and Furlan took notice that the Berry–Klein equation can be derived as an equation of motion from a Lagrangian of the form

$$L(x,\dot{x}) = \frac{1}{2}\left(\dot{x}^2 - \frac{g^2}{x^2}\right) \tag{2.35}$$

and this Lagrangian theory is invariant with respect to (1.21), in the sense of invariance of the *differential, or elementary physical action.* We need to insist on this theory, and we shall follow here the guidance of exquisite presentation of Morton Lutzky, who showed that, in generalizing the adiabatic invariance — as, in fact, the Berry–Klein theory aims to — the Lagrangian (2.35) is instrumental, and is closely connected to the Noether symmetries of the classical harmonic oscillator (Lutzky, 1978). Closer to the spirit of the work of de Alfaro, Fubini and Furlan is the modern work of Joanna Gonera, to which we also shall return, every now and then, for one or two points of physical interpretation, and which seems to us the clearest one among all others, and specifically dedicated to this issue (Gonera, 2013). The first observation based on this last work, is that with respect to the transformation (1.21), where the corresponding group action on space coordinate is defined by considering this coordinate as a function of the corresponding time:

$$x'(t') = \frac{x(t)}{\gamma t + \delta} \tag{2.36}$$

the elementary physical action based on Lagrangian (2.35) is 'quasi-invariant', so to speak:

$$(\alpha\beta - \gamma\delta)\, L\left(x'(t')\frac{dx'(t')}{dt'}\right) dt' = L\left(x(t)\frac{dx(t)}{dt}\right) dt - d\frac{\gamma x^2}{2(\gamma t + \delta)} \tag{2.37}$$

provided, of course, g is a constant. The genuine invariance is conditional, first of all, on the character of the time transformation from equation (1.21): if that transformation is realized by a unit-determinant matrix, then the transformed Lagrangian from (2.37) is just a scaled version of the original, in the sense of the theory presented above [see equation (2.15) ff]. In the case presented here, the exact differential from equation (2.37) replaces the exact differential from equation (2.23). Assuming that this is the case — *i.e.* the time transformations are genuine SL(2,R) transformations — the resulting equation of motion is invariant with respect to transformation (1.21). And this equation of motion is the Berry–Klein equation, which we rewrite here, but expressed in the 'statistics' x:

$$\frac{d^2x(t)}{dt^2} = \frac{g^2}{x^3}, \tag{2.38}$$

where g is considered a *real* constant. We have already touched the idea of solution of this equation on the occasion of Berry–Klein analysis (Mazilu *et al.*, 2019). This was worked out by Eliezer and Gray, and is based on the dynamics of the harmonic oscillator as an 'auxiliary motion' (Eliezer and Gray, 1976). It is time now, to get into deeper details on this issue. Fact is that if, according to Berry–Klein theory, x should be a 'gauge length' for instance, equation (2.38) is instrumental for its definition. Before any other considerations regarding its physical nature, we take it as such: a *length* or a *distance*, either in a plane or in space, helping us in making up our minds as to the meaning of the procedure.

Now, Eliezer and Gray's procedure (*loc. cit. ante*, §7) is actually referring to a proof of what we would like to call *Ermakov–Pinney theorem*, which can be accomplished just by direct calculations. Let us assume that we have a plane situation — the "auxiliary plane motion" in the terminology of Eliezer and Gray — more generally, a two-dimensional case, whereby the square of length x is a generic quadratic form, which may be even taken as a quadratic metric form:

$$x^2 \equiv \alpha u^2 + 2\beta uv + \gamma v^2. \tag{2.39}$$

Then, if u and v are two independent solutions of the second-order linear differential equation

$$\ddot{x} + a(t)\dot{x} + b(t)x = 0, \tag{2.40}$$

the square root of expression (2.39) is a solution of the nonlinear equation

$$\ddot{x} + a(t)\dot{x} + b(t)x = (\alpha\gamma - \beta^2)\frac{(v\dot{u} - u\dot{v})^2}{x^3} \tag{2.41}$$

usually called *Ermakov–Pinney equation* [see Ermakov (2008); see also Pinney (1950)], whence the name of the theorem just proved. Equation (2.38) corresponds to the case where $a \equiv b = 0$ in (2.40), whereby u and v are linear in time, *i.e.* the auxiliary motion of Eliezer and Gray is, in fact, a uniform rectilinear motion, with the constant g to be properly identified right away, of course.

The theory evolved remarkably ever since, for it was realized that it is connected to the idea of adiabatic invariants (Lewis, 1968), from which the whole modern quantum mechanics originated [see Heisenberg (1925); see also Mazilu and Porumbreanu (2018), for detailed explanations on the coming to being of the quantum mechanics]. Meaningfully, and quite naturally we should say, the whole point of the Harold Ralph Lewis' work from 1968 can be connected to that of Berry and Klein's work: what was thought to be just an adiabatic invariant in the case of harmonic oscillator, is actually an *exact* invariant, and therefore can be thought as being *independent of the rate of change of the environment.* The mathematical theory evolved remarkably, as we said, involving the continuous groups, and in the present explanation, as well as in further developments along this work, we shall also use the recent results of Faruk Güngör and Pedro Torres, whereby the geometric approach of the issue at hand is referred to the two expressions that prove so important from a physical point of view (Güngör and Torres, 2017): the quadratic form (2.39) and the Schwarzian derivative, defined, in our case here, by

$$\{x, t\} \equiv \left(\frac{\dddot{x}}{\dot{x}}\right) - \frac{1}{2}\left(\frac{\ddot{x}}{\dot{x}}\right)^2. \tag{2.42}$$

In order to give a hint of the place where the physics is touched by the solution of this problem, let us recall that the physical definition of the frequency involves consideration of the mechanical epitome of periodical variation in time — the very same harmonic oscillator we were discussing previously. The classical definition of the frequency is usually limited to the instantaneous frequency (Mazilu *et al.*, 2019, Chapter 2, equation (2.52) ff) which can be used to properly extract the physical parameters of an oscillator from a general recorded signal, in cases where the phase of this signal is homographic in the time of recording. This means that the Schwarzian derivative of the phase with respect to time is zero. The theory, to be partially reproduced here right away after (Güngör and Torres, 2017), contains an important result which we shall reconsider again further in our development, revealing many other of its properties, inasmuch as it has an important impact in the physical relation between the phase and the amplitude of the signal.

Now, one can prove, by direct calculation, that equation (2.38) is invariant with respect to unimodular transformations (1.21), where the action on space variable x is understood in the sense expressed by equation (2.36). The problem is how to understand the intermediation of the variable from (2.39) within the framework provided by this invariance. One can even declare by now that it represents the general concept of *phase plane*. In order to show this we need to use the properties of the two-dimensional manifold of solutions of the second-order differential equation that naturally describes the harmonic oscillator, be it damped or undamped. We exploit, for this, the properties of **sl**(2,R) algebra, described in the base generated by the action (1.21). The base vectors of this algebra are the operators which we transcribe here, for convenience, in some generic variables, from which we need to extract our previous physical variables:

$$X_1 = \frac{\partial}{\partial \theta}, \quad X_2 = \theta \frac{\partial}{\partial \theta} + \frac{1}{2}\xi \frac{\partial}{\partial \xi}, \quad X_3 = \theta^2 \frac{\partial}{\partial \theta} + \theta \xi \frac{\partial}{\partial \xi}. \tag{2.43}$$

It is important to notice the composition of these operators: they reproduce the basic trait of transformation (1.21), of being rational linear in time and only linear in the space coordinate. As we shall see, this characteristic is reproduced in a specific way for multiple space coordinates. Now, we use two fundamental solutions of equation (2.40) in order to perform the transformation:

$$t' \equiv \frac{dt}{d\theta} = u^2, \quad x = u\xi, \quad W\theta = \frac{v}{u}, \tag{2.44}$$

where W is the Wronskian of the two solutions u and v. This quantity is, in general, not a constant; however, we can manipulate any second-order homogeneous differential equation in such a way as to transform it into one missing the term that contains the first derivative of the unknown function. In this case, the Wronskian of any two solutions of the equation is a constant and one can choose the solutions such that this constant is always 1. Assuming, therefore, the second-order differential equation of the form admitting a constant Wronskian, we can take it as unity, so that

$$\frac{\partial}{\partial t} = \frac{1}{u^2}\left(\frac{\partial}{\partial \theta} - x\dot{u}\frac{\partial}{\partial \xi}\right); \quad \frac{\partial}{\partial x} = \frac{1}{u}\frac{\partial}{\partial \xi}.$$

Thus, we get the following operators satisfying the standard commutation relations of the **sl**(2,R) algebra:

$$\begin{aligned} X_1 &= u^2\frac{\partial}{\partial t} + u\dot{u}x\frac{\partial}{\partial x}, \\ X_2 &= uv\frac{\partial}{\partial t} + \frac{u\dot{v} + v\dot{u}}{2}x\frac{\partial}{\partial x}, \\ X_3 &= v^2\frac{\partial}{\partial t} + v\dot{v}x\frac{\partial}{\partial x}. \end{aligned} \tag{2.45}$$

The bottom line is that using the homogeneous second-order differential equation corresponding to a Ermakov–Pinney equation, we can use infinitely many realization of a **sl**(2,R) algebra associated with this equation. The two-dimensional manifold of solutions of this equation properly generalizes the classical phase plane, and helps in adding to those properties of the harmonic oscillator, revealing still other forces involved in its definition.

2.8 An Idea of Morton Lutzky

The story of the Harold Lewis' discovery has a few interesting connotations. One of them is that already mentioned above: the invariance related to harmonic oscillator is not just adiabatic, it is an exact invariance, mathematically speaking. It is, in a way, of that type sought for by Michael Berry in his quest for generalization of the phase changes. However, with the idea of generalization of the phase plane, another connotation related to harmonic oscillator comes out in the open. Hidden as it were, within the entanglement of calculations, this meaning may be destined to change the emphasis on forces in the dynamics of the harmonic oscillator. We start with the observation that in equation (2.41), considered as a dynamical equation of motion, the right-hand side can be considered as a force driving the harmonic oscillator. Just the same happens for the case of equation (2.38), but it is known that such a force, if central, leads to a spiraling motion. Therefore, in a problem of gauging it would be necessary to assume that the central forces going inversely with the cube of distance are fundamental, and this is an entirely new turn of the situation, which needs a rational explanation.

Fact is that the quadratic form from equation (2.39) is a solution of the third-order differential equation with the coefficients determined exclusively by the coefficients of the second-order differential equation (2.40). This is true independently of any other mathematical or physical considerations, as long as u and v in the construction of the quadratic form x^2 are solutions of the same equation (2.40). The coefficients of that quadratic form can be any triple of numbers or physical magnitudes if we like: one can say that there are a triple infinity of quadratic forms defined this way; we are thus dealing here with a linear span of the three homogeneous quadratics u^2, uv and v^2. Specifically, the third-order differential equation is (see Bellman, 1997, Exercise 3, p. 179)

$$\dddot{q} + 6a\ddot{q} + 2(\dot{a} + 4a^2 + 2b)\dot{q} + 2(\dot{b} + 4ab)q = 0, \tag{2.46}$$

where $a(t)$ and $b(t)$ are the coefficients of the second-order differential equation (2.40). In the case of an undamped harmonic oscillator proper we have $a(t) = 0$ and $b(t) = \omega^2(t)$, so that equation (2.46) becomes

$$\dddot{q} + 4\omega^2 \dot{q} + 4\omega\dot{\omega}q = 0. \tag{2.47}$$

Morton Lutzky has noticed a few important facts related to this equation (Lutzky, 1978b), which we shall bring in from time to time, as we go along with our presentation. The first one of these, to be signaled right away, is that it has a first integral that generates equation (2.41) corresponding to the given coefficients. Indeed, just multiplying by $q(t)$ transforms it in an exact differential giving the integral

$$q\ddot{q} - \frac{1}{2}\dot{q}^2 + 2\omega^2 q^2 = C. \tag{2.48}$$

Now, if we identify q with x^2, and choose the constant C appropriately, this equation becomes

$$\ddot{x} + \omega^2 x = \frac{\alpha\gamma - \beta^2}{x^3}, \tag{2.49}$$

which is, indeed, a version of equation (2.41) corresponding to this situation.

2.9 Conclusion: Asymptotic Freedom as a Manifestation of Scale Transition

The Berry–Klein theory (Berry and Klein, 1984) appears to us as the only theory deliberately sanctioning the concept of scale transition as a criterion in physics. This notion, as we see it, was indeed used before even by Newton, when he invented the concept of forces, however only implicitly, so to speak. Indeed, it is by using the continuous sequence of collision events, with percussions acting toward a unique point in space, that Newton ratified, from a natural philosophical point of view, the existence of centripetal forces responsible for the Kepler motion. Where is the 'implicitness' here?! Ideally, the collision events can be considered as dynamical events involving forces that act within infinitesimal spaces and infinitesimal time intervals — locally and instantaneously, in modern terms — *i.e.* at the *infrafinite scale* in space and time, if it is to use the modern terminology of Nicholas Georgescu-Roegen. Therefore, Newton used forces acting at the infrafinite scale, in order to define field forces — forces acting everywhere and permanently — *i.e.* forces acting at the *transfinite scale* of space and time. Historically speaking, this is the first case of specific *gedanken-experiment* involving the use of a scale transition, at least from the physical point of view. The transition between scales is, indeed, implicitly made *via* an *assumed identity* of the physical objects invented by Newton with the accidental forces of our daily experience. It is by no means the only case of such a transition in history.

Once again, we see the relevance of Berry–Klein theory in two important points issuing from its intervention in the Hamiltonian formalism. First of all, it can be considered as a modern approach of classical *Newton's own cosmology*, an approach that ought to be applied to the modern *Newtonian cosmology* (McCrea and Milne, 1934) at any rate, in order to make it truly 'modern'. Indeed, the hallmark of modern cosmology at large, is the expansion of universe: this seems to be a fact of observation of the universe around us, which can serve as a test for every law of physics. The truly modern example of conformity with such a test is provided by the Wien's displacement law: physically demonstrated for thermal radiation as an adiabatic thermodynamical system, it appears as an exact — *i.e.* independent of the expansion rate of the enclosure containing radiation — universal law, serving as a criterion of selection for any of the laws of radiation. The universality in question can be proved by the *scale invariance*, showing that such a law, true at the *finite* laboratory scale, is also valid at the *infrafinite* microscopic scale, and also at *transfinite* cosmological scale (Mazilu, 2010).

This last case of cosmological scale manifestation of the Wien's displacement law, has been definitely demonstrated by the data on the cosmic background radiation, whose 3K spectrum satisfies the Planck's law of radiation (Fixsen, Cheng, Gales, Mather, Shafer, and Wright, 1996). Now, the foundation of Newton's own cosmology is provided by the forces he invented, an ingredient, if we may say so, which is just as important for that cosmology as the law of radiation is for an expanding cosmology. No one has ever considered, though, that in creating a Newtonian cosmology along the Einsteinian line (McCrea & Milne, 1934), the *field of forces* has to pass the expansion test. This is the second important point of the Berry–Klein theory: it can be taken as ascertaining that the Newtonian force field passes the expansion test, so that from the modern cosmological point of view, provided by the fact of expansion of the universe, Newton's own cosmology is indeed 'modern', and it can be aptly considered as a Newtonian cosmology. This observation has remarkable consequences: the chief among these is the fact that we cannot dispense with forces in physics, we have just to understand them properly. Historically, the first step in this process of understanding of forces is the creation of the theoretical concept of field, independently of the theoretical Kepler problem. It is this conclusion that allows us to build an equilibrium ensemble serving for the interpretation of matter continuum, as in Chapter 1 [see equation (1.7) ff].

This being said, it should be by now easy to understand why we attach a very much importance to the Ermakov–Pinney equation like the one in (2.49): in view of the Berry–Klein theory, it is the key of the scale transition involving forces, inasmuch as it is implied by the existence of Newtonian forces. This may not be too much, but as it turns out, it is the expression of a scale invariance at the infrafinite microscopical scale of the world. This, again, may not mean too much either, but if we twist a little the statement, and express it as meaning that Newton's and Einstein's cosmology involve the same gnoseological criterion of existence, it can appear to mean a lot: *there is no difference between the Newtonian and Einsteinian points of view* in the natural philosophy. Indeed, that equation supports, as we already have said before, the existence of Lewis invariant, which reduces to the Planck's constant in some particular cases, and therefore extends the concept beyond the classical thermodynamical argument. To wit, it is by no means only 'adiabatic' but an exact invariant, therefore universal, just as universal as the Planck's law of radiation, for instance. As a matter of fact, its invariance was first suggested, indeed, by Planck's quantization law (Lewis, 1968), which can be produced by integrating equation (2.49) directly (Rogers and Ramgulam, 1989). Now, as we have shown elsewhere (Mazilu, 2006, 2010), the Planck's constant can be produced itself as a *joint invariant* of two SL(2,R) type actions. No wonder, then, the Lewis invariant in its most general form (Lutzky, 1978b) can be also produced as a *constraint* to be satisfied by a joint invariant of two actions generated through the infinitesimals from equation (2.43). Such a joint invariant is given by the ratio of the solutions of two Ermakov–Pinney equations (2.49), corresponding to two sets of values (α, β, γ). This ratio can be taken as a gauged position involved in the Mariwalla theorem as in equation (1.2), but for the one-dimensional case, so that equation (2.49) is instrumental in the transitions of scale.

Then, by extending an observation of Colin Rogers and Usha Ramgulam, we can give the following theorem (Rogers and Ramgulam, 1989): the ratio of solutions of two equations (2.49), corresponding to different triplets (α, β, γ), is also a solution of an Ermakov–Pinney equation, however, with coefficients determined exclusively by these triplets. The 'time' of this equation is dictated by the equation that gauges the time in the corresponding Mariwalla transformation. However, if we use the Berry–Klein theory as a criterion, the way it was intended initially — *i.e.* dynamically, *via* a Hamiltonian — a problem pops up right away: none of the field of forces involved in equation (2.49) is invariant to expansion, in the very sense of the Berry–Klein theory. Therefore, when it comes to scale transition, the Ermakov–Pinney equation *cannot be taken dynamically*! The right way to consider it is offered by the idea of Lutzky above, involving the group theory: in a word equation (2.49) is a mathematical fact, which must be taken as a universal *kinematic* equation. After all, it essentially involves the microscopic physics and, historically speaking, it is this physics that created the wave mechanics. In its turn, the wave mechanics asks for a concept of interpretation, and this is the whole point at issue: the Ermakov–Pinney equation (2.49) must be used in the *interpretation of a matter continuum as a static set of Hertz material particles*, just as we described in Chapter 1, regardless of scale. It is, for instance, this equation that provides the conditions for a statement like the one of Earnshaw, saying that the ether is composed of 'detached particles' (Earnshaw, 1942). And it should be, in another instance, the very same equation that provides the conditions authorizing the Einstein approach to gravitation. Details on how this is achieved follow right away.

One important observation must be made, though, and a corresponding important notice should be duly taken, before entering into these details: the scale invariance we are talking about here *is not* of the kind involved in scale transitions. The scale invariance we are talking about here is a kind of invariance *inside the same world* — incidentally, we might say, we have here the microscopic world — and is concerned with the size of the fundamental constitutive objects of that world. An appropriate example of *scale transition invariance* would be here one of the sort signaled by Shmuel Sambursky in 1937. The suggestion we take from Sambursky is that while the universe at large is expanding, the atoms should be *shrinking* (see Sambursky, 1937; Sambursky and Schiffer, 1938). Therefore, while at transfinte scale we have expansion *per se*, at the infrafinite scale we should have its contrary, *i.e.* contraction. The truth of this statement would mean the design of a formal mathematical transition between transfinite and infrafinite scales, *physically sustained* by a scale transition invariant, like, for instance, the Planck's spectrum ratifying the expanding cosmology. On the contrary, the scale invariance sought for around Ermakov–Pinney equation here, should be of the nature of the setting that inspired it: a *statistical invariance*, if we may say so, which is guaranteed only for the same scale. And this is, indeed, the case.

According to Lutzky's idea, equation (2.38) is the 'radial' equation of a *free* particle, if we represent the freedom by vanishing of the second-order differentials of Euclidean coordinates. Pending a possible new word for a universal description of the concept of freedom here — the modern *asymptotic freedom* comes to our mind right away — we take equation (2.38) as *kinematic*, suggestive enough at this point of our discourse to mean the elimination of forces from an interpretation scenario. They may exist, but like in the system of fixed stars of Newton, recalled by us here in Chapter 1, the forces acting randomly on each one of them render it free. The point is that using the group theory, this condition can be interpreted as such even in a more general coordinate system, where the measure of a coordinate is a binary quadratic form, and its velocity is to be further defined by a special equation. The wave mechanics asks specifically for a kind of Wigner's principle, providing for instance the transport 'means' for the fluid serving in interpretation, a condition that Poincaré asked for, in the case of Lorentz matter. The fact of the matter is that even the classical mechanics in its Lagrangian form asks for such a definition, by the transition to its Hamiltonian form, in which case the definition of velocity connected with a general coordinate is the only independent constraint that has to be taken into consideration, besides Euler–Lagrange equation (Ray, 1973).

Now, assume such a general coordinate — either a length or a distance — governed by equation (2.38). This equation can be produced by making stationary the action corresponding the Lagrangian from equation (2.35). As we have seen before [equation (2.15) ff], the physical action is simply defined as the time average of the difference of the terms involved in the definition of Lagrangian, and so can be defined the action producing the equation (2.38). This would mean that equation (2.38) is produced by the stationarity of the time average of the difference involved in the expression (2.35). Adopting this reading we have the following theorem of Morton Lutzky that settles the place of the Ermakov–Pinney equation in physics (Lutzky, 1978b): assume two space variables p and q, describing an interpretative ensemble of Hertz particles. Then an Ermakov–Pinney equation for p is produced by making the time average of the difference $(dp/dt)^2 - \omega^2 p^2 - C_1/p^2$ stationary. This stationarity principle produces a constant generalized Lewis invariant connected with the two variables: $C_1(q/p)^2 + C_2(p/q)^2 + p^4[d(q/p)/dt]^2 \equiv C_1(q/p)^2 + C_2(p/q)^2 + q^4[d(p/q)/dt]^2$ as a conserved quantity generalizing the classical Planck's constant, provided q is a solution of the Ermakov–Pinney equation $d^2q/dt^2 + \omega^2 q = C_2/q^3$. As one can see right away the theorem can be formulated for q in exactly the same terms: the Ermakov–Pinney equation for q is produced by making the time average of the difference $(dq/dt)^2 - \omega^2 q^2 - C_2/q^2$ stationary.

The general constant of this procedure is the same generalized Lewis invariant, provided p satisfies the Ermakov–Pinney equation $d^2p/dt^2 + \omega^2 p = C_1/p^3$.

A few observations are now in order. First of all, the Lewis invariant given by Morton Lutzky has a clear parentage with the phase plane of the two variables p and q: it may be taken, for instance, as the product of the two differential forms (2.12), referred to an appropriate time. This is, as a matter of fact, its statistical mechanical origin: a measure of the phase plane 'cell'. However, our physical interpretation of the physical action leading to the Ermakov–Pinney equation, hints toward a somewhat different view, closer, we should say, to a regular geometry, not a geometry of the phase plane. Such a view is motivated by the following, legitimate in context, question: what if the Lagrangian like (2.35) is zero?! In other words, what if the generalized velocity, say (dq/dt), associated with the coordinate q, is *exactly* given by the ratio (g/q), up to a sign, of course. Within the framework of the ideal classical gas, such a condition means sufficiency, as shown before. As it leads, mathematically, to the constancy of the expression (q^2/t), then we should be able to say that the departure of this quantity from this condition is *a measure of 'agglomeration'* if it is to use Compton's expression. This certainly provides an incentive for the physics based on Paul Lévy's approach to the stochastic calculus, presented by us before, but also has a hint as to the kind of geometry we need to follow. Indeed, if this 'exact' coordinate, whatever it is, turns out to be a solution of the linear third-order differential equation in time, then it needs to be the square of another variable x, making, for instance, the transition from the Ermakov–Pinney equation (2.49) to the equation (2.47) which leads to it. Thus, q^2 means x^4, and while q^2 is indicating an Euclidean sphere, of parameter x. It seems that the affine is the physical way in geometry, or vice versa, which is why we shall follow here an *affine foundation* of the physics, to be described in the due time.

And now, for the true point of these conclusions, which turns out to be the kinematical reading of the Ermakov–Pinney equation. Indeed, as we have seen above, there is no room for dynamics here and, therefore, there is no other way to accept the fact that, if exactly zero, the Lagrangian (2.35) leads to a geometrical characterization of the sufficiency, while, if stationary fluctuating, it leads to the 'field equation' (2.38). After all, this last observation is crucial: it means that if a fluctuation of the second derivative of the coordinate q should be provided, it is the Ermakov–Pinney equation that effectively provides it, with respect to the background that, according to the wave-mechanical rules can be called a 'field' (de Alfaro, Fubini & Furlan, 1976). The problem is to rationally explain these fluctuations *as such*, and the explanation is, fortunately, handy as the conclusion of a theorem by H. J. Wagner. This theorem really shows the importance of the physical definition of action as a time average of Lagrangian with respect to an appropriate time distribution (Wagner, 1991), which turns out to be geometrically and statistically inspiring.

Wagner's work is directed to straightening, if we may say so, a "misleading reasoning" of Bhimsen Shivamoggi and Lawrence Muilenburg, that appears as an unfair critique of the work of Morton Lutzky from 1978, just cited by us above (Shivamoggi and Muilenburg, 1991). The work of these two authors uses the continuous group theory too, but in the realization of a group action generated by the vectors listed in our equation (2.43), with θ taken as time and ξ as a general coordinate. The conclusion is that the Lewis invariant is an exact invariant that agrees with the classical theory of adiabatic invariance of the harmonic oscillator — as presented by us right above — only in cases where $\omega t = constant$, *i.e.* the pulsation of harmonic oscillator is defined, classically, as the inverse of a time period. This is in itself a remarkable result, inasmuch as, according to our take on the subject, it shows that the group theory in the action generated by the infinitesimals (2.43) requires a 'phase locking', as it were, of the classical phase of the harmonic oscillator, in order to allow us to perform the quantization. It is, however, the wrong conclusion of Shivamoggi and Muilenburg, namely that Lutzky has not succeeded in his construction, which can only be performed with an action (2.45) that raises objection. After all, as we have shown above, there is a transition between the two actions, no question about that. Shivamoggi and Muilenburg are apparently contending that the construction must only take place for the kind of action from equation (2.43), as if this would be the only group action allowable. They notice, however, that the right construction was indeed realized by Rogers and Ramgulam (*loc. cit.*) with a group action of the type given by us in equation (2.45), but fail to notice that Lutzky did that too, and that apparently motivated the observations of Wagner.

This incident is essential to us, insofar as it touches a few points of fundamental physics. Among these, the most important one is the problem of physical action as defined by a Lagrangian, which is the object of what we like to call *Wagner's theorem*, mentioned by us above. Wagner rightly relegates the observation of Shivamoggi and Muilenburg to the problem of phase as we presented it here for the harmonic oscillator: fact is that the physical action (2.16) for the undamped harmonic oscillator is not invariant with respect to groupal action (2.45). As we have seen above, even the genuine **sl**(2,R) action, results in a relative invariance, whereby the Lagrangian needs a gauging in order to be properly used. What would be, in these conditions, the place of harmonic oscillator, so clearly delineated in a Hamiltonian dynamics by the Berry–Klein theory? The answer is given by Wagner: there is a Lagrangian $(1/p^2)[(pdq/dt - qdp/dt)^2 - (q/p)^2]$, producing the Euler–Lagrange equation $d^2q/dt^2 - (q/p)(d^2p/dt^2 - 1/p^3) = 0$. Thus q is a harmonic oscillator if and only if p is an Ermakov–Pinney particle. Of course, a Lagrangian can be constructed right away, with the roles of q and p switched, but the conclusion is the same. In words, it sounds: kinematically speaking, the reading of Ermakov–Pinney equation is not that suggested in (2.49), whereby the inverse cubic term appears as a perturbation of the harmonic oscillator from the left-hand side, but rather that suggested by (2.38), whereby the right-hand side provides a linear fluctuation of an acceleration field, with respect to a background provided by some inverse cubic accelerations. Notice that the 'exact' acceleration, provided by the 'exact' velocity field that insures vanishing of Lagrangian (2.35), is a significant inverse square acceleration.

In order to assess the situation, it is best to provide a demonstration for the Wagner's theorem. An observation of Morton Lutzky provides the key: applying his theorem above for the particular case $C_2 = 0$, we get the Lewis invariant: $(q/p)^2 + p^4[d(q/p)/dt]^2$ where we chose the constant C_1 as unity, as Wagner did. Now, switching to a new time, θ say, defined by $d\theta = dt/p^2$, the Lewis invariant becomes: $(q/p)^2 + [(q/p)']^2$ where the prime means derivative on θ. This is the energy of a harmonic oscillator working *in the new time* according to an equation of motion produced by the Lagrangian $[(q/p)']^2 - (q/p)^2$, as well known. Switching back to the time t in this new Lagrangian, gives the Wagner's Lagrangian that produces the above kinematical results. The pair $(q/p; dt/p^2)$ is a one-dimensional Mariwalla transformation of the coordinate q and time t, with a scaling factor p. The theory of regularization to be described in the next chapter will provide further clues regarding the position of harmonic oscillator in physics.

Chapter 3

Theory of Regularization of Kepler Problem

In our times, the uniqueness of the universe of natural philosophy of any persuasion, shapes each and everyone of the theories built to explain this universe. Some of these theories, however, carry the imprints of the necessary scale transitions, incidentally imposed by classical dynamics along its way of explaining the physical structures of the universe. As we have shown before [Chapter 1, equations (1.1) ff., and discussion] even the classical dynamics of the Kepler problem itself requires the mixed Mariwalla scale invariance: infrafinite ranges in time transition and finite, or even transfinite ranges, in space transition. This is, in fact, the very nature of the original Newtonian calculus of fluxions. However, one cannot say that, while not recognizing the idea of scale transition of any kind, and therefore categorizing the universe as a unique extant world, our spirit has not encountered similar warning situations in its journey: occasionally it has been forced to remarkable, though sometimes peculiar, attitudes. Such a notorious case, quite rich in lessons for us here, is the *theory of regularization of the Kepler problem*, whereby the harmonic oscillator plays, again, a central part. This is, indeed, a case of intellectual productions specifically related to *scale transitions* in the unique universe of the classical natural philosophy. Involving intimately the harmonic oscillator, these productions help us in further elucidate the nature of results of the Hamiltonian theory of Berry and Klein which, as we have seen, are connected to this physical structure.

In the classical Kepler problem one can talk indeed, of a theory of regularization *per se.* Actually, one is compelled to talk of regularization by the very nature of the forces involved in classical dynamics, a fact noticed even from the times of Newton. The reason stays with the fact that the Newtonian forces are eternal, at least in regards of their actions, but by astrophysics our spirit came to realize that in *close encounters* these forces should coexist with forces ephemeral of different degrees. And, in our opinion, the most important consequence of the theory of regularization of Kepler problem is not one or another among the mathematical consequences it entails, but a hidden reason of the theory, in need to be realized and amplified properly, because it is of importance for a scale relativity. Indeed, even from the beginnings of the modern theory of regularization, a peculiar pattern of it started emerging, without, however, being properly elucidated at the moment's notice. We quote from the introduction of a classical work that established the grounds in the field of regularization:

> In closing, I would like to emphasize the importance of regularization *from a more general point of view.* It suffices for me to point out that the prior *regularization of a differential system* is indispensable in order to intimately penetrate into the *general pattern of solutions* and the *distribution of periodic solutions.* Exactly in this way, opened out by POINCARÉ , it seems now that geometers must preferentially engage their efforts [(Levi-Civita, 1920); *our translation and Italics*]

A few explanatory words for understanding the message of this excerpt seem necessary, in order to better appreciate it. The explanation will be followed by a general theory of regularization of the Kepler problem, presented from the perspective of the idea of scale transition invariance.

The classical Kepler problem, in its dynamical formulation, meant first and foremost a relief of our spirit from the grips of that kind of classicism embodied in the identification of the action at distance with a force. First, it brought into theory the quintessential *concept of field*, and this concept is the one that liberated physics from the Newtonian burden of regularly evaluating the force with the aid of trajectory of motion. However, the idea of trajectory seemed to have been inescapable, insofar as, instead of force, the concept of field has brought in *the necessity of quantization.* This was necessary in order to allow for the theoretical description of the phenomenology at the microscopic scale of the world, where the light — in its electromagnetic stance, of course — would appear to be the defining phenomenon.

One can safely say that the orbit's quantization had a sort of 'undertone', as it were, of radical breakout of physics from the grip of forces invented by Newton. And such a quantization would have, indeed, thoroughly and definitely accomplish this job, were it not for the application of the theory of gases to the problems of astrophysics. For, the application of theory of ideal gases to cosmic matter came in astrophysics with the necessity of a 'friendlier' consideration, if we may say so, of yet another phenomenon specific to this branch of the human knowledge: *the close encounters of astrophysical matter.* Imposing, as we mentioned above, a concurrent consideration of the forces described by a field, and those occurring in collision situations, these close encounters revealed the further necessity of theoretically considering the *infrafinite space range* together and concurrently with the *finite* and *transfinite* space ranges. This necessity, in turn, points out, in our opinion, to the necessity of a theoretical invariance to scale, a problem to which the classical regularization theory answers in an exquisite way in need to be properly realized.

One can say that the forces whose action is done in the infrafinite time range — the forces in their collision instance — were indeed enough to Newton in order to physically validate the forces he invented, viz. the forces in the transfinite time range. As it turned out, however, the experience that these invented forces opened to our knowledge was not sufficiently explained by them in a theoretical manner. More specifically, the application of the theory to cosmic close encounters has shown that Newton's validation was unilateral, inasmuch as it was applied only from the point of view of the *time scale*, while the invented forces revealed the necessity of considering the infrafinite range in the *space scale.* It is this topic which became, by and large, known in celestial mechanics — and in physics, actually — as the *regularization problem*, and the 'transition of space scales' is, in our opinion, what should be understood by that "*more general point of view*" of Tullio Levi-Civita in the excerpt above.

Now, it is not that the idea of field did not mean anything for this very problem. However, it imposed another point of view which can be related to space scale transition: that of the *space expansion* of the very bodies involved in a close encounter. This is a consequence of the fact that the classical dynamical Kepler problem is indeed properly defined only under condition that the dimensions of the bodies involved in it are irrelevant with respect to the distances between these bodies. If anything, it is this property that makes the Kepler problem applicable to any reality. And this asks further for that condition of relative size, which is hardly satisfied in a close encounter. Here, the field theory acted by preserving the idea of potential, and thus proceeded by using the method of perturbations, which was the first, in the modern times, to offer a solution for the theory of satellites (Vinti, 1966). This specific method works, indeed, via a formal series — solution of Laplace equation — whose terms are decided parametrically by the relevant dimensions and the geometrical shape of the body considered as creating the gravitational field. It is to this state of the case that we want to give a turn into a proper description of the genuine matter, and the first move is to show that the Kepler dynamical problem has the necessary capabilities in order to accomplish the task.

3.1 The Distance of Action as a Physical Manifold

Begin by considering the equation of a trajectory obtained as solution of dynamic of the Kepler motion with Newtonian forces. In polar coordinates (r, ϕ), this can be written in the form

$$r = a\frac{1-e^2}{1-e\cos\phi}. \tag{3.1}$$

Here a is the major semiaxis of the orbit, and e its eccentricity. By the transformation

$$\tan\left(\frac{\phi}{2}\right) = \sqrt{\frac{1-e}{1+e}}\tan\left(\frac{E}{2}\right), \tag{3.2}$$

which replaces the polar angle ϕ with the *eccentric anomaly* E, equation (3.1) becomes

$$r = a(1 + e\cos E). \tag{3.3}$$

In this algebraical expresssion one can recognize the modulus of the complex number $x \equiv x_1 + ix_2$, where

$$x_1 = a(e + \cos E), \quad x_2 = a\sqrt{1 - e^2}\sin E, \tag{3.4}$$

which can be considered as a complex representation of the radial coordinate. The time of this 'complex' dynamical problem is given by *Kepler's equation*, which can be obtained according to recipe

$$\frac{dt}{dE} = r\sqrt{\frac{a}{\kappa}} \quad \therefore \quad t = \sqrt{\frac{a^3}{\kappa}}(E + e\sin E). \tag{3.5}$$

The first theoretical instance of the regularization procedure, specifically known as connected with the name of Tullio Levi-Civita, is based on the further observation that the complex number $x \equiv x_1 + ix_2$, with its real components given by equation (3.4), is itself the square of another complex number, $u \equiv u_1 + iu_2$ say, having the components given by

$$u_1 = \sqrt{a(1+e)}\cos\left(\frac{E}{2}\right), \quad u_2 = \sqrt{a(1-e)}\sin\left(\frac{E}{2}\right). \tag{3.6}$$

Therefore, one can say that, in the time given by Kepler's equation (15.5) — that is, a time determined by the eccentric anomaly of the planet's motion described by this Kepler problem — the radial distance itself assumes the physical structure of a field. More specifically, it is a physical structure represented by two harmonic oscillators of different amplitudes and phases. This physical behavior is independent of the magnitude of radial distance — it is thus a regular behavior! — provided the procedure is indeed applied to a Kepler problem dynamically controlled by a Newtonian force of the kind that instated the idea of field in the real of theoretical physics, in the first place, *viz.* a central force with magnitude inversely proportional with the square of distance between the positions of its action.

Again, one can conclude thereby that obligations of cosmology have *reinstated* the idea of forces into this physics 'almost' in the manner in which the quantization procedure *eliminated them.* Indeed, in the quantization of orbits, an orbit is 'frozen', as it were, by a Bohr–Sommerfeld type prescription, while in a theory of cosmological usage it is represented by equation (3.3), derived directly from a Kepler dynamical problem. Only as such *the magnitude of distance* can be represented in terms of two physical oscillators! And, as the dynamical Kepler problem admits a scale invariance by Mariwalla's theorem (Mariwalla, 1982), one can further say that the idea of Keplerian orbit is relevant no matter of scale. To wit, the double oscillator representation of space distance related to this specific action turns out to be valid beyond the electromagnetic image of the atom, which inaugurated the quantum mechanics (Heisenberg, 1925), and therefore might be understood as valid only in the realm of microcosmos. Thus, the physical representation (3.6) can even mean a behavior inside the region covered by the space expansion of the source of field in the Kepler problem, *i.e.* inside the matter *per se*, which is actually what we follow suit. In other words, the Bohr's postulates represent a regular manner of the knowledge in general, for such a behavior of matter is universal!

We must take due notice, however, of two important facts here. First, we need to notice that in describing the concept of action, the geometrical distance itself is, physically at least, a multidimensional quantity, like the magnitude of acceleration on a surface (Mazilu *et al.*, 2019). To wax it lyrical, if Riemann himself would have to rewrite his renowned words about examples of 'multiply extended manifolds' (Riemann, 1867) he should certainly have to add, to his initial 'positions' and 'colors', also the 'geometrical distance perceived in the action of Newtonian forces'. Secondly, this description of the essential argument of the idea of regularization, is based on *a certain* concept of time, which suggests the importance of the time in conformal transformation connected to the idea of regularization. Specifically, this is the time of dynamics measured by the eccentric anomaly through Kepler's equation (3.5). One can say, therefore, that in general not quite any time reflects a motion!

This observation picks up a significant standing if we refer it to the fact that, generally — *i.e.* in the current acceptance in theoretical physics — a certain time is often just a continuity parameter, possibly even differential continuity, arbitrary to a significant degree when it comes to its physical meaning. For a 'dynamical continuity', the time should, nevertheless, be further specified, by the existence of a correlated dynamics, obviously. And equation (3.5) just describes such a 'dynamical' time. Let us take this idea to a more significant level.

Certainly, dot-multiplying by (dr/dt) in dynamical equation of motion describing Kepler problem and integrating the result, we get what the astronomers designate as 'the integral of energy', usually denoted by h (Volk, 1973):

$$\frac{1}{2}\dot{r}^2 - \frac{\kappa}{r} \equiv \frac{1}{2}h. \tag{3.7}$$

On the other hand, dot-multiplying by $\mathbf{r}$ and integrating, we have

$$\mathbf{r}^2\dot{r}^2 = 2\kappa r + C_0 r^2, \tag{3.8}$$

where C_0 is an integration constant, which can be evaluated right away. Indeed, a comparison between equations (3.7) and (3.8) shows immediately that $C_0 \equiv h$. Further on, using the definition of the vector of area variation rate for the area swept by the position vector, written in the form: $\dot{\mathbf{a}} = \mathbf{r} \times d\mathbf{r}/dt$, plus the two well-known geometrical identities: $\mathbf{u}^2\mathbf{v}^2 \equiv (\mathbf{u}\cdot\mathbf{v})^2 + (\mathbf{u}\times\mathbf{v})^2$ and $\mathbf{u}\cdot d\mathbf{u} \equiv udu$, we get the relation

$$r^2(\dot{\mathbf{r}}^2 - \dot{r}^2) = \dot{\mathbf{a}}^2. \tag{3.9}$$

Equations (3.8) and (3.9) can still be rewritten in a somewhat simpler form, if we use as time variable not quite the time given by Kepler's equation, but one linear in the eccentric anomaly: $d\tau \equiv dE\sqrt{(a/\kappa)}$. Then equation (3.5) gives $t' = r$, where the accent means derivative over τ, so that finally

$$\left.\begin{aligned} \mathbf{r}'^2 &= 2\kappa r + hr^2 \\ \mathbf{r}'^2 - r'^2 &= \dot{a}^2 \end{aligned}\right\} \therefore \quad r'^2 + \dot{a}^2 = 2\kappa r + hr^2. \tag{3.10}$$

The differentiation on τ of the last relation here, results in the second-order differential equation:

$$r'' - hr = \kappa. \tag{3.11}$$

The general solution of this equation, when the initial position and initial velocity — in the time τ, obviously — are known, is given by the expression

$$r(\tau) = r_0 + v_0 \frac{\sinh(\tau\sqrt{h})}{\sqrt{h}} + a_0 \frac{\cosh(\tau\sqrt{h}) - 1}{h}, \tag{3.12}$$

where the initial acceleration is offered by equation (3.11) itself: $a_0 \equiv r''(0) = \kappa + hr_0$.

Concluding, the time τ is also well-defined *as a dynamical time*: it represents a 'dynamical continuity just like the time t for the initial dynamics modeling the Kepler motion. However, the two dynamics do not describe the same physical system: what in the time t is the dynamics of the Kepler problem proper — allegedly representing the reality — in the time τ is the dynamics of a harmonic oscillator, forced with a constant force of magnitude given by the constant from the Kepler problem, and representing only *a model* for this reality, as it were. If the energy is positive — the proper astrophysical case of close encounters — in equation (3.12) we have an *inverted harmonic oscillator*. As expected, this type of oscillator describes not only collisions, but yet another type of close encounter as well, subject of study in quantum mechanics: *the tunneling process* (Barton, 1986). It is only for negative energy that we have a regular harmonic oscillator in the time τ. However, no matter of the sign of energy, any solution of this type of regularization has the essential algebraical property of the quadratic conformal transformation of Levi-Civita: there is a system of two harmonic oscillators of different amplitudes and phases, equivalent — in the sense presented above, via the complex structure of the radial coordinate with a Kepler motion in a Newtonian dynamical problem. Indeed, if we denote in equation (3.12)

$$u_\alpha = a_\alpha \cosh(\sqrt{h}\tau/2) + b_\alpha \sinh(\sqrt{h}\tau/2), \quad \alpha = 1, 2, \tag{3.13}$$

we get $r = (u_1)^2 + (u_2)^2$, by an appropriate choice of the coefficients. Here these coefficients are indeed decided by the initial conditions of the problem, as they should classically be indeed, via the relations:

$$\begin{aligned}(1/2)(a_1^2 + a_2^2 - b_1^2 - b_2^2) &= -\kappa/h,\\ (1/2)(a_1^2 + a_2^2 + b_1^2 + b_2^2) &= r_0 + \kappa/h,\\ a_1 b_1 + a_2 b_2 &= v_0/\sqrt{h}.\end{aligned} \tag{3.14}$$

Hence, those initial conditions are simply given by just two vectors **a** and **b**, which can be calculated from the equations:

$$\mathbf{a}^2 = r_0, \quad \mathbf{b}^2 = r_0 + 2\kappa/h, \quad \mathbf{a}\cdot\mathbf{b} = v_0/\sqrt{h}. \tag{3.15}$$

Therefore, as a methodological conclusion, the key point of a procedure of regularization should be a definition of the *regularization time* — a 'fictitious time' in the modern astrophysical phrasing [see Stiefel and Scheifele (1975), for instance]. Afterwards comes an equation like (3.11) governed in that time via the dynamical Kepler problem with Newtonian forces. This equation represents in fact a first integral of the third order differential equation

$$r''' - hr' = 0 \tag{3.16}$$

whereby the integration constant is κ. The fact appears as quite natural from a purely mathematical point of view: equations (3.13) are actually representing two linearly independent solutions of the homogeneous second-order differential equation

$$u'' - (h/4)u = 0. \tag{3.17}$$

Now, any quadratic form constructed with these two solutions is itself a solution of the third-order differential equation (3.16) [see Bellman (1997, Exercise 3), p. 179, or Mazilu and Porumbreanu (2018)].

The square relations (3.14) represent the essentials of *the Levi-Civita regularization*, and they are just an 'assay-sample', so to speak, of a three-dimensional procedure generally known as the *spinorial* or *Kustaanheimo–Stiefel regularization.* This fact becomes obvious if we try to extend the procedure of regularization to the three-dimensional case, in the manner that follows (Volk, 1973). Starting again from the dynamical equation modeling Kepler problem and transforming it directly in the time τ, we get

$$r\mathbf{r}'' + r'\mathbf{r}' + \kappa\mathbf{r} = \mathbf{0} \tag{3.18}$$

which is a three-dimensional damped oscillator. Differentiating once more with respect to τ, and using in the result thus obtained the equation (3.11), we get a three-dimensional replica of equation (3.16)

$$\mathbf{r}''' - h\mathbf{r}' = \mathbf{0}, \tag{3.19}$$

which leads to a corresponding three-dimensional replica of the equation (3.11) itself:

$$\mathbf{r}'' - h\mathbf{r} = -\mathbf{A}. \tag{3.20}$$

Here $\mathbf{A}$ is the *Laplace–Runge–Lenz* vector of this problem. This play of differentiation and integration, first to obtain the third-order differential equation, and then integrate it back to obtain the second-order differential equation, may appear as somehow artificial. In fact, the passage from equation (3.18) to equation (3.19) depends on the equation (3.11), more to the point, on the constant from the right-hand side of this equation. If we would start directly from (3.16), in order to get (3.11) by integration, in the right-hand side of this last equation we should have an arbitrary constant. Now, not knowing anything about the origin of (3.11), the constant could remain arbitrary forever: it is only the fact that we are aware of having originally to deal with the Kepler problem, that allows us to identify our integration constant with that from this problem, and thus to get the previous results. We should therefore assume that the quadratic representation of the radial coordinate when using the time τ, is the only one guaranteeing the fact that this procedure might have some physical background, and is not only a simple trick of formal justification. Yet, the procedure represents much more.

In order to reveal in detail what it represents, we follow, up to a point, the line of reasoning of Otto Volk. To wit, the integration of

(3.20) leads to a three-dimensional analog of the equation (3.12):

$$\mathbf{r}(\tau) = \mathbf{r}_0 + \mathbf{v}_0 \frac{\sinh(\tau\sqrt{h})}{\sqrt{h}} + \mathbf{a}_0 \frac{\cosh(\tau\sqrt{h}) - 1}{h}. \tag{3.21}$$

Here $\mathbf{r}_0, \mathbf{v}_0$ and $\mathbf{a}_0$ are the initial vectors of position, velocity and acceleration, respectively, in the time τ. Notice now that the initial vectors of motion are coplanar, which geometrically means that they can be expressed in the same two-dimensional reference frame from the plane of motion. Let $\hat{\mathbf{e}}_1, \hat{\mathbf{e}}_2$ be that reference frame, chosen as orthonormal. We have a first suggestion on the manner in which we should continue the theory, and that is given *by the form of initial position.* Thus, we choose the initial position in equation (3.21) as

$$\mathbf{r}_0 \equiv (a_1^2 - a_2^2)\hat{\mathbf{e}}_1 + 2a_1a_2\hat{\mathbf{e}}_2 \quad \leftrightarrow \quad r_0^2 = (a_1^2 + a_2^2)^2. \tag{3.22}$$

With this choice, we correctly reproduce the initial condition in position, given by the first equality in equation (3.15). Further on, with *the choices*:

$$\begin{aligned} \mathbf{v}_0 &\equiv (a_1b_1 - a_2b_2)\hat{\mathbf{e}}_1 + (a_1b_2 + a_2b_1)\hat{\mathbf{e}}_2, \\ 2\mathbf{a}_0 &\equiv h\mathbf{r}_0 + (b_1^2 - b_2^2)\hat{\mathbf{e}}_1 + 2b_1b_2\hat{\mathbf{e}}_2, \end{aligned} \tag{3.23}$$

we get from equation (3.22), for any 'moment' τ, a result *formally analogous* to the initial one, *viz.*

$$\mathbf{r}(\tau) = (u_1^2 - u_2^2)\hat{\mathbf{e}}_1 + 2u_1u_2\hat{\mathbf{e}}_2, \tag{3.24}$$

where u_1 and u_2 are two independent solutions of (3.17), defined by the vectors $\mathbf{a}$ and $\mathbf{b}$ according to the initial conditions (3.23).

Otto Volk then proceeds to the further choice of a *one-parameter family of orthonormal frames* in the plane of motion [see also Volk (1976)], expressed by a one-parameter family of two unit orthogonal space vectors of the form:

$$\hat{\mathbf{e}}_1 = \begin{pmatrix} \cos^2\alpha \\ \sqrt{2}\sin\alpha\cos\alpha \\ \sin^2\alpha \end{pmatrix}, \quad \hat{\mathbf{e}}_2 = \begin{pmatrix} \sin^2\alpha \\ -\sqrt{2}\sin\alpha\cos\alpha \\ \cos^2\alpha \end{pmatrix}. \tag{3.25}$$

This gives a corresponding family of position vectors, all in the time τ, which can be rewritten in the form of column matrices having as entries quadratic forms in four variables:

$$\mathbf{r}(\tau) = \begin{pmatrix} m_1^2 - m_2^2 - m_3^2 + m_4^2 \\ \\ 2(m_1 m_2 - m_3 m_4) \\ \\ 2(m_1 m_3 + m_2 m_4) \end{pmatrix}. \tag{3.26}$$

The four components of the vector $\mathbf{m}$ involved in the quadratic forms are defined here by the relations

$$\begin{aligned} m_1 &= u_1 \cos\alpha, \ m_2\sqrt{2} = (u_1 - u_2)\sin\alpha, \\ m_3 &= u_2 \cos\alpha, \ m_4\sqrt{2} = (u_1 + u_2)\sin\alpha \end{aligned} \tag{3.27}$$

and are submitted to the differential condition in the time τ:

$$m_1 m_4' - m_2 m_3' + m_3 m_2' - m_4 m_1' = 0. \tag{3.28}$$

Equation (3.26) is the original *Kustaanheimo–Stiefel transformation* (Kustaanheimo and Stiefel, 1965), whereby the three-dimensional physical space is described by a four-dimensional space, under condition (3.28).

3.2 The Analyticity Condition in Regularization Theory

At this point it is better to give the floor to Paul Kustaanheimo and Eduard Stiefel themselves, because their original approach from the year 1965 has a few points worth noticing from the more appropriate perspective of scale relativity. The basic point to be taken from this moment of our knowledge is, again, that there is an objective dynamics of ideas, inasmuch as, the theory of regularization is connected here with an old mathematical necessity: *the concept of continuity.* At a first sight, one might say that the approach of Kustaanheimo and Stiefel aspires at being altogether just an 'educated guess' for the right procedure of regularization in the three-dimensional case. In fact, this is how they present it themselves. For, indeed, as the idea of *analytical complex transformation* involved in the Levi-Civita procedure allows us to grasp, there cannot be but a guess as to what is going on with the regularization in the three-dimensional case. The basis of this statement of ours is that connection we just mentioned above, with the concept of mathematical continuity: the generalization of analyticity in the multidimensional case, as presented by Kustaanheimo and Stiefel, emerges directly from a natural property of the Cauchy–Riemann conditions of analyticity in the two-dimensional case. And as in the multidimensional case such conditions may very well not be quite as natural as in the typical two-dimensional case — as, in fact, these last ones are not typically the natural continuity conditions from the one-dimensional case — the analyticity conditions thus defined, and, with them, the associated regularization procedure, obviously cannot be but an 'educated guess' indeed.

To wit, in the two-dimensional case, the components of 'physical' vector (3.24) are, obviously, analytic (holomorphic) in the components of the regularizing parameter, insofar as they can be expressed by the quadratic transformation

$$m \equiv m_1 + im_2 = u^2, \quad u = u_1 + iu_2. \tag{3.29}$$

Taken as a property defined over the complex plane of the regularizing parameter u, this analyticity property is quite clear: the Cauchy–Riemann conditions characterizing it are obviously satisfied by the components of the vector (3.24). However, if u is considered itself an analytic function defined on a complex domain, of the variable $x_1 + ix_2$ say, by (3.29) this property is 'transferred' over to the complex plane m. In that case, the complex variable u has to satisfy the Cauchy–Riemann conditions, which we can enunciate in the form of vanishing of the following two differential expressions:

$$I_1 \equiv \frac{\partial u_1}{\partial x_1} - \frac{\partial u_2}{\partial x_2}, \quad I_2 \equiv \frac{\partial u_1}{\partial x_2} - \frac{\partial u_2}{\partial x_1}.$$

These two conditions can be further brought together into a single one, which thus characterizes the general case: the two functions u_1 and u_2 must be two solutions of the same Laplace equation, $\Delta u = 0$. Now, this last condition can be construed as an outcome of the following obvious identities

$$\Delta u_1 \equiv \frac{\partial I_1}{\partial x_1} + \frac{\partial I_2}{\partial x_2}, \quad \Delta u_2 \equiv \frac{\partial I_1}{\partial x_2} - \frac{\partial I_2}{\partial x_1}.$$

Hence a way to generalize the analyticity for higher dimensions by considering that the Laplacian of the multidimensional case is linear in the derivatives expressing the Cauchy–Riemann conditions, as in this last equation for the two-dimensional case. Assuming, of course, that these conditions are still linear in the first derivatives of the components of the analytic application we are referring to. Eduard Stiefel found, and proved rigorously, that such a theory is not possible but only for dimensions 1, 2, 4 and 8 (Stiefel, 1952). This discovery is, actually, making more precise a result obtained earlier by Adolf Hurwitz in a study of composition of quadratic forms (Hurwitz, 1898).

Enter now Kustaanheimo and Stiefel, with the observation that the vector from equation (3.24) can be obtained from the vector **u** by a quasi-orthogonal matrix of the type given by Hurwitz in 1898, with entries linear in the components of this vector:

$$\begin{pmatrix} m_1 \\ m_2 \end{pmatrix} = \begin{pmatrix} u_1 & -u_2 \\ u_2 & u_1 \end{pmatrix} \begin{pmatrix} u_1 \\ u_2 \end{pmatrix} \tag{3.30}$$

and that this relation has a remarkable differential counterpart:

$$\begin{pmatrix} dm_1 \\ dm_2 \end{pmatrix} = 2 \begin{pmatrix} u_1 & -u_2 \\ u_2 & u_1 \end{pmatrix} \begin{pmatrix} du_1 \\ du_2 \end{pmatrix}. \tag{3.31}$$

We find this last relation, indeed, quite remarkable in our terms here, and with very good reasons, on two accounts: first, the linear transformation between the two vectors is here represented by *the same matrix* in the *finite*, as well as *infrafinite scale ranges*, and the entries of this matrix are expressed by *the components of the regularizing vector*. It is like the *space scale transition* is formally identical with the *time scale transition*, with an important emphasis though: the transition in finite range and that in infrafinite range are both accomplished *by the same matrix* with entries linear in the components of the parameters. Secondly, even more remarkable, these properties are preserved for a special case of a Hurwitz matrix of *rank four* in the finite scale range, which was so chosen by Kustaanheimo and Stiefel, as to give:

$$\begin{pmatrix} u_1 & -u_2 & -u_3 & u_4 \\ u_2 & u_1 & -u_4 & -u_3 \\ u_3 & u_4 & u_1 & u_2 \\ u_4 & -u_3 & u_2 & -u_1 \end{pmatrix} \begin{pmatrix} u_1 \\ u_2 \\ u_3 \\ u_4 \end{pmatrix} = \begin{pmatrix} u_1^2 + u_4^2 - u_2^2 - u_3^2 \\ 2(u_1u_2 - u_3u_4) \\ 2(u_1u_3 + u_2u_4) \\ 0 \end{pmatrix}. \tag{3.32}$$

This equation inspired our authors to consider the parameters' space as a four-dimensional space, with the real space 'emergent' so to speak, *i.e.* defined just like in the Einstein's cosmological theory, only by another method. Indeed the first three components of the vector from the right-hand side of equation (3.32) are exactly those from equation (3.26). Therefore, the approach finds thus almost naturally suggested, to consider indeed the real space as emergent from the four-dimensional regularizing parameters' space, with the coordinates expressed in terms of parameters by quadratic forms, according to the 'imbedding' scheme involving quadratic special quadrivectors, and having a vanishing fourth component:

$$\begin{pmatrix} m_1 \\ m_2 \\ m_3 \\ m_4 \end{pmatrix} = \begin{pmatrix} u_1^2 + u_4^2 - u_2^2 - u_3^2 \\ 2(u_1u_2 - u_3u_4) \\ 2(u_1u_3 + u_2u_4) \\ 0 \end{pmatrix}. \tag{3.33}$$

The problem is that the analog of equation (3.31) is no more feasible, because performing the differentiation of equation (3.33), we have to notice that in this scheme $dm_4 = 0$, unconditionally, and a relation in the infrafinite scale range would produce a 4×4 matrix with a line of zeros. However, Kustaanheimo and Stiefel took notice of the fact that an equation like (3.31) can be *enforced* as a condition of definition of the space, representing a certain connection between the infrafinite and finite ranges in regularizing parameters. Indeed, we can write an equation analogous to (3.31):

$$\begin{pmatrix} dm_1 \\ dm_2 \\ dm_3 \\ dm_4 \end{pmatrix} \equiv 2 \begin{pmatrix} u_1 & -u_2 & -u_3 & u_4 \\ u_2 & u_1 & -u_4 & -u_3 \\ u_3 & u_4 & u_1 & u_2 \\ u_4 & -u_3 & u_2 & -u_1 \end{pmatrix} \begin{pmatrix} du_1 \\ du_2 \\ du_3 \\ du_4 \end{pmatrix} \tag{3.34}$$

provided we limit the parameters to those constrained by

$$u_4 du_1 - u_3 du_2 + u_2 du_3 - u_1 du_4 = 0. \tag{3.35}$$

When referred to a continuity parameter, like the time τ before, this is exactly the condition (3.28) above. This incident compels us to a short consideration of the time as a continuity parameter. Certainly, a continuity parameter may not be 'universal', embracing, as it were, the whole change incorporated in a differential, or in a variation proper. Such a variation, taken in its entirety, can be fractal for instance, while the continuity with respect to a parameter reflects just a part of it. For instance, equation (3.35) may not be *equivalent* with equation (3.28). Fact is that equation (3.35) does imply equation (3.28), indeed, but the reciprocal of this statement is not valid. The regularization procedure, therefore, hides here an issue, of which we need for now only to take proper notice.

The essential point in need of clarification here would be a transcendence indeed, however — as the regularization procedure seems to suggest — this transcendence is dictated by the transition between *space scales*. Definitely, the time transformation involved in the Mariwalla invariance involves not the transformation defined by the differential equation $t' = r$, but the transformation defined by $t' = r^2$, in terms of the time scales. Obviously, the procedure described by Volk does not satisfy this demand, inasmuch as it is referring to a transition between infrafinite scales of time involving the first time equation. However, Claude Alain Burdet has shown that there is also a procedure of regularization based on the time definition $t' = r^2$ and this is the one needed in Mariwalla transcendence (Burdet, 1969). Let us describe, in broad strokes, the results of the specific works of Burdet [for a thorough conformity one can also consult (Burdet, 1968), a work where the applicative celestial mechanics terms are clearly defined, even mathematically].

Claude Burdet himself gets equation (3.20) too, but as a result of the *central regularization procedure, i.e.* by two oscillators referred to the center of Kepler's orbit. This regularization procedure is, indeed, related to the time definition $t' = r$, as before. However, there is one more possibility of proceeding in regularization, involving two oscillators *related to the center of force*, whereby the time scale transition is defined by $t' = r^2$. This is the focal regularization procedure. Burdet approaches it as follows: going back to the dynamic equation of the Kepler problem we rewrite it in the form

$$r\ddot{\hat{\mathbf{e}}}_r + 2\dot{r}\dot{\hat{\mathbf{e}}}_r + (\ddot{r} + \kappa/r^2)\hat{\mathbf{e}}_r = \mathbf{0}, \quad \mathbf{r} = r\hat{\mathbf{e}}_r \tag{3.36}$$

and then go over to the time τ defined by $t' = r^2$. The final result is

$$\hat{\mathbf{e}}_r'' + \left[\frac{r''}{r} - 2\left(\frac{r'}{r}\right)^2 + \kappa r\right]\hat{\mathbf{e}}_r = \mathbf{0}. \tag{3.37}$$

In view of the fact that the unit vector $\hat{\mathbf{e}}_r$ is perpendicular on its time derivative, *no matter of this time* — in fact, it is perpendicular to its first differential — we have:

$$\frac{r''}{r} - 2\left(\frac{r'}{r}\right)^2 + \kappa r = \hat{\mathbf{e}}_r' \cdot \hat{\mathbf{e}}_r'. \tag{3.38}$$

This quantity is conserved along the Kepler motion: it is the square of the rate of area swept by the position vector, given by us here in equation (3.9). Then equation (3.38) can be written as

$$\frac{r''}{r} - 2\left(\frac{r'}{r}\right)^2 + \kappa r = \mathbf{a}^2. \tag{3.39}$$

Therefore, the focal regularization portrays the position by an isotropic harmonic oscillator on the unit sphere having the center in the center of force, together with a radial harmonic oscillator, constantly forced with a force having the magnitude equal to the constant from the expression of Newtonian force governing the Kepler dynamics. Both oscillators work in the time τ defined by $t' = r^2$, and have a frequency defined by the area constant specific to the Kepler problem. Summarizing:

$$\begin{aligned} \hat{\mathbf{e}}_r'' + \dot{a}^2 \hat{\mathbf{e}}_r &= 0, \\ u'' + \dot{a}^2 u &= \kappa, \end{aligned} \qquad u \equiv \tfrac{1}{r}, \quad t' = r^2. \tag{3.40}$$

Mention should be made, that Victor Bond obtained the very same results by a more involved independent calculation, which can be taken as an alternative verification of the above theory carried on by Claude Burdet (Bond, 1985).

3.3 A Meaning only Newton would be Able to Perceive

The application of classical dynamics to problems of astrophysics has also revealed a bunch of problems related to the *space shapes* of the celestial objects, profoundly affecting our cosmological ideas. Among these, the problems of the kind of missing mass, dark matter, and such like, occupy a chief place in the modern speculations about the structure of the universe. Rarely are these speculations conducted from the point of view of Newtonian forces, partly because the Einsteinian point of view subjugated the modern positive thinking, partly because the Newtonian theory is considered as completed, not open to any improvement. Even rarer — never, we should say! — was it recognized that this issue is closely correlated with the problem of the first quantization, and that before inventing what seems to be missing, we need to learn the proper lessons from what we already have as firmly established. However, there is an example worth considering, for it has a notable suggestion that can be connected to equation (3.40). It is our opinion that, in view of the natural philosophy illustrated by his *Principia*, only Newton might be able to comprehend such a suggestion, hence the title of the present section.

Indeed, one of the speculations mentioned above, guided by the idea that even the Newtonian theory is open to improvement, and we do not need to invent concepts beyond its area in order to explain our ongoing experience, is a *modified Newtonian dynamics*, MOND in a mnemonic acronym proposed by its author (Milgrom, 1983, 1984). This theory was inspired by facts regarding the complex structure of celestial objects accessible to our observation. One can say that MOND is an example of leap outside the realm of harmonic oscillator as a fundamental physical structure of the world. This theory concerns the Newtonian forces indeed, but not only: Jacob Bekenstein and Mordehai Milgrom revealed (Bekenstein and Milgrom, 1984) that such a theory may also involve naturally the kind of forces eliminated from the theory of ideal gases by Clerk Maxwell [see Chapter 2 of the present work, equations (2.8–2.10) and the discussion there]. In order to better understand what is all about with these forces in a modified Newtonian dynamics, but in some simpler and more understandable terms, let us contemplate the following scenario of a possible generalization of the classical Newtonian forces.

We are, obviously, aware of the fact that in real celestial problems the concept of classical material point does not work properly but only in the limits where the dimensions of bodies involved in interaction are much smaller than the distance between them. Assume then that we want to introduce this condition explicitly in the theory of forces. Consider, therefore, the Kepler problem, *viz.* the archetype of celestial dynamics of any persuasion, be it classical or relativistic. The Newtonian force involved in the corresponding two-body dynamics can be obtained here as the gradient of the potential energy which, up to a sign, momentarily irrelevant for our argument, is:

$$\mathbf{f}(\mathbf{r}) = \nabla W(\mathbf{r}), \quad W(\mathbf{r}) \equiv \frac{k}{r}, \quad \mathbf{r} \equiv r\hat{\mathbf{r}}.$$

According to our 'dimension awareness', as it were, we need a first move in extending the Keplerian experience beyond of the classical condition of material point. This can certainly consist of taking the stand that this Newtonian force should be some approximation of a 'true force', expressed in terms of some specific parameters reflecting the dimensions of bodies involved in the associated dynamics of the Kepler problem. Accordingly, a natural first attitude would be, therefore, to assume that this force, and thus the corresponding potential energy, is just a *quantitative approximation* of that 'true central force', whose algebraical expression depends on the scale where the Kepler motion is represented dynamically. And this scale is obviously decided, first and foremost, by the space extension of the central body considered as creating the field of force, for instance the Sun in astrophysics, or the nucleus for the planetary atom, within microcosm. As this central body always appears as fairly spherical, it can be dimensionally described by a radius, R_0 say, which settles the scale of lengths in the problem. The potential energy of the two-body problem can then be presented as a linear function in the ratio between the dimension of the central body and the distance of the position where we calculate the magnitude of force:

$$W(\mathbf{r}) = \frac{\kappa}{R_0} \cdot \frac{R_0}{r}. \tag{3.41}$$

Now, let us say that instead of a regularization procedure of the corresponding two-body problem (Bond, 1985), we are following the idea of a field-related procedure where the dimensions of the central body enter as parameters for the description of some scale-specific close encounters (Vinti, 1966). According to an established practice, the function (3.41) may be taken as an approximation, for instance as just a linear approximation of the 'true' function. There are, of course, infinitely many such functions yielding (3.41) as linear approximation, especially when its argument (R_0/r) is small, and therefore the Kepler problem can be properly described as a dynamical problem, but we have a sound criterion of choice among these, given by thermodynamics.

Indeed, if it is to maintain the ideal gas as a model for a certain cosmological scenario, this model should contain the long distance logarithmic forces [see Chapter 2, equation (2.7) ff]. For instance, these forces may prevail inside the space of a physically stable structure which, *taken as fundamental constitutive element of the universe*, would be capable to confer it the property of isotropy. To wit, this is the case of galaxies, with respect to which only, the universe around us has the property of isotropy (Tolman, 1934). In such a case, the potential energy of the two-body problem whose approximate expression is the linear function from equation (3.41), should be of the form:

$$W(\mathbf{r}) = \frac{\kappa}{R_0}\ln\left(1 + \frac{R_0}{r}\right). \tag{3.42}$$

If the ratio under logarithm is large enough with respect to unity, the potential energy is practically a logarithmic function that can very well account for an incidental thermodynamics of the fundamental physical unit of this universe. For the theoretical description of such a universe the ideal gas model may have an application secured by the presence of logarithmic forces. This can be, for instance, the case of confinement of matter in the nucleus of a planetary atom (Mazilu, Ioannou, Diakonos, Maintas, and Agop, 2013). Thus, if in our scenario we maintain the idea of force as a gradient, instead of regular Newtonian force we should have from equation (3.42) a central force having the expression

$$\mathbf{f}(\mathbf{r}) = -\frac{\kappa}{r(r+R_0)}\frac{\mathbf{r}}{r}. \tag{3.43}$$

Thus, in this case the Newtonian force appears as a composite force of magnitude given by the (algebraic) sum of the magnitudes of two logarithmic forces. The theory of Bekenstein and Milgrom introduces this kind of forces in a manner which, according to the ideas of Dennis Sciama, reminds us of the inertial forces. Let us, therefore, get into some details of this theory.

Based on our discussion so far, the essential philosophy of MOND can be presented based on the idea that the *inertial mass is different from the gravitational mass*, in a manner that we already displayed before in Chapter 1. Recall equation (1.17), where we alluded to a fundamental difference between the two kinds of masses for a Hertz particle — inertial and gravitational — as due to the presence of two other of its physical attributes besides gravity, namely the charges, electric and magnetic. In that case, the second principle of dynamics for a Hertz material particle would be written as

$$q\,\mathbf{a} = \mathbf{f} \quad \therefore \quad m\,\mu(a/a_0)\,\mathbf{a} = \mathbf{f}, \tag{3.44}$$

where $\mu(x)$ is a continuous function of its argument. For the specific case of Chapter 1, we actually have

$$\mu = \sqrt{1 - \frac{q_E^2 + q_M^2}{m^2}}$$

but we are entitled to consider this as just a special case valid only for a simple material particle: for the more complicated fundamental physical structures of a universe, we can very well assume the stand of Bekenstein and Milgrom. In this case the function $\mu(x)$ may well depend on the ratio between the acceleration a, and a 'gauging' acceleration a_0, specific to the universe in question. In such a case equation (3.44) is formally identical with the fundamental equation of MOND.

We feel that, in order to crystallize our understanding of the issue of forces, we have to frame the problem of inertia in the general positive knowledge related to this subject. And, of course, nothing seems better for this task, than the first move to escape the very principle of inertia, *viz.* the move that led to the theory of general relativity. It involves the obsessive idea that lies down at the basis of the whole physics of the last century, namely to *always recur to measurements* as the last resort in deciding the truth. In specific terms, embodied here in the words of Einstein himself, the possibility of escaping this principle, rests with another principle more firmly engraved in our experience due to the opportunity of measurements:

> ... The possibility of explaining the *numerical equality* of inertia and gravitation by the *unity of their nature* gives to the general theory of relativity, according to my conviction, such *a superiority over the conceptions of classical mechanics*, that all the difficulties encountered in development must be considered as small in comparison with this progress.
>
> What justifies us in *dispensing with the preference for inertial systems* over all other coordinate systems, a preference that seems so securely established by experience? The weakness of the principle of inertia lies in this, that it involves *an argument in a circle*: a mass moves without acceleration if it is sufficiently far from other bodies; we know that it is sufficiently far from other bodies only by the fact that it moves without acceleration. *Are there at all any inertial*

> *systems for very extended portions of the space–time continuum, or, indeed, for the whole universe*? We may look upon the principle of inertia as established, to a high degree of approximation, for the space of our planetary system, *provided that we neglect the perturbations due to the sun and planets.* Stated more exactly, *there are finite regions,* where, with respect to *a suitably chosen space of reference,* material particles move freely without acceleration, and in which the laws of the special theory of relativity, which have been developed above, hold with remarkable accuracy. Such regions we shall call "Galilean regions". [(Einstein, 2004); p. 60; *our Italics*]

Clearly, the experimental quantitative results — the numerical equality of the two kinds of masses — are here taken to prevail over any hypothesis whatsoever: after all, the measurement is, as we said, the new criterion of truth of the whole 20th century. In this specific case, our experience provides numerical values of masses, and only in order to properly use these values in developing our knowledge we need to replace the principle of inertia by a more profound principle. As the story of how the general relativity deals with this problem, and how its results contain those of the classical mechanics, are by and large relatively well known, let us limit our discussion strictly to the very word of the above excerpt.

The 'reference system' of the classical mechanics was always preferential: a Galilean inertial frame. In any other kind of frame we would not be able to interpret physically the mathematical rendition of the experience embodied in the classical laws of motion. Practically speaking, though, such a reference frame does not even exist: that 'neglect of perturbations' of which Einstein is talking about in this excerpt, is a pure fiction, for, in fact, our whole experience is built precisely upon contingencies given by those very perturbations. Consequently, we cannot say what their neglect would mean, but only speculatively, based upon some different kind of experience. As a matter of fact, it is only for delimiting such a feature of our experience that Newton invented the concept of force in its permanent instance! The special relativity takes over the classical reference frame, defining it *under the Newtonian force concept according to the principle of inertia*, and adapting it by the condition of impossibility of uniform rectilinear motions having a speed higher than that of propagation of light in vacuum. Then, again, the rectilinear constant speed motion was suggested by the portion of our experience referring to the *propagation* of light. However, the constant speed rectilinear motion is plainly one of the possible classical characteristics *of motion*, and therefore it should be a classical characteristic of the special relativity just as well.

Now, the second law of Newton, which is the mathematical expression of the principle of inertia in classical dynamics, and which is the root of entire mathematics associated with this rational mechanics — even amended according to special relativity! — represents physically a comparison between the forces acting upon a material point *from exterior* and the *interior force of inertia.* As known, even the motion of free fall can be described by the second principle of dynamics, which is the observation that lured our spirit into including the gravitation among those exterior forces, in the first place. Here is, however, the crux of a problem to be theoretically solved: according to our experience, *the inertia of bodies is only accidentally manifested*, at the moments when an external force is applied on a body. For what is worth, the inertia might not even be a force. But, if considered under this concept, it should be taken as a permanent force; in terms of time it is an eternal force. Gravitation too, if represented by a force, this one should be a permanent force, eternal in time. Therefore, the *principle of inertia should not be applicable to it*, inasmuch as this principle, being based on human experience, involves incidental forces. However, the equation of Galilei for the motion involved in a natural free fall on Earth seems to testify for the applicability of the second law of Newton even in this case: when it comes to action, it is *as if the gravitation acts accidentally* like any other force of human experience. It is this observation that has been raised to the rank of a principle in the classical physics: gravitation *is indeed* a force like any other one!

At this moment, equation (3.40) may come in handy: it can be taken as showing that when the action of the inertia is referred to a time decided by the 'impulses of gravitation', as it were, this action, represented by a force, is a solution of a second-order differential equation. This is how Newton proceeded, anyway, in order to insure, at least for himself we should say, the legitimacy of the principle of inertia for the classical dynamics involved in Kepler's problem: he assimilated the action of gravitation with that of a continuous sequence of collisions (see Brackenridge, 1995, p. 25). And this is why we uphold that 'only Newton would understand'! In hindsight, we can say that he may have realized that the instantaneous action of the gravitation has to be quantitatively specified at any rate. Newton's insistence on the construction of the calculus of fluxions, in order to render his dynamics mathematically, leaves no doubt that such a conclusion is entirely feasible. However, in the system of dynamics he invented, there is no other possibility of proof than via the third principle of dynamics, which is *a priori* taken to transcend the time scales: to Newton it is effective even between permanent forces and collision forces, whose action is accomplished within infrafinite time intervals. It should be quite significant, in our opinion, that two centuries after Newton, Poincaré found the third principle of dynamics in default when introducing the Coulombian forces into dynamical play (see Chapter 1 here). The problem is then to define the ephemeral forces, and this is the point where equation (3.40) comes in handy: the time $\boldsymbol{t}$ is defined in this equation by the measured-time average of the Newtonian force, *i.e.* by a quantity proportional to the impulse of that force. The measured time is here the time involved in the dynamical problem describing the Kepler motion. One can say that this time average of the permanent force is actually a finite version of the infinitesimal time scale impulse. In other words, the average of force over a finite time interval (the quantitative definition of the impulse of a force) determines the new time of a special dynamics to be applied in this case. For, as the inertia *acts permanently* in the measured time, one needs the second of equations (3.40) to describe the accidental behavior of this force, as it appears to our experience, in order to apply the second principle of Newton.

Einstein's idea from the excerpt above, is concerned nevertheless with an amendment brought about by the experience subsequent to the Galilei moment of knowledge: in free fall, the inertia principle seems to be valid in any measured time, but only on *time portions* connected with certain *space portions* with respect to the Earth surface. In other words, the reference frame in which we can apply the inertia principle to the gravitation phenomenon is preferential only in a 'space of reference', as Einstein would say. This space must be extracted from experience and represents a separate hypothesis for which we need different criteria, especially from the measurement perspective. From this point of view the *Wien–Lummer enclosure* and *Einstein elevator* are the two versions of the space of reference that helped effectively build the modern theoretical physics. The first one helped understanding the characteristic *cosmological invariance of the spectrum of light, which thus must be a Planck spectrum*, the last one helped *replacing the principle of inertia by the principle of equivalence*, thereby leading to the identification of inertial and gravitational masses. It is in this well-established quantitative fact of measurement (Eötvös, 2008) that Einstein was able to see that tremendous superiority of general relativity over the classical mechanics, and to build it accordingly, as one can perceive from the above excerpt. Perhaps it is worth detailing this statement a little further, in order to settle our ideas.

Let us reiterate once again that, as a force, the inertia should be a permanent one — at least for the classical natural philosophy — and so should be the action of gravitation. Therefore, in order to apply the second principle of dynamics in describing the gravitation as universal — as a field, in Einstein's expression — one must do some tricks, like, for instance, a regularization that confers some field properties to the coordinates of motion. The Einstein's natural philosophy, guided by Mach's principle, cuts the Gordian knot here: it is all about *a direct relationship* between the two permanent forces. 'Direct relationship' means no 'regularization', no scale transition, no second principle of dynamics — nothing of the kind. In general relativity *the inertia does not occur in accidental manifestations* like in the classical mechanics. Logically, if we can talk of forces here, the relation between inertia and gravitation is one of which the classical mechanics cannot even conceive: a relation between two permanent forces, for none of the two can ever be accidental as such. This is why Einstein himself uses the word 'equivalence', obviously more appropriate in such an instance, and consequently the principle of inertia is replaced with the principle of equivalence! The application of this principle asks, nevertheless, for a subtle physical analysis, that establishes the basis of analogies of general relativity, *viz.* a special phenomenology. In this phenomenology the quantitative identity between inertial and gravitational mass is taken at the level of essential physical identity, and the classical acceleration thus becomes a continuous field, just like the gravitational acceleration. No force as known to the human experience has any place here, and the impression has been created that the force is altogether eliminated from any relativistic scenario. But the most important aspect of this phenomenology is its continuity, bearing on the continuity of the gravitational field. The matter is thus still defined in the classical way, *viz.* by density, but this time the density is 'smeared': the matter is misleadingly extended over space, as if it would fill that space continuously. From time to time one is forced, however, to face the reality, as in the cosmological episode of Einstein, that founded the modern cosmology, reported by us in its essential traits in Chapter 1 above. This cosmological episode, therefore, seems to teach us that even the matter of the general theory should, indeed, *fill the space at its disposal*!

Now, Milgrom's theory is connected to this very issue, in an attempt to put things in order *according to experience.* This experience is embodied here by the results of experimental and theoretical astrophysics, which clearly demonstrate that the matter is accessible to our spirit through *fundamental physical structures* that, as such, should depend on the space scale. However, that experience presents the spirit with a due request: to explain the discrepancy between the observations, for these fundamental structures seem to accept contradicting explanations. Indeed, it is only a matter of cursory notice that in observing these physical structures at the cosmic level, there is always incongruity, even between the different possibilities of humanly perceiving their matter, to say nothing of the possibilities of their theoretical description! There can be no question of continuity in the distribution of matter here: the cosmological solution of Einstein type, for instance, can occur at any scale, not just cosmologically. The spirit is thus reminded one critical fact that escaped to it until now: *the 'permanence' of the inertia is essentially different from the 'permanence' of gravitation*, and that the Mach's principle has to be taken, like we stated before, *cum grano salis.* To wit, assuming a Newtonian stand, while the gravitation acts permanently between material structures separated by distances that make material points out of the fundamental physical structures, the inertia acts permanently only between material structures *situated at infinity* with respect to each other. The inertia principle should thus be reinstated: the acceleration is acceleration again, no matter of force, and the gravity should be treated accordingly. The theory of Jacob Bekenstein and Mordehai Milgrom represents such an approach, referring, however, *only to gravitation* from among the two permanent forces (Bekenstein and Milgrom, 1984). Thus, the force of gravitation is the only permanent force of the theory, and it should be described by field equation, as in the general relativity. But then a necessary twist occurs on reasoning: if we accept the Sciama's thesis that the inertia is represented by logarithmic forces, the forces of inertia are to be found in this theory just naturally, right in the limit where we are searching for them: *at infinity.* Perhaps it is better, though, to finally give the floor to these authors.

The Newtonian point of view is represented, as usually in physics nowadays, by a 'Poisson stand' as it were. And the Poisson equation, taken here in the form $\nabla^2\varphi_N = 4\pi G\rho$, is thought of as being derivable from the Lagrangian:

$$L = -\iiint d^3\mathbf{r}[\rho\varphi_N + (8\pi G)^{-1}(\nabla\varphi_N)^2]. \tag{3.45}$$

The index N stands here for 'Newtonian'. This qualification refers to a potential φ, generated by the matter of density ρ, and describing a field of gravitational constant G. Our authors' attitude is the one already mentioned above, namely of thinking of the Newtonian situation as just of an approximate one in terms of forces. The problem is then to discover the exact field, and this is a matter of assumption. Their essential assumption is that the Lagrangian referring to the *true* field should be of the form

$$L = -\iiint d^3\mathbf{r}\left\{\rho\varphi + (8\pi G)^{-1}a_0^2 F\left[\left(\frac{\nabla\varphi}{a_0}\right)^2\right]\right\}, \tag{3.46}$$

where F is an arbitrary function of its argument. Which argument involves the square of a ratio between the acceleration characteristic to the *true gravitational field*, described by the potential φ, and a reference *scaling acceleration*, a_0. The *field equations* are then the Euler–Lagrange equations corresponding to this Lagrangian:

$$\nabla \cdot [\mu(|\nabla\varphi|/a_0)\nabla\varphi] = 4\pi G\rho, \quad \mu(x) \equiv F'(x^2), \tag{3.47}$$

where the accent means derivative with respect to indicated independent variable.

Now, Bekenstein and Milgrom adopt what we think is a normal standpoint: they assume that there is also a Newtonian field at the same density of matter. This means that the difference between the *true* acceleration and the *Newtonian* one can be described then by the curl of a vectorial field, as follows:

$$\mu(g/a_0)\mathbf{g} = \mathbf{g}_N + \nabla\times\mathbf{h}, \quad \mathbf{g} \equiv -\nabla\varphi; \quad \mathbf{g}_N \equiv -\nabla\varphi_N. \tag{3.48}$$

The immediate conclusions are, in our opinion, staggering: as we mentioned before, the logarithmic forces should indeed be contained in the 'modified' Newtonian theory. Again, the Newtonian forces should be there as well. Their quantitative difference is obvious: it resides in distance as we said — they fade away differently — and this result is quite a clear outcome of the theory, in spite of the fact that it is only qualitatively defined, as a convenient limit. However, the proper qualitative difference obviously depends on the functional form of $\mu(x)$, making out of it an essential tool of the trade. For, indeed, this qualitative definition is to be clearly recognized in the inertial mass. Quoting from the original:

> We shall show that at large distances from a *bound object* of total mass M the curl term in equation (3.48) decreases faster with r than the other two terms. We thus have in this limit
>
> $$\mu(g/a_0)\mathbf{g} = -G\mathbf{r}/r^3 + \mathbf{0}(r^{-3}) \equiv \mathbf{g}_N + \mathbf{0}(r^{-3}) \qquad (3.49)$$
>
> as in equation (3.44).
> For our theory to satisfy the assumptions of MOND we identify μ in the field equations with that of equation (3.44). In particular, as we require $\mu(x) \approx x$ for $x << l$; we find for an *arbitrary bound system* of mass M that
>
> $$\mathbf{g} \xrightarrow{t\to\infty} -(MGa_0)^{1/2}\mathbf{r}/r^2 + \mathbf{0}(r^{-2}) \qquad (3.50)$$
>
> and thus, in this limit,
>
> $$\varphi \to -(MGa_0)^{1/2}\ln(r/r_0) + 0(r^{-1}) \qquad (3.51)$$
>
> where r_0 is an arbitrary radius. This potential leads to an *asymptotically constant* circular velocity
>
> $$V_\infty = (MGa_0)^{1/4}$$
>
> as observed in the *outskirts of spiral galaxies.* [(Bekenstein and Milgrom, 1984), *our Italics, a/n*]

The 'arbitrary bound system' here is what we have called before a 'fundamental physical constitutive unit' of the universe at a certain scale, *i.e.* that unit to be considered in the process of counting when estimating the density of matter. Such a unit is not unconditionally a Hertz material particle, or a harmonic oscillator, for it has a complex physical structure revealed by our observations. The general conclusion is that the inertia [equation (3.50)] can be compared with gravitation [equation (3.49)] only in conditions depending on the function $\mu(x)$, therefore on the constant a_0. As the theory is now, this is an empirical constant, but the last observations, and the equation (3.51) above may be able to indicate where to find it: in a *possibility of wave-mechanical interpretation.* And this possibility only occurs in a special region of space, where the 'principle of equivalence' should be valid.

3.4 Conclusion: The Coordinate Space and the Ordinary Space

It is as if the theory of Bekenstein and Milgrom indicates the necessity of a new constitutive unit of the universe, much in concordance with the 20th century cosmology. Indeed, the potential (3.51) is characteristic to the intensity of a vortex, as the 'spiral galaxies' mentioned by the last remark in the excerpt above. This fundamental constitutive unit of the universe is the one with respect to which the universe is isotropic, according to some modern standards (Tolman, 1934). It replaces the classical Kepler unit in a *gravitovortex model* of the universe, as the author of the model baptized it, and even made out of it the universal model of the world we inhabit (Popescu, 1982, 1988). What are the incentives of such a theory? Start by noticing that, as if in passing, Bekenstein and Milgrom point out a fact that we find particularly important: their field theory *is prone to an interpretation*, just like the original Einsteinian theory that partly inspired it. For, indeed, this field theory is equivalent to a theory referring to an irrotational fluid, and the fluid is the condition *sine qua non* for a proper interpretation in the wave-mechanical approach. Only, once again, this fluid of Bekenstein and Milgrom is *irrotational*, and its stationary flow is not capable of 'carrying too much', as it were: it needs to be properly completed with rotation properties, as in fact the theory demands. Quoting in a free format from the original work:

> It may also be useful to note that our field equation is *equivalent to the stationary flow* equations of an *irrotational fluid* which has a density $\rho \equiv \mu(|\nabla\varphi|/a_0)$ a *negative pressure* $-(1/2)a_0^2F[(\nabla\varphi)^2/a_0^2]$, a *flow velocity* $\nabla\varphi$, and a *source distribution* $4\pi G\rho$. The fluid satisfies an *equation of state* giving the pressure as a function of density as $-(1/2)a_0^2F\{[\mu^{-1}(\rho)]^2\}$ (Bekenstein and Milgrom, 1984), *our Italics, a/n*].

Notice, however, that it is a matter of convenience the fact that φ should be *a potential of velocities* or *a potential of accelerations*: for the field equations, we need the potential of forces, while for interpretation we need the potential of velocities, and the flow velocity of Bekenstein and Milgrom is actually that 'exact' field velocity that makes the Lagrangian (2.35) vanish. From this point of view, the classical dynamical solution of the Kepler problem only provides a reference velocity.

This all means that, operationally, the interpretation is far from being realized by just an ensemble, and this raises a series of major problems. To the extent to which the scale transition is involved, Darwin's labeling of interpretation as 'translation' should be also referring to the very physical structure described by a wave function, which, while mathematically described, in the spirit of initial work of Schrödinger, in the 'system's configuration space', is nevertheless interpreted as an ensemble of positions, which cannot be but in a reference frame, therefore in a 'ordinary space'. This is, again, quite a classical characteristic of a theory of continua for which we shall settle in the present work. It is, obviously, at odds with the orthodox interpretation of the wave function, which is referring, almost exclusively to a weight function *representing probabilities*, not anything else. In thus accepting it the concept of interpretation raises quite a significant concern, in hindsight even critical, which tends to show up especially if we forget about that objective connotation that Schrödinger assigned to his wave function, the one involving the idea of a charge distribution over the space of system.

Incidentally, we have to recognize, though, that this is, in fact, the largest level of acceptance ever possible, inasmuch as it includes, as only a particular case, that experimental level invoked by Darwin in the definition of interpretation, a fact that shall be obvious along our work. The concern we are talking about is conspicuous, if we may say so, and comes with that notable dichotomy regarding the problem of space, which we have mentioned above, but needs to be pinpointed as such. That dichotomy of space actually confounds the whole human knowledge of all times — no matter if natural philosophical, purely philosophical or simply technical in general — being, in fact, unrecognized as such even today. It is the difference between what is *philosophically* accepted as 'the ordinary space', and what is *scientifically* accepted as 'the coordinate space'. However, it was recognized as a problem for the wave mechanics, from the very beginning. Quoting again from Darwin:

> In dealing with the interpretation we have touched on one of the great difficulties which have made it hard to gain physical insight into the wave theory. This is the fact that *the wave equation is not in ordinary space, but in a coordinate space*, and the question arises *how this coordinate space is to be transcribed into ordinary space.* It would appear that most of the difficulty has arisen from *an attempt to apply it illegitimately to enclosed systems, which are really outside the idea of space.* In most of the problems we shall discuss the question hardly arises, but where it does *the correct procedure is so obvious that there is no need to deal with it in advance.* It is tempting to believe *that this will be found to be always the case.* [(Darwin, 1927); *our emphasis*]

It is 'tempting', indeed, to assume that the problem will not pop up, but the evolution of physics proved that the case is quite contrary: Darwin was way too optimistic! We 'need to deal with this issue in advance', indeed: more precisely even before we start anything physical, for the very existence of wave mechanics and quantum mechanics is conditional on the measurement. As it turns out, physically the fact of the matter rests with the *identification* of the *length*, involving the "coordinate space" only, with the *distance*, which involves "the ordinary space" only. And this identification is just a special instance of a universal relationship, fundamental in the description of any universe according to actual human knowledge, which has to be recognized as such.

Indeed, it would appear from the above excerpts from Darwin that the coordinate space is intimately connected with the idea of enclosed physical systems, which is 'outside', so to speak, of the space concept derived from our intuition. However, inasmuch as, physically speaking, we have always to deal only with 'enclosed systems', we need either to bring the 'ordinary space' under this concept of coordinate space, or to bring this last concept under that of ordinary space. This is to be done in physics, as everywhere for that matter, with the aid of two tools: a *clock*, to normalize the perception of *one* physical body, and a *coordinate system*, to normalize the perception of *many* physical bodies. These are the two essential features — or *differentiae*, using a philosophical label — of a general concept of *reference frame*. The starting point of such a construction is the one indicated by the theory of regularization above: *the concept of analyticity*.

Chapter 4

A Natural Condition of Analyticity

As we have seen before, the Kustaanheimo–Stiefel regularization could not be accomplished directly, but via a four-dimensional parameter space. This fact is due to the very definition of analyticity: according to Stiefel's theorem, if the analyticity is expressed by the Cauchy–Riemann linear relations, there is no possibility of regularizing but for dimensions 1, 2, 4 and 8 (Stiefel, 1952). As long as we think of regularization through the property of analyticity, and keep up with the linearity as a property *sine qua non* of the Cauchy–Riemann conditions, this fact is pathologic: the three-dimensional case of regularization has to be considered indirectly, for it is not among those of the Stiefel's list. In fact, as long as it does not concern the theory of fluids, the theoretical physics always used, preferentially, some other conditions, than the Cauchy–Riemann ones, in order to deal with forces in the three-dimensional case. These conditions are closer to the spirit of the potential theory of forces.

Indeed, we have to recognize that the Cauchy–Riemann conditions *per se* are actually purely mathematical in character, and that the true physical conditions, at least regarding the force, are actually the *Helmholtz conditions*, which seem more appropriate for the specific purposes:

$$\nabla \cdot \mathbf{f} = 0, \quad \nabla \times \mathbf{f} = \mathbf{0}. \tag{4.1}$$

These conditions are derived from the practice with Newtonian forces: the first of them represents the force in absence of matter, while the second shows that the force is acting along the closed moving path of a material point. They lead to the Laplace equation *in space* just as naturally as, in the two-dimensional case, Cauchy–Riemann conditions lead to this equation. However, in space the things are more complicated than in the plane. Indeed, in space we may have to do with two potentials, ϕ and $\mathbf{A}$ say, not one as in the two-dimensional case, and the conditions of analyticity in Helmholtz's stand is expressed by the fact that, while ϕ should be a *harmonic scalar function*, $\mathbf{A}$ should satisfy a more involved second-order differential equation:

$$\nabla^2\phi = 0, \quad \nabla \times (\nabla \times \mathbf{A}) = 0. \tag{4.2}$$

Indeed, assuming that the second Helmholtz condition (4.1) *is an identity*, the situation is realized with a function ϕ whose gradient is the force $\mathbf{f}$. Inserting this into the first of Helmholtz's conditions, gives for this function the Laplace equation from (4.2): the function ϕ is not arbitrary, but needs to be a solution of the Laplace equation. On the other hand, if we start from the first condition (4.1) *as an identity*, then this situation may be realized by a vector function $\mathbf{A}$, whose curl is the force $\mathbf{f}$. Inserting this into the second of Helmholtz's conditions, we get the second one of the conditions (4.2). In view of their tremendous importance for physics, mostly for the problem of interpretation, we should obviously stick with the three-dimensional Helmholtz conditions. $\mathbf{A}$ satisfies the Laplace equation only in case it is divergenceless.

In the two-dimensional case, the Helmholtz conditions (4.1), referring to the vector $\mathbf{f}$ having the components (f_1, f_2), can be taken as Cauchy–Riemann conditions for the vector of components $(f_2, -f_1)$ perpendicular to $\mathbf{f}$. Now, as the Cauchy–Riemann conditions in plane are conditions of existence of the Cauchy integral of the complex function corresponding to a vector, one can seek for a generalization of the Cauchy integral in the multi-dimensional case, which would be quite appropriate for building a physics of the fluid interpretative ensembles, for instance. For the three-dimensional case such a formula is (Fulton and Rainich, 1932):

$$4\pi\mathbf{f}(\mathbf{0}) = \oiint \left(\mathbf{f}(\mathbf{r}) \cdot \frac{\mathbf{r}}{r^3}\right)(dS) - \oiint \left[\mathbf{f}(\mathbf{r})\left(d\mathbf{S} \cdot \frac{\mathbf{r}}{r^3}\right) + \frac{\mathbf{r}}{r^3}[d\mathbf{S} \cdot \mathbf{f}(\mathbf{r})]\right], \tag{4.3}$$

where the surface of integration surrounds the position of origin. This formula represents the force $\mathbf{f}(\mathbf{r})$ as calculated in a point — in this specific case the origin — from the values of this force on a surface surrounding that point. It has the particular property of giving a null force, if the point is outside the surface. The formula (4.3), which will be called the *Fulton–Rainich formula* from now on has a particular significance from the point of view of interpretation process. In order to understand this significance, we need the general transport theorem referring to physical characteristics defined by their densities. For an arbitrary differential form the *transport theorem* sounds (Flanders, 1973; Betounes, 1983); see also Mazilu *et al.* (2019)

$$\frac{d}{dt}\int_{\phi_t(V)} \Omega = \int_{\partial\phi_t(V)} \mathbf{v}\cdot\mathbf{\Omega} + \int_{\phi_t(V)} \mathbf{v}\cdot(d\wedge\Omega),$$
$$\mathbf{v} \equiv \frac{d}{dt}\mathbf{\Phi}_t(\mathbf{r}), \quad \mathbf{\Phi}_0(\mathbf{r}) \equiv \mathbf{r}. \tag{4.4}$$

Consider the differential form, representing an *elementary physical quantity* related to a continuum, given by

$$\omega \equiv X \cdot dx^1 \wedge dx^2 \wedge dx^3.$$

Then we shall have, according to transport theorem transcribed here in equation (4.4):

$$\frac{d}{dt}\int_{\phi_t(V)} X\cdot(dx^1\wedge dx^2\wedge dx^3) = \int_{\partial\phi_t(V)} X(\mathbf{v}\cdot d\mathbf{S})$$

because the second term in (4.4) is null for a 3-form in the three-dimensional space. With this result in hand, we can solve the Reynolds' problem (see Mazilu *et al.*, 2019, Chapter 10): calculate the time variation of the volume average of a certain vector. Such a volume average is given by an integral like the left-hand side of the equation above. In the case of a vector, we have three replicas of the integral, representing the components of the vector. Therefore, we can write

$$\frac{d}{dt}\int_{\phi_t(V)} \mathbf{X}\, dV = \int_{\partial\phi_t(V)} \mathbf{X}(\mathbf{v}\cdot d\mathbf{S}) \tag{4.5}$$

which shows that the rate of variation of a vector average in a point, is dictated by the flux of the vector through a surface surrounding that point. If the vector $\mathbf{X}$ is our force from equation (4.3), the terms in square brackets are of the form (4.5). Specifically, the first term under the square bracket of equation (4.3) can be written as the

$$\frac{d}{dt}\int_{\phi_t(V)} \mathbf{f}(\mathbf{r})\, dV$$

provided the rate of variation of the surface is given by a velocity:

$$\frac{d\mathbf{r}}{dt} = \Gamma_1 \frac{\mathbf{r}}{r^3} \tag{4.6}$$

if $\mathbf{r}$ belongs to a position on the surface. Here Γ_1 is a constant necessary for dimensional accordance, having the dimensions of a time rate of a volume (L^3T^{-1}). This way the Newtonian force — or, to be more precise, a vector proportional to it — has the clear interpretation of the rate of variation of the position itself, when this position is on the surface delimiting the volume where we are calculating the average of our force. Inasmuch as this position is on a surface surrounding the origin, we shall have to elaborate further on the variation of that surface. Likewise, the second term in the square brackets of (4.3) can be interpreted as the variation of the mean of Newtonian force itself:

$$\frac{d}{dt}\int_{\phi_t(V)} \frac{\mathbf{r}}{r^3}\, dV$$

provided the variation of the surface delimiting the space region serving for average is calculated in the direction given by the force whose value is supposed to be calculated in origin:

$$\Gamma_2 \frac{d\mathbf{r}}{dt} = \mathbf{f}(\mathbf{r}) \tag{4.7}$$

if $\mathbf{r}$ belongs to the surface. Here Γ_2 is another constant of dimensional adjustment, having the dimensions of a momentum density $(L^{-3})(MLT^{-1})$. This approach provides a natural way to replace the second principle of dynamics by the theory of surfaces, in concordance with Wigner's principle [see Mazilu *et al.*, 2019, Chapter 2, equation (2.19) and discussion], having as expression the equations (4.6) and (4.7). The first term in equation (4.3) is a special kind of *average of the virial* of force to be calculated in origin, involving the components of *solid angle.* Thus, in view of the transport theorem, the Fulton–Rainich formula (4.3) has a very suggestive reading the special virial of a force is given by its value to which we have to add two surface contributions. One of these contributions is generated by the surface variation of our force along the direction of Newtonian forces, the other is generated by the variation of the Newtonian force along the force whose virial we are calculating.

Assume that the host space of the matter has a metric which, in some general coordinates, is given by the quadratic form

$$(ds)^2 = g_{ik}(\mathbf{r})dx^i dx^k.$$

In this case the Laplace's equation takes itself the general form

$$\Delta\psi \equiv \frac{1}{\sqrt{g}}\frac{\partial}{\partial x^k}\left(\sqrt{g}g^{kj}\frac{\partial\psi}{\partial x^j}\right) \tag{4.8}$$

which can be further written as

$$\Delta\psi \equiv g^{ij}\frac{\partial^2\psi}{\partial x^i\partial x^j} + \Gamma^k\frac{\partial\psi}{\partial x^k}.$$

Here we used the notation

$$\Gamma^j \equiv \frac{1}{\sqrt{g}}\frac{\partial}{\partial x^k}\left(\sqrt{g}g^{kj}\right) = \frac{\partial g^{kj}}{\partial x^k} + \frac{1}{2}\frac{\partial(\ln g)}{\partial x^k}g^{kj}. \tag{4.9}$$

Now, equations (4.8) and (4.9) show that we can write

$$\Gamma^j \equiv \Delta x^j \tag{4.10}$$

thus offering mathematical meaning to Γ^j: they are always obtained from the functions representing the coordinates, by applying the Laplace operator. It turns out that these three functions do not represent always a vector. However, if the coordinates are three harmonic functions, *i.e.* if they are three independent solutions of the Laplace equation, these quantities are zero, and this is an essential fact (Diósi and Lukács, 1985). It not only simplifies the calculations, but turns out to have importance of principle in a straightforward generalization of the Cauchy–Riemann conditions.

First of all we need to notice that, according to equation (4.9), the usage the harmonic coordinates actually comes down to a restriction of the *space variation* of the metric tensor. This gives, in fact, the true physical meaning of such a condition. At least in the case of constant curvature, the metric tensor contains the deformation of the space filled with matter, and in such a case it should also contain information on the forces acting inside the matter. Consequently, if we assume that the matter is in a direct relationship with the space it fills, in such a way that a constant density of matter means constant curvature of space, the solutions of Laplace equation corresponding to the metric in matter can, and we claim that they should, be taken as harmonic coordinates. Let us elaborate a little on this statement, in order to explain the motivation of some of our ideas.

From a historical point of view, the harmonic coordinates have been apparently introduced by Théophile de Donder in 1921, in connection with the ideas of general relativity. Vladimir Fock, however, insisted at length on the importance of harmonic coordinates from the point of view of natural philosophy, albeit in the theoretical arena of general relativity (Fock, 1959). There, the harmonic coordinates are four independent solutions of the homogeneous D'Alembert equation, *viz.* the four-dimensional equivalent of Laplace equation from the three-dimensional case. Now, not all the physicists agree with the idea of introducing the harmonic property for coordinates, and not without good reasons, we should say. Indeed, Fock insists on the fact that only the harmonic coordinates can confer legitimacy to some sound ideas about the physical universe, such as Copernicus' idea: in any other coordinates it does not make sense to distinguish between the Copernican and the Ptolemaic systems. If anything, in the three-dimensional case, regularization theory taught us that there is an essential difference between the dynamics enticed by the Kepler problem at different scales of space. And thus, one can say that in the three-dimensional case, while the condition of harmonicity may appear as natural at a certain space scale that suggests harmonicity, marked, for instance, by the dynamics of the oscillators (see Gallavotti, 2001), it might not be quite as natural at some other scale. Therefore, the two systems, Ptolemaic and Copernican should not be equivalent, by any means: usage of the harmonic coordinates seems only particularly justified. On the other hand, though, the general relativity is certainly describing that space scale where the Copernican idea is valid. The common opinion of the opponents of the thesis of privileging the harmonic coordinates, can be simply summarized by the fact that, from the point of view of general relativity, should be immaterial which coordinates we use in the description of the universe. Therefore, Copernicus or Ptolemy — it hardly matters! And if the general relativity is considered as a straight denial of the classical Newtonian theory of gravitation, then it hardly matters, indeed. However, Fock insists upon the point that, on the contrary, the Newtonian theory of gravitation is actually intimately involved in the general relativity as Einstein constructed it, which is, indeed, indisputably the case.

We recalled here this argument in the manner we did, for it has a certain importance, but *in the theory of space* (not of spacetime!), and it takes us, again, back to the founder of modern science — Newton. We start, as Vladimir Fock himself does actually, by noticing the obvious fact that the natural position of a point in space is described in harmonic coordinates, to wit, the Cartesian coordinates. Indeed, these are solutions of the Laplace equations. Trivial solutions, is true — for they are simply first degree polynomials, and these are *by default*, so to speak, solutions of a partial differential equation of second order — but still, we can count them as such solutions:

$$\Delta x = \Delta y = \Delta z = 0. \tag{4.11}$$

This statement may be put in a vector form. Indeed, let us take the identity vector function

$$\mathbf{f}(\mathbf{r}) = \mathbf{r}$$

representing the position vector. In view of equation (4.11), it is a solution of the Laplace equation:

$$\Delta \mathbf{f}(\mathbf{r}) = \mathbf{0}.$$

In such a kind of reading, however, one can say that Newton himself erected the system of his natural philosophy by thinking, implicitly, in harmonic coordinates, for otherwise his system of the world would not be possible. Further on, we are thus entitled to ask ourselves if there is not a natural connection between the Newtonian gravitational forces and the coordinates we associate with material points. And, indeed, it seems that there is such a natural connection after all, which can be exhibited in the following manner.

For an incentive to consider the harmonic coordinates as an essential tool, we need to swing back to the theorem of Thomson (see Mazilu *et al.* (2019); see also Kelvin (1847) and Liouville (1847)], nowadays called the *inversion theorem*, and establishing the fact that if a certain function $\mathbf{f}(\mathbf{r})$ is a solution of Laplace equation in coordinates $\mathbf{r}$, then the function

$$\frac{1}{r}\mathbf{f}\left(\frac{\mathbf{r}}{r^2}\right) \tag{4.12}$$

is also a solution of the Laplace equation, but in coordinates given by inversion with respect to the origin of coordinates:

$$|\xi\rangle = \frac{\mathbf{r}}{r^2}.$$

For a neat proof of this theorem, and pertinent commentaries on it, one can consult the Gaston Darboux's comprehensive work on the geometry of curvilinear coordinates (see Darboux, 1910, Livre II, Chapitre V, §156). It is now pretty obvious that, according to this theorem, *the Newtonian gravitational force is a solution of the Laplace equation*, being the direct transformation of the identity vector function. Indeed it is the vector function proportional to

$$\frac{\mathbf{r}}{r^3} \equiv \frac{1}{r} \cdot \frac{\mathbf{r}}{r^2}$$

and necessarily satisfies the Thomson's inversion theorem, for it is of the form given in equation (4.12) for the particular case of identity vector function.

One can thus state that *the Newtonian forces should be taken as harmonic coordinates.* This certainly gives satisfaction to Vladimir Fock's idea, although only in the classical Newtonian theory. More importantly, though, is the fact that, at least in the classical limits of mechanics, the Laplace equation should be indeed the one that generalizes *the force fields*, by assimilating them with coordinates, or with the fields in general. On one hand, if the Newtonian forces are acting between Hertz material particles inside matter, the geometry of their relative coordinates is thus simply the geometry of matter: there is no need of motion in order to describe the matter, and therefore no need of a *spacetime* geometry. On the other hand, this means that, we now have to enlarge — or restrict, if one likes; it should be a matter of taste — the concept of space coordinates in order to include any three independent solutions of the Laplace equation. Thus, not just the trivial Cartesian ones are harmonic coordinates, so that we can safely say that *the space itself is a particular limit where the fields of forces and the fields of coordinates coincide.* This may even be taken as a definition of space, starting from the concept of matter, as we indicated before. But, most importantly, the property of forces of being coordinates and vice versa, is indeed connected with the property of analyticity as expressed by Laplace equation, as we shall show shortly.

Meanwhile, that idea can be made intuitively even more precise for an incidental operational use, by the following theorem due to P. Weiss: let $\phi_0(\mathbf{r})$ be a harmonic function, regular inside the sphere S of radius a, and on this sphere. Then the function

$$\Phi_1(\mathbf{r}) = \frac{a}{r}\Phi_0\left(a^2\frac{\mathbf{r}}{r^2}\right) - \frac{2}{a\,r}\int_0^a \lambda\Phi_0\left(\lambda^2\frac{\mathbf{r}}{r^2}\right)\,d\lambda \tag{4.13}$$

has the following properties:

(1) it is harmonic and regular outside sphere S;
(2) satisfies the asymptotic condition

$$\Phi_1(\mathbf{r}) \xrightarrow{t\to\infty} 0(r^{-2});$$

(3) on the sphere S it satisfies the continuity condition

$$\frac{\partial}{\partial n}\bigg|_S [\Phi_0(\mathbf{r}) + \Phi_1(\mathbf{r})] = 0, \tag{4.14}$$

where n indicates the derivative in the normal direction of the sphere (Weiss, 1944). In other words, if inside the sphere there is some confined matter *per se*, with Newtonian forces between Hertz material particles, the function from equation (4.13) gives the coordinates in the space continuum outside the sphere as forces.

There is also an interesting extension of this theorem, due to Ludford, Martinek and Yeh, for the case of the interior of sphere S. Namely, in the very same conditions as above for $\Phi_0(\mathbf{r})$, only this time outside the sphere, the function constructed by integral equation:

$$\Phi_1(\mathbf{r}) = \frac{a}{r}\Phi_0\left(a^2\frac{\mathbf{r}}{r^2}\right) + \frac{2}{a\,r}\int_0^\infty \lambda\Phi_0\left(\lambda^2\frac{\mathbf{r}}{r^2}\right)d\lambda$$

is harmonic inside and on the sphere, and satisfies the condition (4.14) (Ludford, Martinek, and Yeh, 1955). These theorems serve in calculating forces exerted by a fluid upon the objects moving through it. In hindsight though, they justify an 'evolutionary theory' for both the gravitational Newtonian force and the space coordinates in general. Specifically, the functions

$$\frac{2}{a\,r}\int_0^a \lambda\Phi_0\left(\lambda^2\frac{\mathbf{r}}{r^2}\right)d\lambda \quad \text{and} \quad \frac{2}{a\,r}\int_a^\infty \lambda\Phi_0\left(\lambda^2\frac{\mathbf{r}}{r^2}\right)d\lambda \tag{4.15}$$

represent, by the Kelvin inversion theorem, solutions of Laplace equation — they are therefore harmonic functions, and can be taken as harmonic coordinates If we are not interested in the limiting and asymptotic conditions, these solutions are just linear combinations of harmonic functions, which have a proper physical interpretation. In order to explain this interpretation, let us recall and describe again one particular aspect of the Newton's invention of forces, in its classical expression in terms of potential.

The *principal* importance of the previous theorems is tremendous for a scale relativity, as may be shown by the phenomenological scenario that follows. Consider, for instance, the case of Sun: certainly, it does not appear the same, at least from the point of view of the size, to the different planets of the solar system. This is why, when we calculate the gravitational potential, and say that it is the work necessary to bring a material point from infinity to the distance r with respect to Sun, we need to consider the fact that at infinity the Sun appears as a point ($a = 0$), like the stars appear to us, while at the stable orbit of the planet in consideration it appears as a sphere of radius $\lambda = a$. This should be the proper interpretation of the function from equation (4.15). Consequently, when calculating the work necessary to bring a material point from infinity to a finite distance r, we need to take somehow into consideration the apparent variation of the dimensions of Sun in the approaching process, in order to give it a physical content. So the Sun is no more a classical material point, *viz.* simply a position in space, but a material point in the sense of Hertz (2003), having a space 'breadth' as it were. The chief trait of Newtonian natural philosophy to be recalled here is that the way we visually perceive the Sun, should be close to the way it is 'gravitationally perceived' by the planets themselves: it is not simply a point in the geometrical sense of position. Consequently, the variation in distance in our imagined physical process of 'bringing from infinity' must be actually accompanied by a natural variation of the apparent dimensions of both the material point creating the field and of the material point moving in the field.

4.1 The Three-Dimensional Light Ray and Fractality

The classical physical ray of light has a spatial conical shape: being a thin pencil of straight lines, it is nevertheless far from the cylinder that Louis de Broglie used in visualizing it. The most appropriate mathematical — in fact, geometrical — description of this spatial shape would be as a *conical congruence of geometrical rays.* At this point some explanation is necessary, concerning the jargon we are using here: it is the jargon of the old algebraic geometry. Thereby, the *concept of congruence of lines* means a family of straight lines depending on two parameters. Such is, for instance, a family of straight lines passing through the same point, usually designated here by the term 'conical congruence'. This is not a surface, but a spatial shape. The qualification 'conical' is used only to suggest that all of the straight lines of this solid have a common point defining a vertex proper. The cone *per se* is only a surface of this solid figure, *a ruled surface of the congruence*, as they say. The nomenclature has the following explanation (see e.g. Jessop, 1903, p. 15 ff): when we build a space geometry based on the straight line as a fundamental element, the space is coordinated by *four real numbers*. Indeed, this many numbers need to be used in order to locate *a straight line in space*, for it can always be taken as an intersection of two planes. One algebraic constraint among the four numbers gives us a *complex of straight lines*. Two such constraints define *a congruence*, as we said. Finally, three algebraic constraints leave only a parameter free among the four, in which case a *ruled surface* of a congruence is defined. The generic line of this ruled surface is a *ruling* or *generator* of the surface.

Now, maintaining the position as the element of space, it is obvious that this geometrical construction of the physical light ray cannot be extended indefinitely in space, but should have a *finite space measure*, as Newton intended for experimental purposes (Newton, 1952). According to his definition the ray should be physically (we should say: *practically*) defined by two successive surfaces in space, cutting the conical congruence transversally. This condition allows us to insert into physical definition of the light phenomenon, those materiality considerations taken by Robert Hooke as physical reasons of the transversality of the light phenomenon (see Hooke, 1665, pp. 55–67). In the position representation of space, this transversality property of the light phenomenon can be realized symbolically in an exquisite mathematical manner, without introducing further arbitrary references. Let us systematically explain this idea, and then go into some historical considerations, the reason of which will become clear as we go on with our story.

Consider, indeed, as Louis de Broglie would claim (Mazilu *et al.*, 2019) some coordinates along the 'axis' of a physical ray: that kind of coordinates allowing us to talk about the 'position along the ray'. If (x, y, z) are the coordinates of such a position in an Euclidean approach, then Hooke's point of view can be kinematically represented by the system of equations:

$$\frac{dx}{y-z} = \frac{dy}{z-x} = \frac{dz}{x-y}. \tag{4.16}$$

Here the proportionality parameter should be simply a differential parameter giving the measure of continuity of the differentials. If, for instance, equations (4.16) represent a motion, then this continuity parameter is the time of the kinematical problem describing this motion. However, equation (4.16) has a much more general interpretation, which can be intuitively understood based on idea of direction of sight from the daily life.

It is indeed an observation of daily experience that we cannot account *directly* for the motion of a remote material point, if this motion is done exclusively along the line of sight. In order to detect such a motion in this case we need still other physical means, like, for instance, changing in the apparent dimensions of the moving body or, if this observation is not possible because of distance, the spectral analysis of the light by the intermediation of which we perceive that body as a distant material point. Otherwise, any such characterization is out of question. In order to be possible to account *directly* — as the classical mechanics would demand — for the motion of a remote material point, this motion must have a component *out of the line of sight.* Equations (4.16) represent such a condition, expressed in pure differentials, for the differential displacement is unconditionally perpendicular to the line of sight:

$$\mathbf{x} \cdot d\mathbf{x} \propto x(y-z) + y(z-x) + z(x-y) \equiv 0.$$

'Unconditionally' has here not only the physical meaning of 'independently of any spectral measurement of light', for instance, but also a geometrical meaning of 'independent of any external parameters'. For instance, the very same property can be expressed by a vector product, involving two or three parameters of an arbitrary direction in space. Moreover, in the form given in equation (4.16), the representation is universal — 'fractal' we should say — being valid even for arbitrary displacements, not necessarily differential, with the association of a parameter of fractal continuity to the system (4.16).

In Cartesian coordinates or, more generally, in coordinates corresponding to a 'Euclidean mentality' in the sense of Cartan (see Mazilu *et al.*, 2019, Chapter 11), equations (4.16) represent the fact that the elementary displacements it models are, for instance, in the plane of the cone basis serving as spatial limit for the light ray in Newton's acceptance. In the general observation of light, that plane is, however, only a plane of orthogonal projection. If the motion is done on a cone proper — a ruled surface of the congruence representing the physical light ray — but somewhere between the light source generating the ray and plane of observation — it is still representable on a circle. In the plane of observation of the displacements, this circle represents nevertheless a fictitious motion if the particle undergoing the motion does not happen to be really in that plane. We will need some reality conditions for this motion, which come down to a certain particular *space gauging* that tells us when the particle is in such and such plane, along such and such direction.

One may appear as quite strange to some, but these ideas have emerged historically, in connection with the first, and only *classical* description ever known of the *the light quanta* (Jeans, 1913). On that occasion people were forced to take the concept of quantum face value — for there was not any other possibility — and exhibit the flaws of quanta in connection with the idea of light ray, in order to finalize the classical concept of *physical* light ray, much in the way it was presented later on, with the occasion of de Broglie's concept. Quoting:

> Now it must, I think, be taken for granted, that *the quanta can have no individual and permanent existence in the ether*, that they cannot be regarded as accumulations of energy in certain minute spaces flying about with the speed of light. This would be in contradiction with many well-known *phenomena of interference and diffraction.* It is clear that, if a beam of light consisted of separate quanta, which, of course, ought to be considered as mutually independent and unconnected, the bright and dark fringes

to which it gives rise could never be sharper than those that would be produced by a single quantum. Hence, if by the use of a source of approximately monochromatic light, we succeed in obtaining distinct interference bands with a difference of phase a great many, say, millions, of wavelengths, we may conclude that *each quantum contains a regular succession of as many waves*, and that it extends therefore *over a quite appreciable length in the direction of propagation.* Similarly, the superiority of a telescope with wide aperture over a smaller instrument, in so far as it consists in a greater sharpness of the image, can only be understood if each individual quantum can fill the whole object-glass.

These considerations show that *a quantum ought at all events to have a size that cannot be called very small.* It may be added that, according to Maxwell's equations of the electromagnetic field, an initial disturbance of equilibrium must always be propagated over a continually increasing space.

We might now suppose that the exchange of energy between a vibrator and the ether can only take place by finite jumps, no quantity less than a quantum being ever transferred to the medium or taken from it. Something may be said, however, in favor of the opposite hypothesis of a gradual action between the ether and the vibrator, governed by the ordinary laws of electromagnetism. Indeed, it has been shown already, in Planck's first treatment of the subject, that by simply adhering to these laws, one is led to a relation between the energy of the vibrator and that of the black radiation, of whose validity we have no reason to doubt. [excerpt from the Lorentz's contribution, rendered on pp. 381–383 and 385–386 of Jeans (1913); *our Italics*]

In order to understand this primeval account for the concept of quantum, we think it is best, on many accounts, to swing back into history, to the times of Newton. For, indeed, this account of quanta is, in fact, an improvement, brought up by quantization, on the Newtonian definition of physical ray (see Newton, 1952, p. 1). The remark above: 'quanta can have no individual and permanent existence in the ether' leaves no doubt about that, for it has a fundamental principled meaning. Perhaps it is better, though, to bring here the originals in our discussion, in order to see what was the gnoseological issue to Newton and his times, and how has its solution evolved in time. The Newtonian definition for the light ray is (*loc. cit. ante*):

> *By Rays of Light I understand its least Parts, and those as well Successive in the same Lines, as Contemporary in several Lines.* For it is manifest that Light consists of Parts, both Successive and Contemporary; because in the same place you may stop that which comes one moment, and let pass that which comes presently after; and in the same time you may stop it in any one place, and let it pass in any other. For that part of Light which is stopp'd cannot be the same to that which is let pass. The least Light or Part of Light, which may be stopp'd alone without the rest of the Light, or propagated alone, or do or suffer any thing alone, which the rest of the Light doth not or suffers not, I call a Ray of Light. (*original Italics*)

It is necessary to add here that 'line', in the phrasing of epoch we are in with Newton, means 'segment' or straight line, may be even direction in the last resort, and that the property of 'contemporaneity' is referring to *simultaneity* in motion of the parts in question. For the interested reader, this meaning is much clearer in the 1787 French translation of the Newton's *Opticks*, and also in a work of our times (Shapiro, 1975). The reading of explanatory Newton's text for this definition makes two further points of the Newtonian natural philosophy. We describe here these two points, as they are instrumental in understanding the modern concept of elementary particle of the theoretical physics, and, more importantly, its gnoseological connection with the modern idea of quantum, as it has been refurbished by starting from the primary image given by Lorentz in that year 1913.

First, Newton referred the definition of the 'light parts' to the *experimental possibility* of realizing them: the 'least Parts' must in no case be understood as particles in the classical sense of the word, as derived from the daily experience with matter. These last ones may exist 'under our own eyes', as it were, independently of us: being stable at the time and space scales of common experience, they do not ask for an experimental definition in the Newtonian manner. The least parts of light, on the other hand, cannot exist this way: they always need to be defined as such, *i.e.* experimentally. And the experimental definition here is dependent, on one hand, on the physical possibility of stopping the light, with opaque screens for instance, and, on the other hand, on the physical possibility of discerning a propagation direction, *e.g.* with the aid of some orifices placed appropriately in such screens. It is only from the interplay of these two experimental procedures — in their ideal possibility, we have to admit, of course — that the 'least parts of light' can be defined the way they were conceived by Newton himself. In hindsight, one can say that the experiments with light told Newton that the simultaneity in defining the wave surface is not generally an actual property: the observations on light are always directional and limited by *a finite space scale*, in close connection with a *finite time scale*. This experimental procedure has evolved into the modern technologies of producing the *elementary particles* of contemporary physics. These particles can by no means be classical material points, or even Hertz material particles, for instance. They have to be more complicated ephemeral structures of the nature of galaxies in a modern isotropic universe (Popescu, 1982, 1988).

In order to better understand this issue, the most appropriate reference, as we see it, is the Maxwell's demon [see Maxwell (1904); see also Chapter 2 above, on the physical idea of sufficiency]. In hindsight, it is, indeed, obvious that the manner in which Maxwell defines his demon, reproduces the manner in which Newton defines 'the least parts' of light, except in one detail that makes all the difference. To wit, for Maxwell the concept of molecule is *ontologically* defined already by the idea of classical material point, while for Newton the equivalent of that very classical material point has to be defined first, therefore gnoseologically, in order to further advance then to the definition of the equivalent of a gas molecule. As it turns out, neither one of the two have been properly defined, but this comparison has the rare virtue of arranging our *gnoseological* priorities; especially if compared with the Lorentz's discussion as excerpted above.

The physics has long forgotten these issues, but this does not mean that the concept of quantum is well-rounded: it is today just as hazy as it was at the time when the quantum was baptized! Lucky for us, that 'contradiction with many well-known phenomena of *interference* and *diffraction*', for instance, was undertaken and analyzed by Louis de Broglie, in order to construct the modern concept of physical ray (Mazilu *et al.*, 2019). This plainly shows that the Maxwell demon cannot work but in the Newtonian way, by first producing physically meaningful particles, and then gathering them into ensembles, for otherwise it has nothing to work with: the classical material point, just like the Hertz material particle, does not exist physically! It is this fact that justifies the second principle of dynamics as a physical law: in layman's words, once the Deity is asked to do *thoroughly* its humanly defined deeds, it is doomed to non-existence!

The second of the two Newtonian points we have in mind with the definition of 'the least parts of light', seems even more important from the perspective of the scale relativity, than the first. This fact should be apparent from the following excerpt, explanatory to the second definition of the Newton's *Opticks*. This definition is referring to the first of the two properties of light, laying at the foundation of the phenomenology of classical theory of light. To wit, these properties are the 'refrangibility' and the 'reflectability', as Newton calls them. Quoting:

> Mathematicians usually consider the Rays of Light *to be Lines reaching from the luminous Body to the body illuminated*, and the *refraction* of those Rays *to be the bending or breaking of those Lines* in their passing out of one Medium into another. And thus may Rays and Refractions be considered, *if Light be propagated in an instant*. But by an Argument taken from the Equations of the times of the Eclipses of Jupiter's Satellites *it seems that Light is propagated in time, spending in its passage from the Sun to us about Seven Minutes of time*: And therefore I have chosen to define Rays and Refractions in such general terms as may agree to Light in both cases. [(Newton, 1952, p. 2), *our emphasis*]

It is not too much to say here, in order to see in this excerpt the essential difference between mathematics and physics when it comes to the definition of a light ray. In fact, it is this very difference that separated Newton from Hooke, irreversibly we should say: this last scientist *invented* a physical light ray in order to create the first physical theory of colors, based exclusively on the idea of *mathematical light rays* (Hooke, 1665, pp. 55–67). This is not to say that the invention is futile: to this very date, Hooke's theory, just like the celebrated Newton's invention of forces, seems to be the only rational theory of the colors of light! So, let us abide with it for a short while.

As we see it, a *Hooke physical ray* is a plane figure: a plane pencil of straight lines delimited by two of them [see Shapiro (1973) for modern details on this and a few other theories of light ray contemporary with Newton]. This light ray can be construed as a longitudinal intersection of a conical congruence of straight lines with a plane through one of them. The simplest of Hooke's physical rays is monochromatic, with a 'light line' having a certain incline on the two delimiting rays of the physical ray, and it is this *incline which defines the color of the ray.* The expression 'light line', is borrowed from Thomas Hobbes phrasing (Hobbes, 1644), and was rediscovered speculatively by Hooke as a periodic motion called 'orbicular pulse', from the assumption of materiality of the light phenomenon. It is the segment, *transversal to the direction of propagation*, which connects the two extremal mathematical rays, advancing in propagation. The *phenomenological* reason is simple: according to Hobbes the light behaves like the heart, more specifically, like an 'orb' that pulsates. Hooke noticed that the continual expansion of light in space cannot support this image, because, in an 'orb', the continuous expansion would cause the matter crumble to pieces according to our experience, a fact that does not occur in any transparent matter, at least not spontaneously. Thus, either the light is not material, or the transversal motion in 'orb' is not continual. Based on his discoveries on elasticity of materials, Robert Hooke goes for this last option: *the light is material* indeed, but the transversal motion in the 'orb' is periodical, 'of short length' according to his own expression, in order not to break the 'orb. The 'light line' of Hobbes thus becomes the 'orbicular pulse' to Hooke, but remains a 'line' nevertheless, whence the planeness of the physical light ray. For a complex physical light ray, the 'orbicular pulse' possesses, according to Hooke, what we would call today *the property of 'fractality'*: it breaks into many *homogeneous* segments. The homogeneity of a ray is thus defined by the property of light *to possess one color*: each one of these segments represents a given color, revealed by refraction, because in such events the color is conspicuous. Thereby, the Hookean manifold of colors naturally reproduces the property of planeness of this geometrical image of physical ray, being a two-dimensional manifold (Mazilu *et al.*, 2019) with the two basic colors, red and blue. According to Hooke's theory, any color can be constructed from these two, by a mixture of them in different proportions. Now, even from this short presentation, it is clear that this kind of physical ray, unlike its Newtonian counterpart — the forces — is an unsecured invention.

Indeed, Newton himself has challenged this *invention* of Hooke [the celebrated *hypotheses non fingo*, reproduced in many contexts but, unfortunately, almost never with its right meaning; see Shapiro (1975) for the details of this issue]. First, in the excerpt above, Newton draws attention to the fact that the 'Rays of Light' can be considered as lines only under the condition that *the light propagates with an infinite velocity*. This is our take on the remark: 'if the Light be propagated in an instant', from the excerpt above. As we see it, the discovery of finiteness of the speed propagation of light (Römer, 1676, 1677), was the first breach into the very issue of the time scales, and only a Newton would have been able to notice it properly! And the proper notice is, we think, that the geometry in this problem is connected with the instantaneous propagation of light, which is *a matter of space scale*, once the finite speed of light is deduced 'from the Equations of the times of the Eclipses of Jupiter's Satellites'.

This, therefore, was the first instance of the Newton's adage *hypotheses non fingo*: the necessity of a proper reference to a specific phenomenology, possibly even to experiments referring to light, the path Newton followed himself. For, Olaf Römer's discovery obviously belongs to such a phenomenology, but it can hardly count as an experiment. The main phenomenological framework of light at the times of Newton was represented by experiments involving refraction and reflection. And Newton has shown that, based on such experiments — *hypotheses non fingo*! — Hooke's theory of a physical light ray is untenable (Shapiro, 1975). Indeed, an experiment like that necessary in validating a model of the physical light ray, can only refer to the colors of light, whence, in particular, the Newton's prism experiments. This subject engages a good part of the Newton's *Opticks*, whereby he, implicitly or explicitly, shows that, based on the phenomenology of refraction, the property of homogeneity connected with the color is, in fact, *a cross-sectional plane property*, not a longitudinal plane one. In modern mathematical terms, this is an invariance property: the color of a properly built homogeneous ray *is invariant with respect to rotations around the ray's axis*, to wit, around the propagation direction. The color of Hooke must be a *one-dimensional property*, so it does not satisfy the phenomenological fact. A little elaboration on these facts may be in order.

According to Hooke's idea, the cross-section of a natural light ray should appear as differently colored along a certain direction in its plane. As it appears, indeed, but only when the construction of the ray is complicated with one more step beyond those implicit in Newton's definition: *passing the light through a prism.* According to Newton's ensemble of experiments that created the specific phenomenology of light, we can then appropriately isolate 'sub-rays', as it were, having *the same color in any cross-section*, *i.e.* homogeneous, with the homogeneity defined by the property of color in any plane intersecting the direction of propagation. The color spectrum appears as directionally elongated, indeed, but the component homogeneous rays are individually invariant the way we just said. The direction of elongation of spectrum can be conveniently changed in the cross-sectional planes, when complicating the construction of rays even further, by adding, for instance, one more prism, perpendicular to the first, as Newton did. Robert Hooke's model ray could not account for this fact. The bottom line is that Newton's experiment-based theory of light suggests a general idea to be useful here, but this idea does not concern *the geometrical shape* of the very physical light ray, as in the case of Hooke's model. Indeed, it is only the *variation of ray color* that has cross- sectional plane extension in preferred directions, not the ray itself, which can even be an axially symmetric solid figure, containing a specific mixture of colors, like the natural Sun light. Probably this observation made such a great impression on Newton that, ignoring the hole which generates the experimental ray to start with, but especially ignoring the prism itself, which decomposes that ray, he charged the light alone with the property of color. This is the key to Newton's experimental theory of colors, and the basis of the subsequent physical theory of light, whereby the color is univocally connected with the frequency. Contrary to the well-established opinion, though, by this Newton *preserved*, in fact, the initial philosophy of Robert Hooke, in a very specific and convenient way, that needs to be pursued further!

Indeed, the main point of Hooke's model is its *fractality*, as we mentioned above. One can say that Newton just added an essential fact: according to the existing phenomenology — defined by the two classical phenomena involving the interaction of light with matter: refraction and reflection — *the fractality is not a one-dimensional property, but a two-dimensional one*! Why is this dimensional property so important? In order to understand why, we need to notice that the light is the only phenomenon *naturally transcending the scales*, be they time scales of space scales. It exists and acts simultaneously at all three scales of the world: infrafinite, finite or transfinite, in any universe. Therefore the fractality, which is mathematically defined by a scale transcendence, has its phenomenological epitome in light. And from this point of view, the defining property of fractals is that of *existing as surface beings*. As a matter of fact this is how they were first noticed with mathematical concerns by Benot Mandelbrot in virtually all material structures (Mandelbrot, 1982). One can say that insofar as the light is material, *the fractality is its defining property* and, moreover, that *the fractality is always a surface property*!

Now, once this bridge has been thrown over the flow of history, the time has come again to continuing our line of ideas here. First, we need to take notice of the fact that there is a much general kinematical interpretation of equations (4.16), concordant with the history of the problem of light, but not only with that: most importantly, it is concordant also with the theory of the Newtonian forces 'at large', so to speak, *i.e.* forces according to their very definition given originally by Isaac Newton. Perhaps it is worth recalling once again, for our reader's benefit, the fact that in describing the forces as 'Newtonian' we mean that these forces are central, and have a magnitude which depends inversely on the square of distance between the positions of their action. However, these are not Newtonian forces 'at large', as we would like to call them now. The definition of Newtonian forces 'at large' can, in our opinion, concentrate around the Corollary 3 of Proposition VII from the Book I of *Principia*, whereby these inventions of Newton are defined by a *procedure of measurement*, which makes a 'secured invention' out of them. In short, this procedure involves, ultimately, motions controlled by those forces. By and large, it asserts that the ratio between the magnitudes of two forces acting upon a moving point toward two different arbitrary points in space, can be recognized in the parameters of the orbit of that moving point by *the ratio of two special volumes* in the space comprising those points [see, *e.g.* Mazilu and Agop (2012), regarding details from a modern perspective, and further historical readings]. Observationally, this definition is sanctioned by the Kepler motion via the axioms of classical dynamics, in a way illustrated, for instance, in Chapter 8 of the work just cited above. It is worth mentioning that in the cases where the orbit is a conic — the proper Kepler problem — the classical dynamics exhibits the two kinds of forces involved in the dynamical Berry–Klein problem: one force acting towards the focus of orbit, the other acting towards the center of the orbit. This last one is an elastic force, anisotropic in character, insofar as its 'elastic modulus' depends on the position along the orbit, therefore on direction (Mazilu and Porumbreanu, 2018). However, in this case the force dynamically controlling the motion, can be reduced to a Newtonian force inversely proportional with the square of the distance to the focus of orbit. To wit, the Newtonian force 'at large', *i.e.* the force according to its physical definition, reduces to the Newtonian force that exclusively entered the physics — through Laplace and Poisson equations — only in cases where the force toward focus, in the Kepler motion defining it, is referred to the force toward the center of orbit. The Poisson or Laplace equations can then be taken, indeed, as means of liberating us from the obligation of defining the force by measurement — according to the original procedure of Newton — but they bring instead the burden of *boundary conditions*, which, as we have shown before, corrupted the whole general relativity theory in Einstein's acceptance. This is one thing. For, on the other hand, it brings also the burden of *initial conditions*, and with them the non-Euclidean description of the matter, the way we have already shown in the present work for the very case of the Kepler motion.

Fact is that a dynamical problem based on equations like (4.16) would involve Newtonian forces 'at large' indeed, and not only in the sense we have just defined according to Newton, but also in an even larger sense: they should be *Wigner type forces*, obtainable from a dynamical principle involving the velocities (Mazilu *et al.*, 2019). In other words, they should be forces defined under that *condition of static equilibrium* necessary, for instance, in a Poincaré interpretation of the electromagnetic field. Let us show that these equations can be integrated, as equations of motions in particular, no matter of their origin. Indeed, the dynamical principle derived from (4.16) is expressed by

$$\dot{x} = \nu(y - z), \quad \dot{y} = \nu(z - x), \quad \dot{z} = \nu(x - y) \tag{4.17}$$

and reveals a motion done always perpendicularly 'to the ray' as it were, as in the case of light, insofar as the speed is always perpendicular to the position vector along the ray. We are searching for constant integrals of this motion, starting from the differential system (4.16) and trying to find exact differentials that could offer us a physical interpretation to the parameter of continuity. The most obvious method involves linear forms in coordinates. For instance, we can derive the exact differential equivalent to the system (4.16):

$$\frac{adx + bdy + cdz}{(c - b)x + (a - c)y + (b - a)z} = \nu dt \tag{4.18}$$

with constant (a, b, c). Such exact differentials can only exist under the conditions

$$c - b = \lambda a, \quad a - c = \lambda b, \quad b - a = \lambda c \tag{4.19}$$

with λ a parameter. This means that the left-hand side of the equation (4.18) is an exact differential only for the cases in which λ has as values the roots of the cubic equation:

$$\lambda(\lambda^2 + 3) = 0$$

representing the condition of compatibility of the system (4.19). In terms of these three roots, to wit, 0 and $\pm i\sqrt{3}$, the following three complex integrals can be constructed:

$$\begin{aligned} x + y + z &= x_0 + y_0 + z_0, \\ x + j^2 y + jz &= e^{i\nu t\sqrt{3}}(x_0 + j^2 y_0 + jz_0), \\ x + jy + j^2 z &= e^{-i\nu t\sqrt{3}}(x_0 + jy_0 + j^2 z_0). \end{aligned} \tag{4.20}$$

Here j is the *cubic root* of unity, as a counterpart of i, which is the *square root* of negative unity. The three complex variables from the left-hand side of this equation are related in many ways to the name of Paul Appell, and they have a tremendous importance of principle, both from physical (Appell, 1893, p. 351) as well as from mathematical point of view (Appell, 1877).

Start by noticing that the first of the integrals (4.20) is a constant of motion. Another constant of motion is quadratic, and can be obtained from the product of the last two of them, *i.e.*:

$$(y - z)^2 + (z - x)^2 + (x - y)^2 = \mathit{const.} \tag{4.21}$$

The trajectory is then to be found in the intersection of this quadric with the real plane given by the first equality from equation (4.20), therefore it also belongs to the quadric

$$yz + zx + xy = \mathit{const.} \tag{4.22}$$

and thus on a sphere. Although formally these coordinates can pass as Cartesian coordinates, they are not such coordinates without further specifications, if it is to judge them from the physical point of view offered by the description of the motion they characterize. Indeed, until further elaboration on this issue, let us just notice that by successive differentiation with respect to the continuity parameter in (4.17), one can find out that the equations of motion 'decouple', so to speak, and each of the coordinates satisfy the very same third-order differential equation:

$$|\dddot{\xi}\rangle + 3\nu^2|\dot{\xi}\rangle = |0\rangle. \tag{4.23}$$

In other words, trying to connect the situation with the principle of inertia in the classical connotation, leads us here to the essential results of the regularization procedure for Kepler problem [see Chapter 3, equations (3.19) and (3.40)]. To wit, the whole 'ket' $|\xi\rangle$, having as components the coordinates from equation (4.17) in an Euclidean frame, is a solution of a specific third-order differential equation. Equation (4.23) is not a harmonic oscillator equation *per se*, as required in a physical theory of ray optics for instance, but it certainly has periodic solutions, as the Hooke's and Fresnel's theory of the light ray would require. By itself, equation (4.23) has to be taken as a condition used in the theory of harmonic oscillators, in order to reduce the third-order derivative with respect to time to the first derivative. It says, however, something of a geometrical importance about the process deriving from a static Wigner principle, represented in a three-dimensional phase space by the equations of motion (4.17). Indeed, equation (4.23) can be integrated once to give something resembling to a second principle of dynamics, just like in Fresnel's theory of light:

$$|\ddot{\xi}\rangle + 3\nu^2|\xi\rangle = |c\rangle. \tag{4.24}$$

Here $|c\rangle$ is a constant vector. Such a phase space obviously generalizes, by dimension at least, the phase plane of a regular harmonic oscillator: as one can easily see, if we settle for a plane of coordinates in (4.17), we get a two-dimensional harmonic oscillator. The solution of (4.24), on the other hand, is offered by a vector:

$$3\nu^2|\xi\rangle = |c\rangle + |a\rangle\cos(\nu t\sqrt{3}) + |b\rangle\sin(\nu t\sqrt{3})$$

with $|a\rangle$ and $|b\rangle$ initial conditions. It is located on the homogeneous quadratic cone in space, having the equation

$$[\boldsymbol{\xi}\cdot(\mathbf{b}\times\mathbf{c}]^2 + [\boldsymbol{\xi}\cdot(\mathbf{c}\times\mathbf{a}]^2 + [\boldsymbol{\xi}\cdot(\mathbf{a}\times\mathbf{b}]^2 = 0, \tag{4.25}$$

where we have adopted the notation $|\xi\rangle \equiv \boldsymbol{\xi}$, $|a\rangle \equiv \mathbf{a}$, $|b\rangle \equiv \mathbf{b}$, $|c\rangle \equiv \mathbf{c}$, in order to be able to use the usual vector products without complicating the writing.

Let us continue along this line for a little while. From the system (4.20) we can get the one-parameter group equations of the very process thus described by equations (4.17). Namely, after taking the inverse of the appropriate matrix in (4.20) we end up with

$$|x\rangle = [\mathbf{1} + (2/3)\mathbf{c}(t)]|x_0\rangle, \tag{4.26}$$

where $\mathbf{1}$ is the 3×3 identity matrix, and $\mathbf{c}(t)$ is the matrix

$$\mathbf{c}(t) = \begin{pmatrix} \cos\psi & \cos(\psi - 2\pi/3) & \cos(\psi + 2\pi/3) \\ \cos(\psi + 2\pi/3) & \cos\psi & \cos(\psi - 2\pi/3) \\ \cos(\psi - 2\pi/3) & \cos(\psi + 2\pi/3) & \cos\psi \end{pmatrix}$$

with $\psi(t) = t(\nu\sqrt{3})$. This matrix is singular: its determinant is

$$\det[\mathbf{c}(t)] = \cos^3\psi + \cos^3(\psi + 2\pi/3) + \cos^3(\psi - 2\pi/3)$$
$$-3\cos\psi\cos(\psi + 2\pi/3)\cos(\psi - 2\pi/3)$$

and in view of the algebraic identity

$$a^3 + b^3 + c^3 - 3abc \equiv (a + b + c)(a^2 + b^2 + c^2 - ab - bc - ca)$$

and trigonometric identity

$$\cos\psi + \cos(\psi + 2\pi/3) + \cos(\psi - 2\pi/3) = 0$$

it is null.

We can even complicate a little the equations of motion (4.17), admitting a gauging where the velocity has also a component of speed along the ray, which would correspond to propagation. This is in the spirit of a unitary description of the light phenomenon, which would include both the propagation — measured always along the ray — as well as the light motion proper — measured orthogonally to the ray — in describing the light. Mention should be made that the situation would correspond to the motion of an electric charge in the field of a magnetic pole (Poincaré, 1896). Then the equations of motion corresponding to those from (4.12) are:

$$\frac{dx}{lx + m(y - z)} = \frac{dy}{ly + m(z - x)} = \frac{dz}{lz + m(x - y)}, \tag{4.27}$$

where l and m are two parameters representing the 'amounts' in which the motion is decomposed along the ray and perpendicular to it, respectively. The integration procedure described above, leads to a differential form a little more complicate

$$\frac{adx + bdy + cdz}{[la + m(c - b)]x + [lb + m(a - c)]y + [lc + m(b - a)]z} = \nu dt$$

which can be considered an exact differential

$$\frac{adx + bdy + cdz}{n(ax + by + cz)} = \nu dt \quad \therefore \quad ax + by + cz = Ae^{n\nu t} \tag{4.28}$$

if and only if a, b, c are solution of the linear algebraic system given by

$$(l - n)a + m(c - b) = 0$$

and its positive permutations. This system has non-trivial solutions only if the constants l, m and n satisfy the algebraic equation:

$$(l - n)[(l - n)^2 + 3m^2] = 0$$

which offers three possibilities of construction of the differentials representing the corresponding kinematics. They are given by the system of values:

$$l = n \;\rightarrow\; a = b = c,$$
$$l - n = im\sqrt{3} \;\rightarrow\; a = jc;\; b = j^2c,$$
$$l - n = -im\sqrt{3} \;\rightarrow\; a = j^2c;\; b = jc.$$

Formally, then, nothing changes with respect to the preceding simpler case: it is just that we have here to do with a *harmonic of the frequency* ν, not quite with the frequency itself.

The preceding might seem a pure kinematics, if we do not consider the Wigner's dynamic principle. However, there is also a 'hidden' dynamics involved here, and this is, we think, the right place to bring up the name of Paul Appell. This dynamics appeared for the first time in 1893, in the known *Traité de Mécanique Rationnelle* of Appell, Tome I, on p. 351, but only as an exercise. Quoting:

> A point is moving in space, under the action of a force whose components X, Y, Z are functions of x, y, z, which verify the relations
>
> $$\frac{\partial X}{\partial x} = \frac{\partial Y}{\partial y} = \frac{\partial Z}{\partial z}; \quad \frac{\partial X}{\partial z} = \frac{\partial Y}{\partial x} = \frac{\partial Z}{\partial y}; \quad \frac{\partial X}{\partial y} = \frac{\partial Y}{\partial z} = \frac{\partial Z}{\partial x}. \tag{4.29}$$
>
> Prove that the integration of the equations of motion is reduced to quadratures. [(Appell, 1893), *Exercise 16, p. 351, our translation*]

The proof is simple: first, one has to define a *complex position vector*, having as components the three complex coordinates as in equation (4.20). Then we need to define a *complex force vector*, having as components three corresponding complex quantities, constructed from the real components of force in the same manner the coordinates are constructed. Obviously, the principles of analysis allow us to infer that, if the real forces are functions of real position, the complex forces must be, likewise, functions of complex positions. Therefore, in the following table constructed by the rules just mentioned:

$$\begin{aligned} &x^1 = x + y + z,\ x^2 = x + jy + j^2 z,\ x^3 = x + j^2 y + jz, \\ &X^1 = X + Y + Z,\ X^2 = X + jY + j^2 Z, \\ &X^3 = X + j^2 Y + jZ, \end{aligned} \tag{4.30}$$

every variable of the second line should be a function of the variables of the first line. Then notice that, under the conditions (4.29), each of the components of complex force — assumed conservative, of course — is a function only of the corresponding complex coordinate from the first line. Therefore, the differential equations of motion can be written as

$$\frac{d^2x^1}{dt^2} = X^1(x^1), \quad \frac{d^2x^2}{dt^2} = X^2(x^2), \quad \frac{d^2x^3}{dt^2} = X^3(x^3) \tag{4.31}$$

and can be solved by integrating twice, indeed. The property is transmitted as such over to the real corresponding quantities, because the transformations (4.30) are always non-singular. So the Appell's result is proved. However, there is more to be read in this problem than can be perceived at the first sight. For, with the dynamics (4.31), like with the dynamics of the harmonic oscillator, we have a genuine *procedure of regularization*, which turns out to be universal for the three-dimensional case, without any appeal to the four-dimensional case, as in the Kustaanheimo–Stiefel theory of regularization.

4.2 A Three-Dimensional Regularization Procedure and its Physical Meaning

The observations right above turn our reasoning in an entirely different direction when it comes to conceiving the theory of regularization: it should be *a natural step imposed by the necessity of scale transition* in physics. This conclusion comes out itself just as naturally from the fact that the regularization was always conceived as a necessary consequence of the *analyticity conditions*, expressed by a *meaningful dynamics* describing the Kepler problem. One can say that the first dynamical description of the Kepler problem — the one that led to the invention of the central forces of Newtonian kind — was only incidental among many other dynamical descriptions, one of which can very well be the Appell's dynamics (4.31). It certainly qualifies as a 'meaningful dynamics' for a regularization procedure, a dynamics that does not even need a change of the time scale. Provided, of course, to be the expression of an accompanying analyticity condition.

This problem takes an interesting turn here: in the three-dimensional case as presented above, the conditions of analyticity, unlike those of Kustaanheimo and Stiefel, do not necessarily need a definition of the Cauchy–Riemann conditions based on Stiefel's property of linearity. Consequently, once we dispense with the linearity condition, we can also dispense with the four-dimensional regularizing parameters' manifold, and proceed directly for the three-dimensional case. All it takes is to carry the analogy between the three-dimensional case and the two-dimensional case from its very physical roots: *the concept of vector*. For instance, Fulton and

Rainich were able to generalize the two-dimensional Cauchy *integral* formula to the multi-dimensional case, and thus to represent a vector that satisfies the Helmholtz conditions (4.1), *only in vector form* [(Fulton and Rainich, 1932); see equation (4.3) here, for the three-dimensional case]: there is no complex multi-dimensional Fulton–Rainich formula. Their procedure takes advantage of the fact that, in the two-dimensional case the Cauchy integral formula has a vectorial counterpart, with meaningful operations between the vectors under the integral sign. As these operations are meaningful no matter of the dimensions of vectors, the formula from the two-dimensional case could be simply transliterated in any dimension. The key point here is that the complex plane Cauchy integral has a clear vectorial counterpart, which is not the case in any other dimension. However, the formulas in equation (4.30) allow us to conceive *a complex three-dimensional space*, constructed based on the *cubic root* of unity as complex fundamental unit in space, instead of the usual *square root* of negative unity which is the complex fundamental unit in plane. Thus, we can carry the analogy between plane and space from its very roots, as we said before: in space just as in the plane, any vector is equivalent to a complex number, with the equivalence defined by equation (4.30). And thus, the Appell's conditions from equation (4.29) give us a set of *three-dimensional Cauchy–Riemann conditions*, manifestly equivalent to those from the two-dimensional case (Miles, 1954).

The conditions (4.29) represent, indeed, *genuine conditions of analyticity*, as intuitively envisaged. For, in the two-dimensional case, the Cauchy–Riemann conditions are only a guarantee of analyticity, which ensues from the classical definition of the derivative in the one-dimensional case. To wit, in the two-dimensional case, unlike the one-dimensional case, there is no guarantee that the limit

$$\lim_{x \to x_0} \frac{\mathbf{X}(\mathbf{x}) - \mathbf{X}(\mathbf{x_0})}{\mathbf{x} - \mathbf{x_0}} \tag{4.32}$$

would not depend on the path we are following in reaching from the complex point $\mathbf{x}$ to the complex point $\mathbf{x_0}$. But if the two-dimensional Cauchy–Riemann conditions are satisfied, this limit, in case it exists, is independent of the path followed, a condition which is almost *naturally* satisfied in the one-dimensional case; in this case we have almost no alternative direction. Now, equation (4.29) secures the very same path independence, *but in space.* Indeed, in the three-dimensional complex space we can declare (Miles, 1954): if the conditions (4.29) are satisfied, then the *complex force is an analytic complex function of the complex position*, in the sense that the limit (4.32) is independent of the path followed in space for reaching $\mathbf{x_0}$ when starting from $\mathbf{x}$. Thus, if we accept the Wigner's dynamic principle, the denominators in (4.27) are the components of some complex force given by a complex three-dimensional number representing an analytic function in the sense of Miles. Indeed, in this case equation (4.29) gives

$$\frac{\partial X}{\partial x} = \frac{\partial Y}{\partial y} = \frac{\partial Z}{\partial z} = 1,$$
$$\frac{\partial X}{\partial z} = \frac{\partial Y}{\partial x} = \frac{\partial Z}{\partial y} = -m,$$
$$\frac{\partial X}{\partial y} = \frac{\partial Y}{\partial z} = \frac{\partial Z}{\partial x} = m.$$

Obviously, (4.17) is just a particular case here, for $l = 0$ and $m = \pm\nu$.

There is, nevertheless, a significant difference between the cases of two-dimensional and three-dimensional analyticity thus defined, besides the form of the equations of condition, mostly from the physical point of view.

First of all, the two-dimensional analyticity is referring to a *potential*, unlike the three-dimensional analyticity which is referring *directly to forces.* Again, everything revolves around Laplace equation. In the two-dimensional case, the analytic forces are gradients of a potential which is solution of Laplace equation. Correspondingly, in the three-dimensional case, the analytic forces are solutions of the same Poisson-like equation. Indeed, by handling the equalities in equation (4.29) one can come up with the conclusion that

$$\Delta X = \Delta Y = \Delta Z, \tag{4.33}$$

where Δ means the Laplacian, as usual. Which proves our statement: *any three solutions of a given Poisson-like equation can represent an analytic force*, of complex components given as in equation (4.30). In particular any three solutions of the same Poisson equation proper — *i.e.* the equation having the density of matter in the right hand side of Laplace operator — describe the matter of a given density as a function of position in space, and therefore these solutions represent the forces in matter.

Equation (4.33) is universal: it can define the forces in free space, as well as in the matter *per se*, depending on the density function that stays in the right-hand side of the Poisson equation taken into consideration. As noticed before, the Newtonian force satisfies the conditions (4.33) both in matter and in the free space [see equation (4.12) and discussion]. But there is more to this condition than meets the eye. Assume, indeed, the canal surface representing de Broglie tube, around the central matter of our planetary model. Consider the matter *per se*, interpreted as an ensemble of Hertz material particles, described in the manner of Larmor for his 'cloud of meteors' (Larmor, 1900). Just in order to settle our ideas, we describe the motion of the matter inside this de Broglie tube by a classical fluid of Hertz material particles, along lines of current given by the Kepler orbits of the particles with respect to each and every one of the corresponding particles of the central matter from the nucleus of the planetary atom. Such a line of current is characterized by a velocity, having just two components in its plane. The problem is that, when referred to the whole stream of particles inside de Broglie tube, this velocity should be described as a three-dimensional vector. The dynamical Wigner problem just accomplishes this task. Let us elaborate on this issue.

4.3 Coordinates as Measures

The sphere of inversion used before [see equations (4.13)–(4.15)] is one case leading to a significant example of such complex coordinates. It plays the part of a gauging device for a certain expanse of space, being therefore 'empowered', so to speak, to gauge somehow that region. *Mathematically speaking*, the precise definition of such a device is inessential: one can simply assume that the radius of the gauging sphere is unity, and thus proceed to the general construction of the inversion with no problem. However, *physically* the things are quite complicated, at least for the decision upon validity of the point approximation of the position of a celestial body, if not for anything else. We repeatedly signaled before that the Newtonian gravitation cannot be used in explaining the Kepler motion but in the limit of very small dimensions of Sun and planets with respect to the distance between them. Also, at the cosmological level, the universe is not isotropic but at a certain scale, specifically that scale where the galactic nebulae can be considered material points with respect to the distance between them. We are therefore bound to assume a certain homogeneity of the physical structure of the universe with respect to the dimensions of the bodies populating a region of space at a certain space scale, and this homogeneity asks, from physical point of view, for an additional statistical characterization. Specifically, we shall assume that those bodies are a *normal population of spatially extended physical particles*, of extension to be statistically specified.

That specification comes in handy on a population of a spherical shapes, as surely is the case with most of the celestial *bodies*, but 'moderately spherical' we should say, *i.e.* with incidental departures from the condition of sphericity. The practical engineering example, to be considered as a guiding case here, is the one of the agglomerates with a high degree of erosion, due for instance to the friction between particles in sedimentology (Bloore, 1977). It is far from a collection of stars, is true, but at this statistical stage the distance between material components of the agglomerate hardly matters. What matters is the manner of construction of the statistics. Thus, in such a population we assume that the relevant dimension of the spatially extended particles (the radius of a generic sphere enclosing the matter for instance) should be normally distributed with the probability density

$$W_R(r|R_0, \sigma) = \frac{1}{\sigma\sqrt{2\pi}} \exp\left(-\frac{1}{2}\frac{(r - R_0)^2}{\sigma^2}\right). \tag{4.34}$$

The two parameters of this statistics: the *mean* R_0 and the *variance* σ^2, quantitatively describe what we mean here by 'moderately spherical shape'. This is not the ideal geometrical shape, but what is currently termed, in sedimentology, for instance as a particle with 'advanced sphericity'. More precisely, the particles, taken individually, may have rather irregular surfaces. Nevertheless, their global shape, which can be perceived at a certain space scale where the surface irregularities can appear as 'smoothed away' as it were, should also appear as having no exaggerately flattened or elongated aspect.

This definition is more convenient in the case of planets and stars, rather than anywhere else: even though these bodies do have details of surface irregularities, we perceive them as round, very close to spherical shapes. On this note, we can get into some illuminating details. The Earth, for instance, has an advanced sphericity, however not quite at any space scale. This has been noticed first in relation with the idea of horizon. In the latter times it was also confirmed as a general impression of the views from space ships in circumterrestrial trips. In order to consolidate the idea even further, it helps noticing that in the case of Sun and Moon, the reasoning went the other way around: in the beginning we did not have but the 'general impression' of sphericity and perfect roundness of those celestial bodies, and only in the more recent times the technology allowed us to distinguish details of shape irregularity of their surface. In the case of Sun these details are downright dynamical. The bottom line here is that such a population can be dimensionally characterized by a mean radius R_0 as in equation (4.34), and a statistical spreading of the dimensions having a standard deviation σ. *Irregularities are to be fractally describable over such a generic spherical shape.*

Leaving the technical details to already existing publications (Nedeff, Lazăr, Agop, Eva, Ochiuz, Dimitriu, Vrăjitoriu and Popa, 2015), we need to highlight a few important points on this issue. To the extent where the shape can be described as a tensor (Winkelmolen, 1982), it can be estimated by sampling: a sample of three measurements from the normal population characterized by the density (4.34) represents the three eigenvalues of the shape tensor at a given instance. These measurements can be either measurements of the space extensions of a particle in three different directions in space, or measurements of the extensions in space of three different particles of the population along the same direction. From each three such measurements, taken as giving the eigenvalues of the shape tensor, this very tensor is defined up to a rotation in space. Let s_1, s_2, s_3 be the results of the three *lengths measurements* in sampling. Then the parameters R_0 and σ of the population of spherical particles are given by

$$3R_0 = s_1 + s_2 + s_3, \quad \sigma^2 = (s_2 - s_3)^2 + (s_3 - s_1)^2 + (s_1 - s_2)^2. \tag{4.35}$$

The measured values s_1, s_2, s_3 can practically be estimated considering for instance the intersections of the particles of population with an arbitrary straight line, as we mentioned above.

In our opinion, the conditions of dimensional homogeneity of a population of fundamental physical structures in a universe must be always observed, before we take the further step of neglecting the dimensions of these structures, in order to consider them as material points. So much the more such prerequisites are to be considered in the counting moment of estimating the density of matter. Quoting from Larmor (see Mazilu *et al.*, 2019, Chapter 6):

> It results from a very general proposition in dynamics that as the central meteor moves along its path the region occupied by the group of its neighbors multiplied by the corresponding region in their velocity diagram remains constant. Or we may say that the *density* at the group considered, *estimated by mere numbers, not by size*, varies during its motion proportionally to the extent of the region on the velocity diagram which corresponds to it.
>
> This is true *whether mutual attractions of the meteors are sensibly effective or not*; in fact, the generalized form of this proposition, together with a set of similar ones relating to the various partial groups of coordinates and velocity components, forms an equivalent of the fundamental law of Action which is the unique basis of dynamics theory [(Larmor, 1900); *our Italics*].

The estimation 'by mere numbers' is not quite independent of 'size': in physics one should take some precautions of homogeneity of that 'size'. And there is even a positive, Einsteinian aspect of the problem, if we may say so, as only by a material point can one indicate a position in space, for there is no real possibility that man may work with Hertz material particles. A generic material point may even be made into an affine reference frame, quantitatively described by a minimal sampling of the populations in space. The volume of such a reference frame can be taken as defining an *a priori* repartition density of the position of a point in space by a Hertz material particle: the position is equally probably distributed in the volume of a fundamental physical structure of the universe. This fact can be algebraically expressed by an inverse proportionality in the sense of Larmor. To wit, this density would then be given by the reciprocal of

$$\lambda_1^3 + \lambda_2^3 + \lambda_3^3 - 3\lambda_1\lambda_2\lambda_3 = \frac{3}{2}R_0\sigma^2, \tag{4.36}$$

where R_0 and σ are the parameters estimated by sampling from a Gaussian population of spherical particles, for samples of volume three $(\lambda_1, \lambda_2, \lambda_3)$ (Mazilu and Porumbreanu, 2018). Thus, in order to settle the ideas: when Joseph Larmor talks of considering 'large and small meteors alike', in order to define 'their density at the group considered... estimated by mere numbers not by size' as 'inversely as the volume occupied by them', he suppresses a space statistics well-known and necessary. This is the statistics representing qualitatively the precision with which a body from the 'cloud of meteors following an orbit' — as a material point in the sense of Hertz — indicates a position in space. This statistics defines an *affine reference frame* by its volume, and such a reference frame can be made into an 'Einstein elevator' pursuing the orbit in space. But this is not all of it, by any means.

Taken as such, the reference frame itself is characterized by a metric, playing an essential part in the description of a Schwarzschild type structure. Let us transcribe here a few formulas in the parameters $(\lambda_1, \lambda_2, \lambda_3)$. In the complex variables defined by equation (4.26) through relations:

$$x^1 = \lambda_1 + \lambda_2 + \lambda_3; \quad x^2 = \lambda_1 + j\lambda_2 + j^2\lambda_3; \quad x^3 = \lambda_1 + j^2\lambda_2 + j\lambda_3$$

the Barbilian metric [see Barbilian (1937); see also Mazilu *et al.* (2019, equation 4.30)] corresponding to the ternary cubic (4.36), and measuring the logarithmic statistical variation of the volume of population, as estimated by this cubic, is given by

$$(ds)^2 = \frac{K^2}{9}\left[3(d\phi)^2 - \left(d\ln\frac{x}{x^1}\right)^2\right] \tag{4.37}$$

where we used the notation:

$$x^2 \equiv x\,e^{i\phi} = (x^3)^* \tag{4.38}$$

with a star meaning the complex conjugate. The parameters $(\lambda_1, \lambda_2, \lambda_3)$ define therefore an affine reference frame (see Mazilu *et al.*, 2019, equations (4.33) and (4.34)) having the characteristic angle, θ say, in terms of which we have

$$\begin{aligned} \left(\frac{x}{x^1}\right)^2 &= \frac{1-\cos\theta}{1+2\cos\theta}, \\ (ds)^2 &= \frac{K^2}{9}\left[3(d\phi)^2 - \frac{1}{4}\left(d\,ln\frac{1-\cos\theta}{1+2\cos\theta}\right)^2\right]. \end{aligned} \tag{4.39}$$

If the reference frame is Euclidean, then $\cos\theta = 0$, and the metric thus constructed is the square of an exact differential.

4.4 On a Statistics Prompted by Appell's Coordinates

The previous statistical results on size of a population of regular shapes like spheres are obtained based on the theory of small sample estimation, specifically samples of size three. The three measurement in a sample are then interpreted as coordinates of a point. Of course, according to what we just said above, these coordinates need to be taken as lengths in the classical sense of the word: they are obtained by measuring pieces of material. Speaking of the sample size in deciding these coordinates, an interesting circumstance arises here, to give an interpretation to the parameter of a family of quadratic variance probability densities, in connection with the fractional linear transformations. As we have seen in Chapter 2 [equation (2.33) and discussion] the linear fractional transformations of time can warrant the use of time intervals as variants for a Gaussian distribution of the distances in space in view of the theorem of Paul Lévy. Then the problem of simultaneity is referring to an ensemble of time moments characterized by a Riccati equation. Along the geodesics of the **sl**(2,R) Lie algebra, this equation becomes an ordinary Riccati equation. That equation can be written in the form

$$d\theta = \frac{dt}{a^1t^2 + 2a^2t + a^3} \tag{4.40}$$

allowing an exquisite interpretation for the parameter θ of the geode- sics: *it can be taken as the density of probability of a Cauchy distribution of the simultaneous times.* The parameters of geodesics can then be taken as maximum likelihood estimates from samples of size three, taken from the ensemble of simultaneous times. Indeed, the probability density corresponding to the elementary probability (4.40) can be written in the form

$$\frac{1}{\pi}\frac{\sqrt{\Delta}}{a^1t^2 + 2a^2t + a^3} \equiv \frac{1}{\pi}\frac{\sigma}{(x-\mu)^2 + \sigma^2}, \tag{4.41}$$

where $\Delta \equiv a^1a^3 - (a^2)^2$, assumed positive, μ is a location parameter of the ensemble and σ is a scale parameter. Applying maximum likelihood considerations to this last form of the probability density, Thomas Ferguson finds the following estimates of the location and scale parameters for samples of size three (Ferguson, 1978):

$$\mu = \frac{\sum x_1(x_2 - x_3)^2}{\sum (x_2 - x_3)^2}, \quad \sigma = \frac{\sqrt{3}(x_2 - x_3)(x_3 - x_1)(x_1 - x_2)}{\sum (x_2 - x_3)^2}, \tag{4.42}$$

where the summation runs over all positive permutations of the three measurements values x_1, x_2, x_3 of a sample. These estimates have quite an important connotation.

Peter McCullagh has noticed an important property of the one-dimensional Cauchy distribution, which is obtained as a benefit of a complex parametrization of this distribution (McCullagh, 1996). The parameters of a statistical distribution density are usually taken as real, but McCullagh shows a clear advantage of representing them in a complex form, at least when it comes to the Cauchy distribution.

He starts with the fact that the density of this distribution for a single variate X can be written in the form

$$f_x(x|\theta) = \frac{|\theta_2|}{\pi|x-\theta|^2}, \quad \theta \equiv \theta_1 + i\theta_2 \tag{4.43}$$

where θ continues to signify the parameter of distribution as it did before, but this parameter now becomes a complex number. The real part of this parameter gives the 'location' of data, while the imaginary part roughly characterizes the 'spread' of their distribution, *i.e.* it is a scale parameter as before. Now, one knows that this class of distributions is closed with respect to the homographic transformation of the variable: any linear fractional transform of the variate X has also a Cauchy distribution. However, the complex representation of the parameter brings to light one of the most important consequences of this theorem: if X belongs to the Cauchy class with the complex parameter θ, *i.e.* writing symbolically $X \approx C(\theta)$, then we have

$$\frac{aX+b}{cX+d} \approx C\left(\frac{a\theta+b}{c\theta+d}\right). \tag{4.44}$$

This property allows us to give *efficient estimators* for the complex parameter θ, based on the principle of maximum likelihood, as Ferguson did before, but in a more expedite way and, what is more important, with significant theoretical consequences.

As a rule, the likelihood function used in estimations is simply the product of the values of the probability density for the different measured values of X. In taking the maximum likelihood with respect to parameters, it would be therefore appropriate to work with the logarithm of the likelihood, and this is what practically happens. For instance if one measures two values of X having the probability density (4.43), say x_1 and x_2, the Cauchy likelihood function constructed based on this information is simply:

$$L(\theta|x_1, x_2) = \frac{\theta_2^2}{\pi^2|x_1-\theta|^2|x_2-\theta|^2}.$$

The likelihood is maximum with respect to θ when the derivatives of this function with respect to θ_1 and θ_2 are null. In terms of the log-likelihood, which is a lot easier to handle, we then have:

$$\frac{\partial}{\partial \theta_1} \ln L(\theta|x_1, x_2) = \frac{\partial}{\partial \theta_2} \ln L(\theta|x_1, x_2) = 0. \tag{4.45}$$

In view of the fact that

$$\ln L(\theta|x_1, x_2) = -2\ln\pi + 2\ln|\theta_2| - \sum \ln(x_k - \theta) - \sum \ln(x_k - \theta^*),$$

where the summation extends over the two measured values and a star denotes complex conjugation as usual. Here we took an exception from the adopted rule of summation in this work, whereby the Latin indices always run over three values, in view of what will be seen immediately. The two equations (4.45) become:

$$\begin{aligned} &\sum \frac{1}{x_k - \theta} + \sum \frac{1}{x_k - \theta^*} = 0, \\ &\sum \frac{1}{x_k - \theta} - \sum \frac{1}{x_k - \theta^*} + \frac{2}{i\theta_2} = 0. \end{aligned} \tag{4.46}$$

If we sum up these equalities, results in

$$\sum \frac{1}{x_k - \theta} + \frac{1}{i\theta_2} = 0. \tag{4.47}$$

Clearing the denominators, and solving for θ_1, we get

$$\theta_1 = \frac{x_1 + x_2}{2} \pm \sqrt{\left(\frac{x_1 - x_2}{2}\right)^2 + \theta_2^2}$$

which shows what one already knows well about the Cauchy distribution. First, with the information of only two measured values we cannot have an estimation for the mean; it can be any value between the two measured ones. As to the variance estimator, it is also indeterminate, but this is quite a natural characteristic, so to speak, of this type of repartition, because it has no finite moments of second order.

At this point we can easily see the advantage of equation (4.44): it shows that the best estimation from the data of the Cauchy distribution involves just as many measured values of X, as the evaluation of a real linear-fractional, or Möbius transformation, in McCullagh's terms. Therefore, we need to have *three measurements* of the statistical variable X, in order to characterize a Cauchy distribution the best possible way. The general estimator will then be calculated from a particularly convenient Cauchy distribution through a well-defined transformation. Let us do some calculations, which, by the way, will now involve a summation over three values of the index, thus making our use of Latin k 'legal', as it were.

In equations (4.46) and (4.47) nothing changes, except the fact that, as we said, the sum should be now performed on three values of X, say x_1, x_2, x_3 in the samples of size three, instead of two values as above. So, instead of (4.46) we have

$$\sum \frac{1}{x_k - \theta} + \sum \frac{1}{x_k - \theta^*} = 0, \quad \frac{3}{i\theta_2} + \sum \frac{1}{x_k - \theta} - \sum \frac{1}{x_k - \theta^*} = 0$$

and instead of (4.47), we have

$$\frac{3}{2i\theta_2} + \sum \frac{1}{x_k - \theta} = 0 \tag{4.48}$$

as well as the complex conjugate of this equation. Now, the direct calculation of the estimators for θ_1 and θ_2 is rather tedious, anyway not quite as simple as in the case of samples of size two, however with more reliable results. This makes the property (4.44) worth using in simplifying the procedure of solution.

Indeed, the procedure amounts to choosing three particular values for X, say $-1, 0, 1$, and calculate the estimator of θ for them; then take the homographic transform of this estimator through the homography that carries $-1, 0, 1$, into the values x_1, x_2, x_3 of X, and the problem is solved, for such a real homography is well defined. Let us consider that the values (x_1, x_2, x_3) do correspond to the values $(-1, 0, 1)$ in this order. If the matrix of this homography has the entries a, b, c, d, then we can find it, up to a normalization factor, from the system of equations

$$x_1 = \frac{-a + b}{-c + d}; \quad x_2 = \frac{b}{d}; \quad x_3 = \frac{a + b}{c + d}.$$

This gives

$$\frac{a}{x_2x_3 + x_1x_2 - 2x_1x_3} = \frac{b}{x_2(x_3 - x_1)} = \frac{c}{2x_2 - x_3 - x_1} = \frac{d}{x_3 - x_1}. \tag{4.49}$$

The problem is now to find the estimator θ for the particular values $(-1, 0, 1)$. This can be easily done from equation (4.48) and its complex conjugate, which give the system

$$\theta_1 = 0, \quad 3\theta_2^2 = 1.$$

Therefore, in this particular case we have simply $i/\sqrt{(3)}$ as an estimator for the parameter: it is purely imaginary. The estimator according to arbitrary data of the sample (x_1, x_2, x_3) will then be obtained through the homography given by equations (4.49):

$$\theta = \frac{(x_2x_3 + x_1x_2 - 2x_1x_3)(i/\sqrt{3}) + x_2(x_3 - x_1)}{(2x_2 - x_3 - x_1)(i/\sqrt{3}) + (x_3 - x_1)}.$$

In real terms we have:

$$\theta_1 = \frac{\sum x_1(x_2 - x_3)^2}{\sum (x_2 - x_3)^2}, \quad \theta_2 = \sqrt{3}\frac{(x_2 - x_3)(x_3 - x_1)(x_1 - x_2)}{\sum (x_2 - x_3)^2}, \tag{4.50}$$

where the summation runs over the three positive permutations of the values in a sample. As equation (4.42) suggests, the parameter θ should be one of the roots of a second degree equation with complex roots. By a more involved calculation, starting from the values (16.50) and using the Viète relations, one can restore this equation, so that it is:

$$\left\{\sum (x_2 - x_3)^2\right\}\theta^2 - 2\left\{x_1(x_2 - x_3)^2\right\}\theta + \left\{\sum x_1^2(x_2 - x_3)^2\right\} = 0. \tag{4.51}$$

Therefore, we have the significant result which may be assigned to Thomas Ferguson (*loc. cit. ante*): the complex parameter of a Cauchy distribution is the root of a second degree equation whose coefficients are the first and second moments of the data from samples of size three, calculated by using the three weights:

$$p_1 = \frac{(x_2 - x_3)^2}{\sum(x_2 - x_3)^2}, \; p_2 = \frac{(x_3 - x_1)^2}{\sum(x_2 - x_3)^2}, \; p_3 = \frac{(x_1 - x_2)^2}{\sum(x_2 - x_3)^2}. \tag{4.52}$$

In this system of weights, θ_1 is the sample mean, while θ_2 is the sample variance, as it should be.

The property of fractional linear class of Cauchy variates, of being also Cauchy variates has been a subject of study for a long time in theoretical statistics (Pitman and Williams, 1967). The main statistical interest of these studies is in results regarding the directional statistics and time series (Jammalamadaka and Sengupta, 2001). Typical in such studies is the *problem of phase* and the specific functions related to it, like the imaginary exponential and the tangent functions, from which Pitman and Williams started their observations. As the reader might have guessed by now, the problem of phase is also one of our tasks here, but from a more general point of view, which we relegate to Stoka theorem (Mazilu, 2006). Namely, from a structural algebraic point of view one can say that the univariate Cauchy distribution, just like the univariate and bivariate Gaussians, are distributions of the same class. The underlining algebraic structure is a **sl**(2,R), realized by group actions in one, two or three variables with three parameters [see Chapter 2, equations (2.41), (2.42) and Chapter 8, equations (8.35), (8.36), Mazilu *et al.*, 2019]. It is from this, larger point of view, that one might think that the Cauchy and Gaussian distributions are of the same class, the 'sl(2,R) class', as it were. From such a point of view, the above observations of Peter McCullagh, are about to unravel fundamental properties of space, time and matter, related to the property of simultaneity via the concept of phase.

4.5 An Example of Yang–Mills Fields Related to the Three-Dimensional Regularization

Just to give our reader a taste of what is involved here, notice that the probabilities from equation (4.52) are related to the invariants of the equations of motion (4.17) related to Wigner principle. The common denominator in (4.52) is given by the quadratic invariant of motion from equation (4.21), and the condition of normalization of the probability vector of components (p_1, p_2, p_3) is represented by the projection of this vector along diagonal of the first octant in a Euclidean reference frame. Let us determine this terminology a little further, in relation to the complex variables from equation (4.20).

Now, if we are going to continue to use the Euclidean reference frame given by the eigenvectors of the symmetric tensor, for purposes of orientation of the space containing continuous matter, we are bound to consider that the plane of residence of the Novozhilov's averages (see Mazilu *et al.*, 2019, Chapter 4, equation (4.35)] is not quite any plane: it is an *octahedral plane* of this reference frame. In order to give our presentation a necessary geometrical vision, we refer the reader to the exquisite presentation of Silviu Olariu, for the so-called *tricomplex numbers*, as defined above in equation (4.20), and their properties (Olariu, 2002, Chapter 2). A first connection is worth mentioning here, though: an octahedral plane of ours, is any plane parallel to Olariu's nodal plane Π, and we take it as such in *any octant* of the Euclidean reference frame. Such a plane is always normal to the *trisector direction* of the octant (*loc. cit. ante*, Figure 2.1). This is the volume diagonal direction of the octant.

When referred to the light phenomenon, equations (4.17) represent the local motion transversal to the line of sight as it were, represented by the position vector in an Euclidean reference frame. Therefore, the constant quadratic integral of equations (4.17) given in equation (4.22) should have at least a geometrical meaning, if not even a physical one. And it has indeed, even an exquisite geometric meaning: like the quadratic form (4.21), it is the projection of the vector of components (yz, zx, xy) along the *trisector direction* of the first octant of reference frame, having the components (1, 1, 1). This can certainly qualify as a gauging condition property of the integral (4.17). The vector in question has still another gauging virtue: its projection along the position vector is proportional to the volume whose diagonal is this vector. However, one of the most important gauging procedures connected to it, is that it is connected with some specific Yang–Mills fields (Uy, 1976). We go a little ahead of ourselves here, in saying that this gauging procedure is closely associated with a *reverse interpretation*: an interpretation of the discrete matter by a continuum. The specifics follow.

Zenaida Uy's scaling procedure, presented in the work just cited (*loc. cit. ante*) can be described geometrically according to the general theory of what we have called before 'transition quantities' [see Chapter 8, equations (8.17) to (8.22), and discussion, Mazilu *et al.*, 2019], or dual quantities, as follows. Define the skew-symmetric matrices for the case of structure constants of the rotation group, *i.e.* according to the rule: $(\mathbf{h}_k)_{ij} = \varepsilon_{kij}$, where the total skew-symmetric Levi-Civita tensor provides the structure constants of the rotation group, as usual. This definition results in the following three matrices forming a closed algebraic system, generating that part of the three-dimensional rotation group in the Euclidean space connected with the identity matrix:

$$\mathbf{h}_1 = \begin{pmatrix} 0 & 0 & 0 \\ 0 & 0 & 1 \\ 0 & -1 & 0 \end{pmatrix}, \quad \mathbf{h}_2 = \begin{pmatrix} 0 & 0 & -1 \\ 0 & 0 & 0 \\ 1 & 0 & 0 \end{pmatrix},$$
$$\mathbf{h}_2 = \begin{pmatrix} 0 & 1 & 0 \\ -1 & 0 & 0 \\ 0 & 0 & 0 \end{pmatrix}. \tag{4.53}$$

These matrices represent a three-dimensional linear basis in the space of skew-symmetric 3×3 matrices. Choose now the vector having components of second degree in some generic coordinates, x_k say:

$$\mathbf{b}_0 = \begin{pmatrix} x_2 x_3 \\ x_3 x_1 \\ x_1 x_2 \end{pmatrix}. \tag{4.54}$$

The action of the base matrices (4.53) on this vector defines three new vectors

$$\mathbf{h}_1 \cdot \mathbf{b}_0 = x_1 \begin{pmatrix} 0 \\ x_2 \\ -x_3 \end{pmatrix}, \quad \mathbf{h}_2 \cdot \mathbf{b}_0 = x_2 \begin{pmatrix} -x_1 \\ 0 \\ x_3 \end{pmatrix},$$
$$\mathbf{h}_3 \cdot \mathbf{b}_0 = x_3 \begin{pmatrix} x_1 \\ -x_2 \\ 0 \end{pmatrix}. \tag{4.55}$$

Zenaida Uy builds then the following four vectors

$$\begin{aligned} &\mathbf{b}_1 \equiv f(x_1, x_2, x_3)(\mathbf{h}_1 \cdot \mathbf{b}_0), \quad \mathbf{b}_2 \equiv f(x_1, x_2, x_3)(\mathbf{h}_2 \cdot \mathbf{b}_0), \\ &\mathbf{b}_3 \equiv f(x_1, x_2, x_3)(\mathbf{h}_3 \cdot \mathbf{b}_0), \quad \mathbf{b}_4 \equiv g(x_1, x_2, x_3)(\mathbf{1} \cdot \mathbf{b}_0) \end{aligned} \tag{4.56}$$

where f and g are two functions, arbitrary for the moment. Then she takes notice of the important fact that the tetrad $(\mathbf{b}_1, \mathbf{b}_2, \mathbf{b}_3, \mathbf{b}_4)$ can be so chosen as to represent *static Yang–Mills* SU(2) fields. This can happen under following conditions: the gauge field intensities should be given by the classical electrodynamics relations, modified according to Yang–Mills nonabelian prescription [see Yang and Mills (1954); see also Wu and Yang (1969)]

$$\mathbf{f}_{\mu\nu} \equiv \partial_\nu \mathbf{b}_\mu - \partial_\mu \mathbf{b}_\nu - \mathbf{b}_\mu \times \mathbf{b}_\nu \tag{4.57}$$

under the following 'equations of motion'

$$\partial_\nu \mathbf{f}_{\mu\nu} + \mathbf{b}_\nu \times \mathbf{f}_{\mu\nu} = 0, \quad \partial_\nu \mathbf{b}_\nu = 0. \tag{4.58}$$

Here, the usual summation over repeated indices is assumed. The static feature of the field is explicitly recognized in the fact that the tetrad $|\mathbf{b}\rangle$, and the corresponding field intensities, do not depend explicitly on any 'time' coordinate, x_4 say, that might incidentally complete the position $\mathbf{x}$ to an event. Zenaida Uy further assumes that the functions f and g depend on coordinates via the volume $\omega = x_1x_2x_3$ of the cuboid whose diagonal is the position vector $\mathbf{x}$, and, additionally, that $g = \pm f$. One can then calculate the 'electric' and 'magnetic' gauge field intensities associated to the tetrad $|\mathbf{b}\rangle$, by the following prescriptions, replicating the well-known classical definition of electromagnetic field intensities (Wu and Yang, 1969):

$$\mathbf{E}_k \equiv i\mathbf{f}_{k4}, \quad \mathbf{H}_k \equiv \frac{1}{2}\varepsilon_{kij}\mathbf{f}_{ij}. \tag{4.59}$$

Calculating effectively the electric field here, with (4.56) and the definition (4.57) of field intensities, we have the result:

$$\mathbf{f}_{k4} = -\partial_k \mathbf{b}_4 - \mathbf{b}_k \times \mathbf{b}_4 \therefore \mathbf{E}_k = -i\left[g^{-1}g'\left(\sum x_2x_3\right)\mathbf{b}_4 + g\mathbf{e}_k\right] \tag{4.60}$$

where the summation runs over the positive permutations of the numerical indices, and a prime means derivative with respect to the unique variable — in this case ω — as usual. For the magnetic fields, using again (4.56), (4.57), and the second of (4.59), we further have

$$\mathbf{f}_{ij} = \omega f^2(\omega)\varepsilon_{ijk}x_k \begin{pmatrix} x_1 \\ x_2 \\ x_3 \end{pmatrix} \therefore \mathbf{H}_k = \omega f^2(\omega)x_k \begin{pmatrix} x_1 \\ x_2 \\ x_3 \end{pmatrix}. \tag{4.61}$$

Here we have to notice that the second condition from equation (4.58) is an identity, for we have

$$\partial_\nu \mathbf{b}_\nu \equiv \partial_k \mathbf{b}_k = [\omega f'(\omega) + f] \sum \mathbf{u}_k$$

where $\mathbf{u}_k$ are the columns from equation (4.55). In view of the fact that $\sum \mathbf{u}_k = 0$ the result is automatically the null vector:

$$\partial_k \mathbf{b}_k = [\omega f'(\omega) + f] \left[\begin{pmatrix} 0 \\ x_2 \\ -x_3 \end{pmatrix} + \begin{pmatrix} -x_1 \\ 0 \\ x_3 \end{pmatrix} + \begin{pmatrix} x_1 \\ -x_2 \\ 0 \end{pmatrix} \right] \equiv \begin{pmatrix} 0 \\ 0 \\ 0 \end{pmatrix}.$$

The *geometric* interpretation of this situation is as follows. Take the elementary vector surface 1-forms:

$$\begin{aligned} ds^1 &= x_2 dx_3 - x_3 dx_2, \\ ds^2 &= x_3 dx_1 - x_1 dx_3, \\ ds^3 &= x_1 dx_2 - x_2 dx_1. \end{aligned} \tag{4.62}$$

They can be written as the bilinear forms, using the $\mathbf{h}$ matrices from equation (4.53):

$$ds^k \equiv \langle x|\mathbf{h}^k|dx\rangle \tag{4.63}$$

so that the column matrices used in the definitions of the vectors from (4.55) are, in fact, defined by identities $\mathbf{u}_k \equiv \mathbf{h}_k|x\rangle$. The problem now, is the presence in our theory of the vector from equation (4.54). As one can see, however, a justification is not altogether out of hand: it is the essential vector for the absolute geometry based upon classical idea of volume [see Mazilu *et al.*, 2019, Chapter 4, equations (4.30) to (4.32) and discussion]. For instance, we get the important gauging relation mentioned above, by noticing that

$$\mathbf{x} \cdot \mathbf{b}_0 = 3x_1x_2x_3 \tag{4.64}$$

which allows for a statistical discussion of stresses as fluxes, as we shall show here pretty soon.

Physically speaking the coordinates can be anything that helps us in locating an event in space. The inversion theorem, for instance, allows us to consider forces as coordinates. They can be lengths just as well as distances, or the eigenvalues of a matrix. In general, we can say that the coordinates are field quantities, and this is the point of this last, concluding we should say, part of the present chapter. The coordinates here are static Yang–Mills fields, to wit, fields of the kind that concerns electrodynamics or chromodynamics. The important message of this story is the fact that the coordinates in general, just as the fields considered here, are connected to a mandatory three-dimensional 'regularization' — imposed, of course, via the necessity of scale transition in physics — by the concept of three-dimensional analyticity, as detailed in this chapter. The remaining ones will extend this connection in its essential physical aspects: the transport, the so-called reverse interpretation, the idea of propagation.

Chapter 5

The Physical Structure of Rotating Matter According to Gödel

Kurt Gödel insists on the suggestive concept of a *compass*, in order to emphasize that the general relativity is physically built upon the observation of motion, and as long as we give up the idea of force, the motion is accomplished in a setting like the one of a ship on the waves. In this context, Einstein's natural philosophy is based on the idea that the *compass of inertia* has the same indications as the *compass of gravitation.* The Newtonian principles of classical dynamics may be taken, indeed, as suggesting this identity (Einstein, 2004) but, in fact, the concept of force as defined by Newton according to Kepler laws, certainly does not suggest it. In order to sustain this point of view, and to give alongside an example of the difference between the two compasses, we shall try to cast into analytic form the Corollary 3 of the Proposition VII from Book I of Newton's *Principia*, which we reproduce here for the benefit of our reader:

> The force by which the body P *in any orbit* revolves about *the center of force* S, is to the force by which the same body *may revolve in the same orbit*, and *the same periodic time*, about *any other center of force* R, as the solid SPRP 2, contained under the distance of the body from the first center of forces, and the square of its distance from the second center of force R, to the cube of the right line SG, drawn from the first center of force S parallel to the distance RP of the body from the second center of force R, meeting the tangent PG of the orbit in G. For the force in this orbit at any point P is the same as in the *circle of the same curvature.* [(Newton, 1974, p. 51); *our Italics*]

As we see it, this is actually a definition of the force based on a precise *idea of measurement.* The only real and, in fact, physically necessary ingredients here, are *the orbit* and *the periodic time.* Of the two points, S and R, helping in defining the force, one is *our arbitrary choice* — the point toward which we want to calculate the acting force — while the other one can be *arbitrary at random*, depending on the subjective chance of whether we know the force acting towards it. Apparently, both of these points should be chosen inside the orbit. However, with the analytical development of the theory, even this condition can become unnecessary, allowing, for instance, the Newtonian definition of the force correlated with the light phenomenon, the way Newton defines it in his Opticks [(Newton, 1952, pp. 79 ff.); see also (Mazilu *et al.*, 2019)].

In order to best describe the principle of measurement used by Newton in defining the forces, notice that the last mention in the excerpt above, referring to 'the circle of the same curvature', seems to send our reasoning to a practical tool accomplishing the 'measurement' of two forces, acting along *different directions*, by each other: the *slingshot.* And what we mean here by this, is not the modern materialization of such a device, created since the emergence of rubber into the life of humanity, but the classical thing used by little David to kill the giant Goliath. The strings of this device allow putting to work, concurrently, two forces acting clearly in different directions: the force of *inertia*, acting horizontally, and that of *gravity*, acting vertically. Newton uses the idea of a sling only once in his *Principia* (Newton, 1974, pp. 2–3) in order to illustrate the definition of centripetal force, as being "that force, *whatever it is*, by which the planets are continually drawn aside from the rectilinear motions". However, we feel that one needs to insist a little more on this issue, from the modern point of view inspired by Gödel's work. We even venture to get here a bit ahead of ourselves, just for the sake of clarifying present and future issues.

The difference between inertia and gravitation is best illustrated if the stone is 'whirled', to use Newton's own word, in a horizontal plane, above the head, in a classically illustrated posture when shooting the sling. It is in this case that, as we said, the force of inertia acts horizontally as a centrifugal force, while the force of gravitation acts clearly vertically. Therefore, in this ideal case, the device — which obviously works based on the equilibrium of forces — allows a 'measurement' proper, of the force of gravitation by the force of inertia — or vice versa, of course — in any instantaneous vertical plane. In our opinion, this is a perfect illustration of a manifest difference between the *compass of inertia* and the *compass of gravity* within the very classical theory of forces. This time though, contrary to the main modern illustration of the concept, related to the second principle of dynamics in the Kepler problem, the compass of gravitation points always in the same direction, while the compass of inertia changes continuously the direction of pointing. And, we might as well add, this is the way Ernst Mach has understood what later came to be known as the 'Mach principle'. Starting from this sling shooting image, and considering only the concept of force, without any further specification of its nature — *i.e.* accepting, with Newton, the force as 'whatever it is' — we can very well dispense with the strings, and imagine a planet in motion, like in the case of a Kepler problem, as Newton himself did. Thus, the attraction *per se* can be considered as the action of a force which is 'measured' against another force that may or may not be a force representing inertia. Newton's case in the above excerpt is referring to *two attraction forces*. However, historically, the things evolved in such a way that the difference between inertia and gravitation has been brought to bear again on our knowledge. It is for this reason that we shall follow, by and large, this history a little closer, in order to properly introduce Gödel's ideas.

Still, before this 'main course', as it were, we may be allowed to offer an 'appetizer', by showing the importance of that idea of 'compass', used extensively, even as a working concept lately, we might say, by the general relativity, in its capacity of a cosmological theory. Recognition of the fact that the manner of working of the age-old sling involves *two compasses*, one of inertia and the other of gravitation, could not be achieved otherwise. Nor could one be able to comprehend the more evolved idea that, in a Kepler motion, the manner of working of these two compasses is formally the same as in the sling shooting. With a little different detail though, which, nevertheless, makes a big difference: the daily compass of gravitation on Earth aims not in a fixed direction, the vertical, but to a fixed point in space, namely the Newtonian center of force in the modern mathematical rendition of the problem. Of course, the compass of inertia changes its orientation exactly like in the case of the old sling.

Speaking of inertia and gravitation, one is naturally led to think about the *Mach's principle*, mentioned above quite a few times in a different context, according to which the inertia would be determined by the action of the distant masses only, ideally those masses located at infinity with respect to the body to be described by inertia. In spite of its name, this principle was introduced to the theoretical physics of 20th century by Albert Einstein, and the idea stays at the foundations of general relativity. One can safely say that *there is not* a formulation of Mach's principle due to Ernst Mach himself. [See the collection of dedicated works (Barbour and Pfister, 1995), especially the contribution (Norton, 1995) in that collection]. However, browsing the Mach's celebrated *Science of Mechanics*, we have been able to find a part which seems to us as being the closest to a concept of principle, while concomitantly suggesting the structure of a inertia compass *per se*. Here is, therefore, the following excerpt explaining Mach's own position with respect to the main Newtonian current of cosmological ideas:

> The expression "absolute motion of translation", Streintz (in *Physikalische Grundlangen der Mechanik, Leip- zig 1883, a/n*) correctly pronounces as devoid of meaning and consequently declares certain analytical deductions, to which he refers, superfluous. On the other hand, with respect to *rotation* (*original emphasis here, a/n*), Streintz accepts Newton's position, that *absolute rotation can be distinguished from relative rotation*. In this point of view, therefore, one can select *every body not affected with absolute rotation as a body of reference* for the expression of the law of inertia.
>
> I cannot share this view. For me *only the relative motions exist*, and I can see, in this regard, *no distinction between rotation and translation*. When *a body moves relatively to the fixed stars, centrifugal forces are produced*; when *it moves relatively to some*

> *different body,* and not relatively to the fixed stars, *no centrifugal forces are produced.* I have no objection to calling the first rotation "absolute" rotation, if it be remembered that nothing is meant by such a designation except *relative rotation with respect to the fixed stars* (*original emphasis, a/n*). Can we fix Newton's bucket, *rotate the fixed stars,* and *then* (*original emphasis here, again, a/n*) prove the absence of centrifugal forces? [(Mach, 1919, Appendix XX, pp. 542–543); *our Italics, except as indicated*]

First, one can notice that Mach was against the idea of any absolutes, as it were, mostly when it comes to describing the motion. Newton's bucket was intended to prove the inertia as an effect of rotation with respect to absolute space (Newton, 1974, Volume I, pp. 10–11). Apparently, for Ernst Mach it was hard to understand the existence of such a thing that can act "but cannot be acted upon" (Einstein, 2004, p. 58) like the absolute space. So, according to Einstein, "he was led to make the attempt to eliminate the space as an active cause in the system of mechanics". Thus, when it comes to action, Mach contended that *only matter can act upon matter*, and thus the inertia should not be due to the absolute space, but to distant matter. Because at that time the matter could not be thought of but as what we have called before as 'authentic matter', the distant matter could not be represented but as the matter contained in the distant stars. Hence, in a way, Mach reinstated, if we may say so, the fixed stars in their own rights, according to the very Newtonian idea of action at a distance. But then, the proof or disproof of the existence of centrifugal forces in the case of Newton's bucket experiment, would be, according to Mach, pending *not on the rotation of the bucket*, but *on the rotation of the distant stars.* Which, of course, is not to be accomplished within the capabilities of man!

In fact, it was astrophysically proved that the distant stars seem to move only along the visual direction, specifically to depart from us (Hubble, 1929). In the language of general relativity, this would mean that the compass of inertia points, indeed, exactly along the direction of the compass of gravitation, and thus both these directions are to be settled, somehow, with the assistance of the *compass of light.* This is, actually, the whole meaning of the idea of propagation fostered by Poincaré, as shown previously in Chapter 1. And, as we have seen, he has not succeeded in establishing an explicit independent equation of propagation for gravitation. In fact, nobody did, so that even today, in the general relativity, people treat the compass of gravitation as being identical with the compass of light. However, the compass of inertia is still suspected of being different, as it was to Newton initially, and the first sound suspicion was indeed raised, in the modern terms of Einstein's general relativity, by Kurt Gödel.

Fact is, that the compass of inertia was always thought of as being realizable as a regular magnetic compass used in seafaring navigation, pointing out in a fixed direction, exactly in the manner described by Ernst Mach in the excerpt above: a piece of elongated material suspended like a magnetic needle, in the interior of a material sphere for instance [(Harrison, 2000), p. 243]. This way, the inertia acts to maintain the direction of the needle, but in spirit of the first law of classical dynamics — *i.e.* as long as no other forces act, it maintains its state, represented by orientation — only with a little 'twist' of imagination, brought up by Gödel's idea. Namely, the compass points in the very same direction during the motion within a 'substratum' of the universe, exactly like a floating ship at sea. This substratum would represent the matter at a cosmic scale according to Einstein's ideas, and should be recognized in a specific rotation of the sphere of device.

But then, in order to validate the Machian concept of a compass of inertia, two essential ingredients would be necessary, at least from a theoretical point of view. First, we need to prove that the fundamental property of this compass — *viz.* that of remaining constantly oriented while the sphere rotates in any possible way — is *independent of the space extension of the sphere.* This is exactly the point that Gödel's idea has gradually brought up and, we believe, it was also the reason of a high appreciation of Einstein for Gödel's achievement. However, according to our knowledge, such a space scale independence was never an object of independent study like, for instance, its celebrated counterpart, the thermal radiation, which has been independently studied, in order to finally realize that it has a spectral distribution independent of the dimensions of the 'hohlraum' [the Wien's displacement law; see Mazilu (2010)]. In this respect, the Machian compass of inertia shares the fate of another essential ingredient of speculative thinking: *the Einstein elevator.*

The second, and by far more important ingredient necessary for the validation of the Machian concept, is a way to describe the action of the rotating matter upon our compass: it is this interaction that has to be invariant with respect to the change in the dimensions of that material sphere containing the compass of inertia. The classical theory of forces has no possibility of describing it. However, it is describable in a general relativistic setting, as the *Lense–Thirring effect*, of course, when one does not consider the change in dimensions of the hollow sphere [for a critical account, updates and English translations of the original articles from 1918 of Hans Thirring and Joseph Lense, see (Mashhoon, Hehl and Theiss, 1984)].

5.1 Classical Forces and Compass Indications

Now, after this digression, let's get back to our present point, which, according to our discussion here, can be simply reduced to a rather suggestive statement: *the classical Kepler motion is a compass defining forces.* It is from this perspective of the modern general relativity, that one can state that Newton used the classical Kepler motion as a tool for defining the very concept of force. In order to properly illustrate this statement, we use the algebra involved in an argument by James Whitbread Lee Glaisher on the excerpt above from Newton (Glaisher, 1878). This argument is valuable for us on two accounts: first it is, according to our knowledge, the only one which has an analytical form expressly related to the formulation of Corollary 3 of Newton. Secondly, it was triggered by a problem that explicitly connects the classical dynamics, therefore the inertia, to the Kepler problem, bringing about genuine issues on this connection. So, it may be worth our while in needlepointing our full story of these issues on the canvas provided by such a valuable work.

To start with, using the geometry of triangle in order to handle the concepts in that excerpt, Glaisher reduces the ratio of those two forces involved in the problem, as it was defined by Newton himself, to the more manageable expression

$$\frac{FORCE\ to\ R}{FORCE\ to\ S} = \left(\frac{SN}{RM}\right)^3 \frac{RP}{SP}. \tag{5.1}$$

This expression involves, on one hand, the perpendiculars SN and RM of the two centers of force onto tangent in P to the orbit, and, on the other hand, the distances RP and SP of the moving point to the two centers of force. This expression is particularly prone to an analytical form and, further on, to a differential calculus. Let us reproduce here just the analytical part, in order to get a feeling of the necessary calculations and their results, in order to be compared with those based on the second principle of dynamics. Choose S, in case it is a privileged position in space occupied by a material point, as the origin of a reference frame. This would mean that a force characterization the gravitation acts between the two material points S and P. Furthermore, this choice would mean a reference frame and a coordinate system fit to gravitational force. Then, referring the generic coordinates, (ξ, η) say, in the plane of motion, to such a frame, the equation of orbit in a Kepler motion, can be written as

$$C(\xi, \eta) \equiv a_{11}\xi^2 + 2a_{12}\xi\eta + a_{22}\eta^2 + 2a_{13}\xi + 2a_{23}\eta + a_{33} = 0, \tag{5.2}$$

where C here is taken as meaning a 'Conic'. The tangent to this conic is one essential ingredient in Newton's procedure. In an arbitrary point P, of coordinates (x, y) say, its equation is

$$(a_{11}x + a_{12}y + a_{13})\xi + (a_{12}x + a_{22}y + a_{23})\eta + a_{13}x + a_{23}y + a_{33} = 0. \tag{5.3}$$

Thus one can calculate the distances from the two centers of force to this straight line, by a well-known analytical procedure. One gets expressions dependent on the coordinates of the point of application of force:

$$\overline{\mathbf{SN}} = \frac{1}{\sqrt{\mathbf{z}^2}}, \quad \overline{\mathbf{RM}} = \frac{1 + \mathbf{z} \cdot \overline{\mathbf{SR}}}{\sqrt{\mathbf{z}^2}}, \tag{5.4}$$

where $\mathbf{z}$ is the vector of components

$$\frac{a_{11}x + a_{12}y + a_{13}}{a_{13}x + a_{23}y + a_{33}}, \quad \frac{a_{12}x + a_{22}y + a_{23}}{a_{13}x + a_{23}y + a_{33}}. \tag{5.5}$$

Equation (5.1) can then be written in the form

$$\textit{Force to } S = (1 + \mathbf{z} \cdot \overline{\mathbf{SR}})^3 \frac{r}{\sqrt{r^2 + \overline{\mathbf{SR}}^2 - 2\mathbf{r} \cdot \mathbf{SR}}} \textit{Force to } R, \tag{5.6}$$

where $\mathbf{r}$ is the position of the moving point with respect to S. Therefore, *if we know* the force toward R, we can calculate the force toward S. For such an occasion, Newton analyzed a number of particular cases in order to be used appropriately. One of these cases shows that, if R is in *the center of conic*, then the force between P and R is proportional to distance between them (Newton, 1974, Corollary 1 of Proposition X, Problem V, p. 54). Therefore, if we choose R as the center of the conic section, then (5.6) simplifies to

$$\textit{Force to } S = \mu r\, (1 + \mathbf{z} \cdot \overline{\mathbf{SR}})^3. \tag{5.7}$$

Here μ is a constant coming from the law of force towards center of conic, which is known to be proportional with the distance. Using (5.5) in order to calculate the dot product from the parenthesis of this equation gives

$$\textit{FORCE to } S = \mu a_{33}^3 \frac{r}{(a_{13}x + a_{23}y + a_{33})^3} \tag{5.8}$$

which is the main of Glaisher's analytic results. It was, in fact, established earlier by William Rowan Hamilton, not analytically though, but in geometrical synthetic terms, which is why it is also known under his name (Hamilton, 1847). Based on equation (5.3) we can formulate this result in words: the force acting on P toward the center S is proportional to the distance from P to S, and inversely proportional to the cube of distance from P to the polar of S with respect to the orbit. Using the equation of the orbit (5.2), the expression (5.8) can still be written as

$$FORCE\ to\ S = \mu a_{33}^3 \frac{r}{(a_{11}x^2 + 2a_{12}xy + a_{22}y^2 - a_{33})^{3/2}}. \tag{5.9}$$

Consequently, (5.8) and (5.9) are the only expressions of force, allowed by the geometry of the Kepler problem. Or so it seems, for there is more to it along the lines of physics.

The principles of dynamics were, apparently, not involved here, so we cannot make this way a comparison between inertia and gravitation. However, to be fair, we should say that the dynamics *per se* was certainly involved, *but only to an infrafinite time scale.* Indeed, as we repeatedly stressed, Newton used the idea of impulse of a force in order to physically assure that a central force acting continuously, obeys *the second of the Kepler laws* (Newton, 1974, p. 40). We know now, by the Berry–Klein theory of scaling, that this physical warranty is invariant even with respect to the expansion of a 'space' of fairly arbitrary shape (Klein, 1984), just like the Planck spectrum, and even more so: while the Planck spectrum is an adiabatic invariant, the Berry–Klein invariance is independent of the rate of enclosure expansion. However, until we do not use the second principle of dynamics in finding the expression of force, we cannot say anything about the compass of inertia at the current finite scale, not even if that 'fairly arbitrary shape' of Klein, which is in fact an ellipsoid (Klein, 1984), is somehow connected with a dynamics. Fact is, that when it occurred for the first time (Bertrand, 1877), the modern problem of finding the *possible expressions* of the central force acting upon planets in motion, was not conceived otherwise, but only in connection with the second principle of dynamics. The Paul Appell's method described by us (Mazilu *et al.*, 2019) is related to this historical incident.

5.2 Bertrand's Problem, Binary Stars and the Compass of Inertia

Glaisher's geometrical argument was, in fact, triggered by the same incident (Glaisher, 1878): a problem, launched in the scientific literature by Joseph Bertrand. As we mentioned above, the Glaisher's approach is unique, though, among all solutions presented to this problem, by the fact that it uses only geometry and the original definition of Newton for forces. Bertrand was only interested in eliminating the second and the third laws of Kepler from the logic related to the procedure of obtaining the expression of central force that determines the motion of celestial bodies (Bertrand, 1877), and we think we can even articulate *his* reason for this. In the second of the works we have in mind here (Bertrand, 1877b), he mentioned the name of Yvon Villarceau who "in his beautiful research on binary stars" would have already found the expression of the central force which, pulling toward a given point in a plane, "can determine an elliptical pathway". We have located this, indeed beautiful, work of Villarceau in the collection (Villarceau, 1850), on p. 71, and largely commented on its meaning within Newtonian theory of forces (Mazilu and Porumbreanu, 2018). Here, and for now, we are merely interested in the reason for which Joseph Bertrand has chosen only the first law of Kepler in order to select the analytic expression of the magnitude of central force. Even though not mentioned by Bertrand explicitly, the reference to Villarceau seems to elucidate here: it stays in those *arbitrary points in plane* from the Corollary 3 of Newton excerpted above, toward which the forces are supposed to point during their actions.

Fact is, indeed, that the binary stars do not obey the Kepler laws — or, more to the point, the first of them — without tweaking a little their original formulation. They are systems of two neighboring stars, out of which one seems to be stationary, while the other moves around it. The pathway of this motion is always an ellipse: this is the only true resemblance with the original Kepler motion, for the center of force is, indeed, in the stationary star of the system, but that star *is never located in the focal point of the ellipse* as in the planetary system. That star can, however, be recognized as the center of force in the very manner Newton described the procedure, *i.e.* by the second of Kepler laws: the position vector of the companion with respect to it sweeps equal areas in equal times. This is the second law of the Keplerian system, indeed, but *not* with respect to the focus of the orbit. One can suspect that in defining the centripetal forces as in the Corollary 3 reproduced above, Newton might have been aware of the existence of such systems, which, by the way, started being signaled in the sky just about that epoch [specifically, in 1650; see Aitken (1918)]. People, especially those inclined to a conspiration theory, may also theorize that Newton had some information on these systems, judging by the formulation of that corollary. However, knowing the possibilities of Newton's imagination, this does not seem likely: if he was able to invent forces based on the generalization of the idea of a sling in shooting, that very feat leaves no doubt that, knowing that there are real binary systems whose center of force can be anywhere inside the ellipse, Newton would have been able to improve on the very Hooke's idea of physical light ray, in order to give a concept of modern physical light ray that was, by and large, obvious to physics only in the times of Louis de Broglie, as we have shown in Mazilu *et al.* (2019, Chapter 3). But, let us come back to our present subject-matter, in order make some comments and draw some conclusions on this issue — and some other important collaterals in fact — as we go along.

Joseph Bertrand must have had realized how much of the first law of Kepler is of essence and, by his reference to Villarceau, we feel entitled to assume that this is due to his knowledge about the binary stars. Anyway, this seems to be indicated by his continual dedication to the Kepler problem, on which he even has a fundamental theorem — of cosmological extraction, we should say — on forces (Bertrand, 1873): the only central forces capable of generating closed orbits according to the classical laws of dynamics are the central force with magnitude *proportional to distance* and the force with magnitude *inversely proportional to the square of distance.* Exactly those forces involved in the proof of Berry–Klein invariance principle! (see Mazilu *et al.*, 2019, Chapter 9, equations (9.34) ff.).

The first one of these two laws of forces was given by Newton himself, *directly* we might say, as signaled before (Newton, 1974, Corollary 1 of Proposition X, Problem V, p. 54). The second one can only be given *indirectly*: one has to assume the first law of force, in order to give the second one by the procedure of measurement indicated in the Corollary 3, as shown above. This way, however, becomes obvious that, with respect to the Kepler motion, the first kind of these laws of force is not 'real' in the classical sense, for *there is nothing 'authentically' material in the center of a Keplerian orbit.* This non-existence of something material in the center of force became a characteristic of the physics of the light phenomenon only after Augustin Fresnel, and has been pushed to its extreme by the quantum mechanics (Heisenberg, 1925). Probably as a result of this continual preoccupation with the relation between trajectory and the associated dynamics, in 1877 Joseph Bertrand set out to prove the theorem that:

> If Kepler would not have deduced from observations but a single one of his laws: *the planets describe ellipses whose focus is occupied by the Sun*, one could conclude, from this result alone taken as general principle, that the force governing them is directed toward the Sun and is inversely proportional with the square of distance [(Bertrand, 1877a); *our translation, original Italics*].

And he even proves it, indeed. One can notice here the identity of the general principle adopted with the first of the Kepler laws in its initial formulation. Then, probably realizing, as we said, the necessity of a specific generalization as indicated by the existence of binary stars, Bertrand proposes, in the final of the article just quoted, such a generalization:

> *Knowing that the planets describe conic sections, and assuming nothing else ever, find the expression of the components of the force that drives them, expressed as functions of the coordinates of their application point.* [(Bertrand, 1877a); *our translation, original Italics*].

There was a wealth of responses to this challenge from different points of view, and with different manners of attack of the problem. As we said, Glaisher's work was enticed by this challenge. A significant one of these rejoinders was that of Paul Appell, also described by us (Mazilu *et al.*, 2019, Chapter 7), which we take as instrumental in defining the *inertial properties of the physical surfaces.* As we go along, we shall mention some other contributions. However, on the present occasion we merely insist on the contributions of Gaston Darboux from France (Darboux, 1877, 1884) and Ugo Dainelli from Italy (Dainelli, 1880), as being of essence in describing the difference between the inertia and gravitation.

The repercussions of Newton's initial theory on the class of centripetal forces responsible for the Keplerian motion are by far much larger than it would seem at the first sight. The most important among these is that such a force, even though central, can have a magnitude depending on the position in a general manner, not just through the distance between the center of force and the moving point. The equation of Keplerian trajectory, the orbit, depends on five parameters, giving the position, the orientation and the size of the orbit in its plane. These parameters can be aptly called *physical parameters*. Two kinds of physical parameters are then involved in the implicit equation of a Keplerian trajectory: the ones provided by the expression of force, and the ones accidentally coming, as *initial values* for instance, from the specific problem of motion taken into consideration. Let's get into a few details of the problem.

A particular setting of the dynamical problem involving central forces, usually chosen by the symmetry of the problem, as it were, entails the polar coordinates, (r, ϕ) say, of the plane of motion referred to the center of force. Particularly, because the force is central, assuming that its magnitude depends in a certain way only on the coordinates of the application point of its action, it has no other components except the one along the direction to the center of force. As a consequence, we can choose a special coordinate system: here r is the distance *from the center of force to the moving point*, while ϕ is the polar angle with respect to an arbitrarily chosen reference direction passing through the center of force. Also, one considers a unit mass moving classical material point, because the mass is, for the moment, only an unimportant constant. The classical equation of motion then takes the form

$$F(\mathbf{r}) = -\ddot{r}(t) + r\dot{\phi}^2, \tag{5.10}$$

where a dot over symbol means differentiation with respect to time, as usual, and $F(\mathbf{r})$ is the magnitude of force, assumed to depend only on the position of its action. Taking here the area constant with respect to the center of force into consideration, denoted suggestively $\dot{a}$ as usual, and passing from the derivative on time, to that on the polar angle, we get *the Binet's equation* in the form:

$$u^2\left(\frac{d^2u}{d\phi^2}+u\right)=\frac{1}{\dot{a}^2}F(\mathbf{r}),\quad ur=1. \tag{5.11}$$

This equation is the basis for Darboux's analysis. As far as we can reason on this statement of the second principle of dynamics, the mathematical rendition from the equation above, can be used just as well for finding the most general permissible form of the magnitude of force, as function of the coordinates of its point of application. One can write (5.11) in an obvious form that gives *the expression of the magnitude of force* explicitly, in terms of the inverse of the radius:

$$F(\mathbf{r})=\dot{a}^2u^2\left(\frac{d^2u}{d\phi^2}+u\right). \tag{5.12}$$

Now, if we know the function $u(\phi)$, which is given by the equation of the orbit, the expression of force is only a matter of calculus. As it happens, the Kepler's first law offers a particularly easy and significant solution in the system of polar coordinates chosen to express the force. Namely, except for the cases where the conic passes through the center of force, which, in our case, is the origin of coordinates, the general equation of this conic is of the form

$$u(\phi)=m\cos\phi+n\sin\phi+\sqrt{M\cos 2\phi+N\sin 2\phi+P},$$

where m, n, M, N, P are the *five real parameters* which can define the conic *in its plane.* No mention about some special position of the center of force — the origin of coordinates in this case — is ever to be made here: as in the original case of Newton, that origin can be arbitrary in the plane of motion. Differentiating here with respect to the polar angle, we get

$$\frac{d^2u}{d\phi^2}+u=\frac{P^2-M^2-N^2}{(M\cos 2\phi+N\sin 2\phi+P)^{3/2}},$$

which gives the force from equation (5.12) in the form

$$F(\mathbf{r}) = \frac{\dot{a}^2(P^2 - M^2 - N^2)}{r^2(M\cos 2\phi + N\sin 2\phi + P)^{3/2}}. \tag{5.13}$$

Consequently, the equation of orbit has a different number of physical parameters than the force itself, which means that, for a given force expression, there are a multiple infinity of possible orbits for every moving material point, just as Newton assumed in the beginning. In order to see exactly how many from among the physical parameters of the orbit are also involved in the expression of force, let us notice that the expression of force from equation (5.13) can be cast into the form

$$F(\mathbf{r}) = \frac{\mu}{r^2(\alpha\cos 2\phi + \beta\sin 2\phi + \gamma)^{3/2}}, \tag{5.14}$$

where we used exclusively Greek symbols for the constants, in order to mark the fact that they are connected with the expression of force. In this case the corresponding orbit has the form:

$$u(\phi) = m\cos\phi + n\sin\phi + \theta\sqrt{\alpha\cos 2\phi + \beta\sin 2\phi + \gamma}. \tag{5.15}$$

It depends therefore on three more parameters, over those introduced by the expression of force: m, n and θ, not all of them essential though. Therefore to a given central force, in a position in space there is *a priori* a triple infinity of orbits, having the parameters m, n and θ as 'coordinates', so to speak.

In order to better grasp what is really involved here, we actually need to solve the dynamical problem for the case of force given by equation (5.14). Using the very same Binet's equation (5.12), this time as an equation of motion whereby

$$\frac{F(\mathbf{r})}{u^2} = \frac{\mu}{(\alpha\cos 2\phi + \beta\sin 2\phi + \gamma)^{3/2}}$$

equation (5.12) thus becomes

$$\frac{d^2u}{d\phi^2} + u = \frac{(\mu/\dot{a}^2)}{(\alpha\cos 2\phi + \beta\sin 2\phi + \gamma)^{3/2}}. \tag{5.16}$$

As well known, the general solution of this is a sum between the general solution of the corresponding homogeneous equation and a particular solution of the non-homogeneous one. The general solution of the homogeneous equation is of the form

$$u(\phi) = m\cos\phi + n\sin\phi$$

while a particular solution of the nonhomogeneous one is simply

$$\Phi(\phi) = \frac{(\mu/\dot{a}^2)}{\gamma^2 - \alpha^2 - \beta^2}\sqrt{\alpha\cos 2\phi + \beta\sin 2\phi + \gamma}.$$

By equations (5.13) and (5.14) the fraction before square root sign is θ, so that the general solution of equation (5.16) is indeed given by equation (5.15), as expected. But here it becomes quite clear that only the parameters α, β and γ do not depend on the initial conditions of equation (5.16), so that we would have *a priori a triple infinity* of orbits corresponding to a central force of given analytical expression. However, there is more to it than what we have just showed.

The family of conics from equation (5.15) assumes the special property of having two common tangent directions controlled by the force itself. Indeed, in cartesian coordinates equation (5.15) can be written as

$$(1 - mx - ny)^2 = \theta^2[(\alpha + \gamma)x^2 + 2\beta xy + (\gamma - \alpha)y^2]. \tag{5.17}$$

These conics are tangent to the straight lines given by equation

$$(\alpha + \gamma)x^2 + 2\beta xy + (\gamma - \alpha)y^2 = 0$$

intersecting each other in the center of force, which is thereby the pole of the straight line

$$mx + ny - 1 = 0$$

with respect to the orbit. Consequently, in this approach, a force having the analytical expression given by equation (5.14), controls *the tangent directions to the trajectory*, specifically the two tangent lines that pass through the center of force. If $\beta^2 + \alpha^2 - \gamma^2 > 0$, the two tangent directions are real. Otherwise they are complex. In both cases, though, their point of intersection, *i.e.* the center of force, is a real point, for two complex lines in a plane always intersect in a real point. Consequently, the force from equation (5.14) characterizes both the situation of the binary stars — in which case the asymptotic directions meet in the region inside the orbit — and that of the light ray as defined by Robert Hooke — when the center of force, identified with the source of light, is in the exterior of the trajectory Hooke (1665, pp. 55–67) and Hooke (1705); see also previous Chapter 4 here]. The first case is that of the complex tangents, while the second corresponds to real tangents — the geometrical rays delimiting the Hooke's physical ray, for instance. This last situation would certainly have been grasped by Newton, had he have some information on the geometrical structure of the binary star systems. In which case, we would certainly have today a different image of the ray in general, and of the light ray especially!

Now, as long as we have at our disposal only the equation of orbit in order to get the force, the problem has a twofold solution. Indeed, if we take into consideration the equation of the trajectory in the expression of the force given by equation (5.14), then this equation takes the form

$$F(\mathbf{r}) = \frac{\mu r}{(\mathbf{a} \cdot \mathbf{r} + b)^3}, \tag{5.18}$$

where $\mathbf{a}$ and b are three arbitrary constants. Therefore, both the equations (5.14) and (5.18) can equally represent the solutions of the problem thus formulated, as Gaston Darboux himself noticed. This is in accordance with the purely geometrical results of Glaisher, from equations (5.8) and (5.9).

Equation (5.18) has, however, just like its geometrical counterpart (5.9), a somewhat post factum appearance, so to speak. Indeed, while equation (5.14) can be obtained directly, as a consequence of the second principle of Newton, equation (5.18) asks further that, after we have already applied that principle and obtained the equation of the orbit, to consider it once again explicitly in order to eliminate the quadratic term from the expression of the force, and thereby obtain (5.18). This would mean that the expression (5.18) cannot be considered independently of that already given in equation (5.14). In other words, the three constants **a** and b cannot be arbitrary parameters: they represent precisely the position of a certain orbit with respect to the focus of the orbit. The problem then arises, if, and in what conditions, an elliptic trajectory is the expression of a central force given by equation (5.18) as such, *i.e.* independently of the general equation (5.14) for the magnitude of force. In order to solve this problem we use again the second principle of classical dynamics, with the force given by equation (5.18).

The Binet's equation (5.12), corresponding to an 'independent' force like that from equation (5.18), is

$$\frac{d^2u}{d\phi^2}+u=\frac{(\mu/\dot{a}^2)}{(bu+a\cos\phi)^3}, \tag{5.19}$$

where a is, this time, the magnitude of the vector **a**. This equation can be solved by transforming it into

$$\frac{d^2v}{d\phi^2}+v=\frac{b(\mu/\dot{a}^2)}{v^3}, \quad v\equiv bu+a\cos\phi$$

as a result of the fact that $\cos\phi$ is automatically a solution of the homogeneous differential equation corresponding to (5.19). Thus $\nu(\phi)$ is the magnitude of the variable vector having as components the solutions of the homogeneous differential system

$$\nu^2=\xi^2+\eta^2, \quad \frac{d^2\xi}{d\phi^2}+\xi=0, \quad \frac{d^2\eta}{d\phi^2}+\eta=0.$$

The general solution of this system depends on three constants, representing the components of two arbitrary vectors whose vector product is fixed:

$$\xi(\phi)=\alpha_1\cos\phi+\beta_1\sin\phi, \quad \eta(\phi)=\alpha_2\cos\phi+\beta_2\sin\phi,$$
$$(\alpha_1\beta_2-\alpha_2\beta_1)^2=b(\mu/\dot{a}^2).$$

The trajectory is therefore a conic, indeed:

$$\boldsymbol{\alpha}^2x^2+2(\boldsymbol{\alpha}\cdot\boldsymbol{\beta})xy+\boldsymbol{\beta}^2y^2=(ax+b)^2. \tag{5.20}$$

This time, though, the force expression determines *the straight line conjugated to the center of force*, while the initial conditions determine the two directions tangent to the trajectory, and passing through the center of force. Notice, however, that the two tangent directions are in this case *unconditionally complex*. Therefore, the center of force should always be inside the orbit, as in the case of the binary stars.

The Darboux's original conclusion (Darboux, 1877) should be, therefore, slightly amended. The most rational way to do it is to start from a definition of the orbit, based on its conjugated triangle. Indeed, a certain conic can be defined in its plane through three distinguished directions: two straight lines, tangent to the conic, crossing each other in the center of force, which is the pole of a third straight line with respect to the conic itself. This last line passes through the two points where the tangents meet the conic. The corresponding analytic expression of the orbit in its plane is, either that from equation (5.17) or that from equation (5.20). The two tangents are given by the quadratic form — specifically they are the two linear factors of the quadratic form — while the linear form under the square power in the orbit's equation gives the straight line polar to the center of force, where the two tangents meet.

The forces from equations (5.14) or (5.18), are therefore, indeed, the only solutions of the problem as set by Bertrand, but they 'act', if we may say so, in quite different manners in determining the analytic form of the orbit of motion, in duality one might say. First, let us say it once more: regardless of the manner of this 'action', *the point of intersection of tangents to the orbit is the center of force.* It is always a real geometrical position no matter of the nature of the two tangent directions. Therefore, it follows, mathematically, the physical nature of the problem: there is a *real* position to be occupied by a material point generating the force field, in case there is such a point. In the first case, *i.e.* that of equation (5.14), *the force determines a set of trajectories having common tangents.* These tangents are decided by the quadratic form from the expression of force, while *the polar of the center of force is given by the initial conditions* of the motion. In the second case, *i.e.* that of equation (5.18) *the force determines a set of trajectories having the same polar of the center of force*, given by the linear form from the expression of force. The quadratic form is here given by the initial conditions of the motion. In this case, the center of force is always inside the trajectory, while its polar is external. It is therefore clear that only a force like that from equation (5.14) can characterize a physical light ray, while the force from equation (5.18) characterizes the binary star systems, or the like, in particular the original Keplerian motion. However, physically speaking, there also exist, among these last systems, some that can be characterized by the force from equation (5.14).

Rounding up the results, we have therefore two systems of central forces, as a solution of the problem as given by Gaston Darboux, which we write in cartesian coordinates of the motion plane as

$$F_1(\mathbf{r}) = \frac{\mu r}{(\alpha x^2 + 2\beta xy + \gamma y^2)^{3/2}}, \quad F_2(\mathbf{r}) = \frac{\mu r}{(\mathbf{a}\cdot\mathbf{r} + b)^3}. \tag{5.21}$$

The trajectories corresponding to these forces, written in the implicit form are, respectively

$$\begin{aligned} \alpha x^2 + 2\beta xy + \gamma y^2 &= (ax + by + c)^2, \\ \boldsymbol{\alpha}^2 x^2 + 2(\boldsymbol{\alpha}\cdot\boldsymbol{\beta})xy + \boldsymbol{\beta}^2 y^2 &= (\mathbf{a}\cdot\mathbf{r} + b)^2. \end{aligned} \tag{5.22}$$

Here the parameters a, b, c, as well as the vectors $\boldsymbol{\alpha}$, $\boldsymbol{\beta}$, are given by the initial conditions of the motion. The vectors satisfy to a geometrical condition showing that *their vector product is fixed.* The center of force is the point of intersection of the lines given by the quadratic forms from the left-hand side of the equations, which is the pole of the straight line given by the right-hand side. The Newtonian force having the magnitude inversely proportional with the square of distance, is obtained in the first case of (5.21) for $\alpha = \gamma$ and $\beta = 0$, while the isotropic elastic force comes from the second case (5.21) for the vector $\mathbf{a}$ null.

Therefore, knowing only the shape of the orbit, in order to find, by means of Newtonian dynamics, the true algebraical expression of the central force which maintains it, we have to decide which one from the terms in 'canonical' equation of the orbit is specified by the initial conditions of the motion. If the parameters of the polar of the center of force with respect to orbit are specified by the initial conditions, then the magnitude of force is given by the first expression from equation (5.21). On the other hand, if the two tangent lines to the orbit, determined by the polar of the center of force are specified by the initial conditions of the motion, then the magnitude of force is given by the second expression from equation (5.21).

The right-hand side of the equation (5.10), even though referring only to the magnitude of force, contains the whole story: geometrically, the inertia is not neatly oriented. There should be a component of acceleration along the direction to the center of force, and one along the normal to the orbit. This fact can be expressed in a number of ways. However, in connection with the incident raised by Joseph Bertrand, we find significant the intervention of the Italian school of geometry (Battaglini, 1878; Dainelli, 1880). According to Ugo Dainelli, Giuseppe Battaglini, who took into consideration the problem *along a conic*, as originally posed by Bertrand, has a result that, for us at least, seems exquisite and deserves a generalization, undertaken by Dainelli himself in his contribution to the solution of Bertrand problem. Quoting:

> Professor Battaglini found that: ≪ in general, when a mobile traverses a conic, *the accelerating force* can be considered as *the resultant of two forces*, one *directed along the position vector with respect to an arbitrary fixed point*, the other *directed along the tangent to the curve*;...≫ [(Dainelli, 1880); *our translation and emphasis, a/n*]

Consequently, pending the third principle of dynamics, the inertial force of classical Kepler motion can never be directed toward a fixed point, be it the focus of the conic, or any other point, as, for instance, in the case of binary stars. In other words, in the classical dynamics of the Kepler problem *the compass of inertia* and *the compass of gravitation* can never point in the same direction. From a general perspective, the very second principle may be in danger, so that the Battaglini's result should be worth generalizing for an arbitrary trajectory, a task that Dainelli undertook in his contribution.

Let us, therefore, present the demonstration given by Ugo Dainelli (Dainelli, 1880), in order to properly illustrate this case.

Assume that the motion is plane, and its trajectory is given in the implicit form

$$\Phi(x, y) = 0 \tag{5.23}$$

without momentarily bothering with the center of force: Φ is just a result of observations. The velocity of the motion has the direction of the tangent to trajectory, which, by the theorem of implicit functions, comes to

$$\frac{dy}{dx} = -\frac{\partial \Phi}{\partial x} \Big/ \frac{\partial \Phi}{\partial y} \Rightarrow \frac{dx}{dt} = \pm k \frac{\partial \Phi}{\partial y}; \quad \frac{dy}{dt} = \mp k \frac{\partial \Phi}{\partial x}, \tag{5.24}$$

where k is a constant here. Notice that the time t is a continuity parameter, that may or may not be the time *per se*. In these conditions even the velocity is actually much more general, in the sense that what we take as the constant k may actually be an arbitrary function of the position along the trajectory of motion. In order to account for this situation, we will rewrite the equation (5.24) as

$$\frac{dx}{dt} = \pm k f \frac{\partial \Phi}{\partial y}, \quad \frac{dy}{dt} = \mp k f \frac{\partial \Phi}{\partial x} \tag{5.25}$$

where $f(x, y)$ is an arbitrary function of the coordinates of the moving point. We can calculate now the components of the acceleration vector, and from these, using the Newtonian recipe, which involves the second law of dynamics, we can calculate the components of force. Up to a factor, these are:

$$\begin{aligned} X = &-k^2 \left(\frac{\partial^2 \Phi}{\partial y^2} \frac{\partial \Phi}{\partial x} - \frac{\partial^2 \Phi}{\partial x \partial y} \frac{\partial \Phi}{\partial y} \right) f^2 \\ &+ k^2 \frac{\partial \Phi}{\partial y} \left(\frac{\partial f}{\partial x} \frac{\partial \Phi}{\partial y} - \frac{\partial f}{\partial y} \frac{\partial \Phi}{\partial x} \right) f, \\ Y = &-k^2 \left(\frac{\partial^2 \Phi}{\partial x^2} \frac{\partial \Phi}{\partial y} - \frac{\partial^2 \Phi}{\partial x \partial y} \frac{\partial \Phi}{\partial x} \right) f^2 \\ &- k^2 \frac{\partial \Phi}{\partial x} \left(\frac{\partial f}{\partial x} \frac{\partial \Phi}{\partial y} - \frac{\partial f}{\partial y} \frac{\partial \Phi}{\partial x} \right) f. \end{aligned} \tag{5.26}$$

One can thus see explicitly that the force has, indeed, *two vector contributions*, thus generalizing the case of Battaglini to an arbitrary trajectory. One of these contributions is quadratic in the arbitrary function defining the velocity, the other is linear. This linear contribution is tangent to the orbit, thus being perpendicular to the normal, no matter of the functional form of Φ. As for the quadratic contribution one can say nothing just yet: we do not know where it pulls. We can say where it pulls, though, in case we know the form of the function Φ. This, in particular, is the case of the laws of Kepler, specifically, of the first of those laws. Indeed, if we assume that the motion obeys the first of the Kepler's laws, but without mentioning the center of force, as in Bertrand's formulation of the problem, then the trajectory from equation (5.23) is given by equation:

$$2\Phi(x,y) \equiv a_{11}x^2 + 2a_{12}xy + a_{22}y^2 + 2a_{13}x + 2a_{23}y + a_{33} = 0, \tag{5.27}$$

then the quadratic component can be expressed as

$$-k^2 f^2 \Delta(x - x_c); \quad -k^2 f^2 \Delta(y - y_c); \quad \Delta \equiv a_{11}a_{22} - a_{12}^2, \tag{5.28}$$

where x_c and y_c are the coordinates of the center of conic (5.27). Therefore, when the trajectory is a conic, the quadratic contribution from the expression of force is an *elastic force*, which, in a Kepler motion, is known to be always directed towards the center of the trajectory. It has, however, an 'elastic constant' which is variable with the direction in the plane of motion and this elastic constant is decided, on one hand, by the parameters of the trajectory in its plane, and, on the other hand, by the square of the arbitrary function from the definition of the velocity of motion. If that function is a constant, we have a proper elastic force, as known from the case of the harmonic oscillator.

The Newtonian standard in measuring the forces responsible for the Keplerian motion, has a precise connotation within the framework of classical dynamics as represented by Newton's laws. Specifically we have in mind the second law, therefore material systems, because only these possess inertia. The light seems to be something else entirely. Namely, if we admit that the equation of trajectory is presented in an implicit form as before, it necessarily results the existence of two forces, which this time are vectors. These behave exactly as those stipulated in Corollary 3 of the Proposition VII from *Principia*. Aside from these specific issues, the Dainelli's theory points out, on one hand, to the importance of a Wigner dynamical principle, in the form of equation (5.25), whereby the potential receives a precise meaning. On the other hand, this theory shows that equations (4.27) represent, in fact, a general scheme for describing the inertia forces: they always have two components, no matter of the reference frame chosen to represent the the finite space scale. One of these components should act along the position vector, the other perpendicularly to it.

The arbitrary function f of this theory still remains undecided, even in cases where the motion satisfies to the first law of Kepler. In order to say something about that function we have to go deeper, considering, for instance, the second of the Kepler laws. For, in general, the motion described by equation (5.23) obeys the second of the Kepler laws only in special conditions. Admitting that we know the position of the center of force, and taking the origin of the coordinates in that position, from equation (5.25) we have:

$$y\frac{dx}{dt} - x\frac{dy}{dt} = \pm kf\left(x\frac{\partial\Phi}{\partial x} + y\frac{\partial\Phi}{\partial y}\right).$$

Thus, in order to have the second law of Kepler satisfied, we must have

$$f\left(x\frac{\partial\Phi}{\partial x} + y\frac{\partial\Phi}{\partial y}\right) = c,$$

where c is a constant. Therefore, we are now able to find the arbitrary function, and by this the variable elasticity constant from equation (5.28). Using equation (5.27) gives

$$f(x, y) = -\frac{c}{a_{13}x + a_{23}y + a_{33}}. \tag{5.29}$$

The components of force are now the ones given by Battaglini:

$$
\begin{aligned}
X &= -k^2c^2\Delta\frac{x-x_c}{(a_{13}x+a_{23}y+a_{33})^2} \\
&\quad - ck^2(y_cx-x_cy)\frac{a_{12}x+a_{22}y+a_{23}}{(a_{13}x+a_{23}y+a_{33})^3}, \\
Y &= -k^2c^2\Delta\frac{y-y_c}{(a_{13}x+a_{23}y+a_{33})^2} \\
&\quad + ck^2(y_cx-x_cy)\frac{a_{11}x+a_{12}y+a_{13}}{(a_{13}x+a_{23}y+a_{33})^3}.
\end{aligned} \tag{5.30}
$$

The unknown function f is always well defined when the second of Kepler laws is obeyed, no matter of the algebraical nature of the function Φ. Notice that the components of force of 'non-elastic' nature depend on the magnitude of the area swept by the position vector with respect to the center of conic.

5.3 Berry–Klein Equation of Motion: The Physical Meaning of Regularization

What we just presented above, speaks by itself of the idea we wanted to promote: the classical dynamics, the Kepler motion simply describes the structure of a compass based on gravitation; that compass indicates not a fixed direction, though, but a fixed point. However, when it comes to scale transition, we need to consider the Berry–Klein 'equation of motion' (2.38), whose solution is a gauge length, and which is, in fact, a condition selected by a specific definition of invariance of forces, no matter if oriented in direction or to a point. As we have seen in Chapter 2, that condition cannot be interpreted dynamically, but only geometrically, or kinematically at the most. It is then worth our while to analyze one of its most important geometrical meanings, for, as we have shown before, it is instrumental in the transition between scales. In a word, the following analysis shows that the idea of angle must transcend the scale: it is the same, in the infrafinite and transfinite scales, as it is at the finite scale.

The geometry of three-dimensional space can be written in the language of position vectors of the form

$$\mathbf{r} = r\,\hat{\mathbf{e}}_r. \tag{5.31}$$

Here r is a length of the position vector in a certain reference frame, whereby $\hat{\mathbf{e}}_r$ is the unit vector orienting this length. For instance in equation (5.31) we denoted:

$$\mathbf{r} \equiv |x\rangle = \begin{pmatrix} x \\ y \\ z \end{pmatrix}, \quad \hat{\mathbf{e}}_r \equiv \begin{pmatrix} \sin\theta\cos\varphi \\ \sin\theta\sin\varphi \\ \cos\theta \end{pmatrix}. \tag{5.32}$$

In most of the applications connected to the physics of matter in space only the first and second symmetric differentials are of importance. The first-order differential, which is always 'symmetric' by its very nature, can be calculated directly from (5.31), giving:

$$d\mathbf{r} = (dr)\,\hat{\mathbf{e}}_r + r\,d\hat{\mathbf{e}}_r. \tag{5.33}$$

In order to calculate the differential of the unit vector, we use the second of the equalities in (5.31), with the result

$$d\hat{\mathbf{e}}_r = (d\theta)\hat{\mathbf{e}}_\theta + (\sin\theta d\varphi)\,\hat{\mathbf{e}}_\varphi \tag{5.34}$$

where the two unit vectors from the right-hand side here are given by matrices

$$\hat{\mathbf{e}}_\theta \equiv \begin{pmatrix} \cos\theta\cos\varphi \\ \cos\theta\sin\varphi \\ -\sin\theta \end{pmatrix}; \quad \hat{\mathbf{e}}_\varphi \equiv \begin{pmatrix} -\sin\varphi \\ \cos\varphi \\ 0 \end{pmatrix}. \tag{5.35}$$

Therefore, the differential of the position vector is

$$d\mathbf{r} = (dr)\,\hat{\mathbf{e}}_r + (rd\theta)\hat{\mathbf{e}}_\theta + (r\sin\theta d\varphi)\hat{\mathbf{e}}_\varphi. \tag{5.36}$$

The square of the differential of the position vector is given by the inner product:

$$d\mathbf{r}\cdot d\mathbf{r} = (dr)^2 + r^2(d\theta^2 + \sin^2\theta\,d\varphi^2) \equiv (dr)^2 + r^2 d\Omega^2 \tag{5.37}$$

with an obvious definition for $d\Omega$, where we recognize the Euclidean metric in spherical coordinates. One can verify the Frenet–Serret equations, describing the frame of unit vectors associated with the spherical coordinate system:

$$\begin{pmatrix} d\hat{\mathbf{e}}_r \\ d\hat{\mathbf{e}}_\theta \\ d\hat{\mathbf{e}}_\varphi \end{pmatrix} = \begin{pmatrix} 0 & d\theta & \sin\theta d\varphi \\ -d\theta & 0 & \cos\theta d\varphi \\ -\sin\theta d\varphi & -\cos\theta d\varphi & 0 \end{pmatrix} \begin{pmatrix} \hat{\mathbf{e}}_r \\ \hat{\mathbf{e}}_\theta \\ \hat{\mathbf{e}}_\varphi \end{pmatrix}. \tag{5.38}$$

This equation helps in establishing the second symmetric differential of the position vector. Indeed, from equations (5.36) and (5.38) we have

$$\begin{aligned} d^2\mathbf{r} = {} & (d^2r - rd\Omega^2)\hat{\mathbf{e}}_r \\ & + (rd^2\theta + 2drd\theta - r\sin\theta\cos\theta d\varphi^2)\hat{\mathbf{e}}_\theta \\ & + (r\sin\theta d^2\varphi + 2\sin\theta drd\varphi + 2r\cos\theta d\theta d\varphi)\hat{\mathbf{e}}_\varphi. \end{aligned} \tag{5.39}$$

In the system of Newtonian dynamics, this vector represents the acceleration, of course when it is referred to an adequate continuity parameter, playing the part of the time of problem. Momentarily, we do not proceed like that, but will discuss the general case of the differentials. However, by abusing a little of the classical nomenclature, the components of the vector (5.39) will still be designated as 'accelerations', just as the components of vector (5.36) will be called 'velocities'.

In the case of classical free particle, the components of acceleration vanish, a condition that comes down to a system of three differential equations:

$$\begin{aligned} d^2r - rd\Omega^2 &= 0, \\ rd^2\theta + 2drd\theta - r\sin\theta\cos\theta d\varphi^2 &= 0, \\ r\sin\theta d^2\varphi + 2\sin\theta drd\varphi + 2r\cos\theta d\theta d\varphi &= 0. \end{aligned} \tag{5.40}$$

This system can be solved by starting with the quadratic form $d\Omega$ depending only on angles in the first of the equations. It satisfies to a simple differential equation. First, we have by direct differentiation:

$$d(d\Omega^2) = 2(d\theta d^2\theta + \sin\theta\cos\theta d\varphi^2 + \sin^2\theta d\varphi d^2\varphi). \tag{5.41}$$

Now, using here the last two equations from (5.40) results in

$$rd(d\Omega^2) + 4dr(d\Omega^2) = 0$$

and thus

$$d\Omega^2 = \frac{R^4}{r^4}dt^2, \tag{5.42}$$

where R^2 is a constant having the dimensions of a rate of area, if the continuity parameter t is time. Therefore, we just have introduced the 'time parameter' t, to be measured by the metric of the unit sphere. This equation defines Ω itself as a continuity parameter on the unit sphere. Inserting the result (5.42) into the first equation (5.40) we get

$$d^2r = \frac{R^4}{r^3}dt^2. \tag{5.43}$$

This equation can be solved to give the known solution:

$$r^2 = At^2 + 2Bt + C \tag{5.44}$$

with A, B and C constants satisfying to the constraint:

$$R^4 \equiv AC - B^2. \tag{5.45}$$

Now, a little digression on the previous results seems necessary, in order to justify our further proceedings. Notice that the equation (5.42) is a direct consequence of the last two equations (5.40). They can be formally integrated as follows: first we have them properly arranged in the form of a homogeneous differential system having a skew-symmetric matrix:

$$d\begin{pmatrix} r^2 d\theta \\ r^2 \sin\theta d\varphi \end{pmatrix} = \begin{pmatrix} 0 & \cos\theta d\varphi \\ -\cos\theta d\varphi & 0 \end{pmatrix} \begin{pmatrix} r^2 d\theta \\ r^2 \sin\theta d\varphi \end{pmatrix}, \tag{5.46}$$

which is easier to solve. Indeed, (5.46) can be solved by matrix exponentiation, with the result

$$\begin{pmatrix} r^2 d\theta \\ r^2 \sin\theta d\varphi \end{pmatrix} = \begin{pmatrix} \cos(\cos\theta d\varphi) & -\sin(\cos\theta d\varphi) \\ \sin(\cos\theta d\varphi) & \cos(\cos\theta d\varphi) \end{pmatrix} \begin{pmatrix} a \\ b \end{pmatrix}, \tag{5.47}$$

where the differential constants a and b are constrained by the condition:

$$R^4 \equiv a^2 + b^2.$$

The two differentials in the left-hand side of equation (5.47) have also the meaning of elementary area components on the surface of the unit sphere. Indeed, from equations (5.31) and (5.36) we get:

$$\mathbf{r} \times d\mathbf{r} = (r^2 d\theta)(\hat{\mathbf{e}}_r \times \hat{\mathbf{e}}_\theta) + (r^2 \sin\theta d\varphi)(\hat{\mathbf{e}}_r \times \hat{\mathbf{e}}_\varphi).$$

Therefore, the elementary area of the unit sphere is actually a vector in the plane (θ, φ), having as components the differentials from equation (5.47)

$$\mathbf{r} \times d\mathbf{r} = -(r^2 \sin\theta d\varphi)\hat{\mathbf{e}}_\theta + (r^2 d\theta)\hat{\mathbf{e}}_\varphi. \tag{5.48}$$

Now it is time for introducing the dynamics, and this will be done here in the classical way, based on the fact that this vector represents the differential of the area swept by the position vector of motion.

However, just for our security, as it were, we stop here for a little while, in order to check one important identity, namely the identity (3.38), essential in the theory of regularization. Notice first, that the vector (5.48) is perpendicular to the vector $d\hat{\mathbf{e}}_r$, as defined by the first of the Frenet–Serret formulas from equation (5.38). The magnitude of the vector (5.48) is the same as the magnitude of $r^2 d\hat{\mathbf{e}}_r$. In terms of time rates, this means $r^2 d\hat{\mathbf{e}}_r/dt$, where t is the time of motion. Thus, in terms of the 'fictitious time' of the focal regularization, this quantity means $\hat{\mathbf{e}}_r'$, where a prime means derivative with respect to that time, so that equation (5.48) implies $(\mathbf{r} \times d\mathbf{r})^2 = (\hat{\mathbf{e}}_r' \cdot \hat{\mathbf{e}}_r')dt^2$, a result that we used already in proving the identity given in equation (3.38).

Coming back to our present analysis, notice that, as long as we have to do with central forces, the last two equations (5.40) are not affected. Consequently, the integral (5.42) persists, but the equation (5.43) gets, in its right-hand side, an expression depending on the magnitude of the force. So that instead of (5.43) we shall have:

$$d^2r - r(d\Omega^2) = f(\mathbf{r})dt^2. \tag{5.49}$$

Here t is the time of the problem, as defined before, and $f(\mathbf{r})$ is the magnitude of force, up to a sign. This is purely a second-order differential equation for the magnitude of the position vector, to be solved once we know the magnitude of the force. In order to go over to the new time here, we usually take notice that equation (5.48), offers the 'area constant' of the motion by relation

$$r^2\dot{\Omega} = \dot{a} \equiv R^2$$

with an obvious notation for the area constant, and R^2 given by (5.45), or by the norm of (5.47). This relation shows that, no matter of orientation of the plane of motion — the force is central! — the area rate in a general dynamical problem is apparently given by the arclength on the unit sphere, which has the same algebraic expression as the magnitude of the elementary area of that sphere. Now, using this conclusion into equation (5.49), brings this equation to the classical Binet's form (5.11), which we transcribe here, for convenience, as:

$$u^2\left(\frac{d^2u}{d\Omega^2} + u\right) = \frac{1}{\dot{a}^2}f(\mathbf{r}), \quad u\,r = 1.$$

Many particular cases of solution of this equation are known, depending on the expression of magnitude of force, out of which for our present interest only two of them are in order: the case of Newtonian forces, and the case of isotropic elastic forces.

For the case of classical Newtonian forces, including the gravitational force and electric Coulombian forces, we have $f(\mathbf{r}) = k\,u^2$, where k is a constant, and therefore equation (5.11) can be written as

$$\frac{d^2u}{d\Omega^2} + u = \frac{k}{\dot{a}^2}. \tag{5.50}$$

The general solution of this equation has the form

$$u(\Omega) = \frac{k}{\dot{a}^2} + u_0 \cos(\Omega + \Omega_0) \tag{5.51}$$

with u_0 and Ω_0 integration constants, to be assigned according to some initial conditions. Obviously, these trajectories are conics described with reference to the focus as the center of force.

In the case of elastic forces we have $f(\mathbf{r}) = k/u$, and the equation (5.11) becomes

$$\frac{d^2u}{d\Omega^2} + u = \frac{k}{\dot{a}^2}\frac{1}{u^3}. \tag{5.52}$$

This is a regular two-dimensional harmonic oscillator of unit frequency with area constant given by k/R^4. The trajectory is therefore a conic, whereby the geometrical center is also a center of force:

$$u^2(\Omega) = \frac{\sum a^2}{2} + \sqrt{\frac{(\sum a^2)^2}{4} - (ad - bc)^2 \cos 2(\Omega + \Omega_0)}. \tag{5.53}$$

Here the summations run through values a, b, c, d, which are calculated in order to account for the initial conditions of the two components of motion, connected with our notations here by equations:

$$ad - bc = \frac{k}{\dot{a}^2}, \quad \tan 2\Omega_0 = \frac{2(ac + bd)}{a^2 + b^2 - c^2 - d^2}. \tag{5.54}$$

These results are to be compared with the results of Claude Burdet from equation (3.40) referring to the focal regularization procedure. The Berry–Klein scaling procedure (see Mazilu *et al.*, 2019, Chapter 9), results in equation [(9.42), *ibid.*] which we transcribe here in the form

$$m\mathbf{x}'' + k\mathbf{x} = -\nabla_x V(\mathbf{x}). \tag{5.55}$$

In view of our discussion thus far, this equation says basically that the constitutive particles of a physical system in an expanding field of forces must be an isotropic harmonic oscillator. Seems just natural: starting from reflection and refraction phenomenology, and assuming that the light is material, Hooke described it as such a periodic process. Two centuries later, the diffraction has been added to phenomenology of light, whose specific materiality helped Fresnel to explicitly assert the mathematical form of periodicity of the light phenomenon. This finally established the modern physical optics as we know it today. However, we need to keep in mind the *sine qua non* condition for equation (5.55): the forces represented by the potential $V(\mathbf{x})$ are *Newtonian forces.* Only these forces are invariant to Berry–Klein scaling procedure from equation [(9.35), Mazilu *et al.*, 2019] therefore only these forces allow the condition [(9.41), *ibid*] defining the gauge. Therefore, to be fair, equation (5.55) should actually read

$$\mathbf{x}'' + \Omega_0^2 \mathbf{x} = k \frac{\mathbf{x}}{x^3}, \quad \omega_0^2 \equiv \frac{K}{m} \tag{5.56}$$

where k is a constant, including the ratio between gravitational and inertial mass of the moving material point. Equation (5.56) can be solved by the previous method, and leads to equation (5.50) with solution (5.51), for the magnitude of the vector $\mathbf{u} \equiv \mathbf{x}/x^2$.

However, equation (5.53) seems to tell us another story, if contemplated through the optics opened to our reason by the Bertrand's problem [see equations (5.15) and (5.16)]. What may naturally puzzle one when analyzing the Newtonian theory of forces is a certain apparently natural philosophical inconsistency in their definition; they are defined by a measurement procedure, no question about that. There is a problem though, with their reality: while, for instance, the force between sun and a planet is real, with the physical reality ratified by the presence of the two material bodies in the sky, the force with which it has to be 'compared' in order to be 'measured' is a fictitious force, inasmuch as it has to pull the planet towards the center of the orbit, *i.e.* toward a point where *matter does not exist.* Only when compared to this force, the attraction between sun and planet is described by a genuine Newtonian force of the kind that generated the modern concept of field, *i.e.* central, with a magnitude inversely proportional to the distance between bodies. One can say that the general relativity dispensed the natural philosophy with this kind of criterion for the reality of forces, a dispensation which, unfortunately, took a turn into the idea that the general relativity eliminated altogether the forces from our knowledge.

The truth of the matter is that, as Berry–Klein theory shows, according to the idea of field derived from the very Newtonian forces, the scale transition actually involves a kinematics, and thus the contradiction disappears, at least formally, even in the classical physics. One can rightfully say that the concept of field is not complete when derived disregarding its true origin, to which our knowledge has to turn, forced, as it were, by the necessity of scale transition. According to this, only the 'exact' acceleration is Newtonian; however, the fluctuations of the acceleration with respect to the background of 'inverse cubic power' are linear.

Chapter 6

A Geometry of Reverse Interpretation of a Fundamental Physical Structure

Before anything else, let us assess the Planck's fundamental hypothesis represented by equation (5.41) in a little more detail. We take it as representing the statement: *the energy of the fundamental physical structure of a universe* — whatever this may be — *should be itself a statistical variable, characterized by an exponential distribution with quadratic variance function.* Insofar as that fundamental physical structure of a universe has to be used in the process of counting for establishing the matter density, we need to know what is its possible physical structure, and how is it related to the concept of scale. In Planck's original case, it was the harmonic oscillator, on which we need to dwell a little longer, for it has the essential properties needed in characterization of that concept.

The second-order differential equation characterizing a damped harmonic oscillator, can be written in terms of physical parameters from Chapter 2 [equation (2.11) ff.] as

$$M\ddot{q} + 2R\dot{q} + Kq = 0. \tag{6.1}$$

As known, the complete set of solutions of this equation unfold over a two-dimensional manifold, depending on three parameters. In terms of parameters playing the part of coefficients of this equation, they can be written in the complete form as

$$\begin{aligned} q(t) &= e^{-\lambda t}\left(z\, e^{i(\omega t+\phi)} + z^*\, e^{-i(\omega t+\phi)}\right), \\ M^2\omega^2 &\equiv MK - R^2 = M^2(\omega_0^2 - \lambda^2). \end{aligned} \tag{6.2}$$

The manifold of these solutions represents an ensemble of oscillators, whose element is identified by three parameters in the complex form: z, its complex conjugate z^*, and the unit modulus factor $e^{i\phi}$. One might say that, by equation of motion of the harmonic oscillator, *the frequency* is somehow statistically related to this ensemble of oscillators, and thus, such a parameter describes an ensemble of harmonic oscillators. Assuming, of course, that it is possible to find such a statistics, ideally a sufficient statistics — like the temperature in the case of molecular kinetic energy of an ideal gas — whose representative should be the frequency. We witnessed such a statistical connection, but related to the real part of the frequency as it were: the parameter λ is related to the statistics involved in transition between the two kinds of energies of harmonic oscillator [see Chapter 2, equations (2.15) ff., and comments]. The classical scenario we can take in consideration for an analogy here is that involved in the statistics of thermal radiation, as revealed by Wien's type of radiation law. That statistic is also referring to the kinetic energy of a molecular gas, whose representative parameter is the temperature. One can think of the fact that the dependence of the law of radiation on the ratio between frequency and temperature has its physical reasons in this purely statistical fact. Taking the case of temperature definition as a standard in our analogy, one reproducible feature pops up right away, that needs to be assessed for the case of frequency: the statistic to which the parameter temperature is related concerns invariants with respect to rotation group, to wit, the magnitudes of velocity vectors of a molecular swarm defining the classical temperature as absolute in the manner of Lord Kelvin. This feature can be replicated, indeed, even though in a specific way, in the case of frequency, for the ensemble of oscillators, having its elements described by the parameters z, z^* and $e^{i\phi}$.

Our assessment of this situation starts with the observation that there is a group that 'sweeps out' the ensemble of oscillators of the same frequency, this time understanding by 'frequency' its *imaginary* part. Indeed, the ratio of any two linearly independent solutions of the differential equation (6.1), τ say, is a solution of the following differential equation:

$$\{\tau, t\} = 2\omega^2, \tag{6.3}$$

where the curly brackets denote the *Schwarzian derivative* of τ with respect to the time of dynamical problem (6.1). Such a ratio has the general connotation of a phase variable of the kind characterized statistically by the Cauchy class of distributions [see equation (4.44), and the discussion of Peter McCullagh's theory]. As known (Needham, 2001), the Schwarzian derivative is defined by the third-order nonlinear operation

$$\{\tau, t\} \equiv \left(\frac{\ddot{\tau}}{\dot{\tau}}\right)^{\cdot} - \frac{1}{2}\left(\frac{\ddot{\tau}}{\dot{\tau}}\right)^2,$$

where the dot is shorthand for the time derivative, as usual. According to properties of this nonlinear operation, it is invariant with respect to the fractional linear transformation

$$\tau \leftrightarrow \tau' = \frac{a\tau + b}{c\tau + d} \tag{6.4}$$

with a, b, c, d four parameters, which *we take as real* here. Then the left-hand side of equation (6.3) is invariant to linear fractional transformation, and therefore the ensemble of oscillators of the same frequency can be 'scanned' continuously by this action in one variable with three parameters. For, indeed, the set of all transformations (6.4) corresponding to all the possible values of these parameters represents an SL(2, R) type of action.

Thus the ensemble of all oscillators corresponding to the same frequency is in a one-to-one correspondence with the finite transformations of SL(2, R). This allows us to construct an 'individual' parameter τ for each oscillator of the ensemble, guided by the form of the general solution of equation (6.3). This solution can indeed be written as

$$\tau' = u + v\tan(\omega t + \phi), \tag{6.5}$$

where u, v and ϕ are constants, characterizing a given oscillator from the ensemble as before. Identifying the phase from (6.5) with that from (6.2), we can write the 'individual' parameter of an oscillator in terms of the complex initial conditions $z = u + iv$ and $e^{i\phi}$, in the form

$$\tau' = \frac{z + z^*\tau}{1 + \tau}, \quad z^* \equiv u - iv, \quad \tau \equiv e^{2i(\omega t + \phi)}. \tag{6.6}$$

Our results right away can be summarized as follows:

(1) *each oscillator* represents a *continuous family* of cubic equations depending on time; this is the statistical property describing an element — the oscillator — as in the case of historical Planck's statistics, only in a geometrical setting, allowing a scale transition.
(2) parametrically, the complex initial conditions of the oscillator — the amplitudes — are given by the roots of the second-order equation (4.51), where x_k are the roots of a cubic of that family.

In physical terms this means that the ensemble of elements corresponding to a given frequency represents a family of matrices giving a measured field. The physical quantities measured are the *normal* and *shear components* of the field. The time of physical evolution of the field is given by phase angle of orientation of the octahedral vector in the local octahedral plane. Let us document these statements.

6.1 The Case of Cubic Equation with Real Roots

As one can grasp from the above results, in the case of oscillators, the gauge freedom should be way richer than the arbitrariness of phase lets it to be comprehended for the case of the hyperbolic plane in the Kepler problem (see Mazilu *et al.*, 2019). Its group has been revealed for the first time, in the context of hyperbolic geometry, by Dan Barbilian on the occasion of a study of the Riemannian space associated with a family of cubics (Barbilian, 1938). We shall briefly reconsider Barbilian's theory, insisting on some particular technical points necessary for our own reference. The basis of approach is the fact that the simply transitive SL(2, R) type action [see Baker (1901, Example 3, pp. 116–119)]

$$r_k \leftrightarrow \frac{ar_k + b}{cr_k + d}, \tag{6.7}$$

where r_k are the roots of a cubic equation and a, b, c, d are real parameters, induces a still simply transitive group for the quantities h, h^* and $k \equiv e^{i\phi}$, whose action is represented by the correspondences:

$$h \leftrightarrow \frac{ah + b}{ch + d}, \quad h^* \leftrightarrow \frac{ah^* + b}{ch^* + d}, \quad k \leftrightarrow \frac{ch + d}{ch^* + d}k, \tag{6.8}$$

which will be called, from now on, the *Barbilian group*. Perhaps it is best to tell the story of passing from (6.7) to (6.8) with details on the parameters h, h^* and the phase ϕ, for these are related to a variety of fundamental ways in which the nature allows us to 'harvest' our primary space and time knowledge. We already took a glimpse at these quantities in Chapter 4, and hinted to the statistics incorporated in their algebraical structure. It is now time for some further details on those statistics and, more importantly, for the mathematics of handling them.

We will take a general cubic equation in the so-called *binomial form*, with explicit *binomial coefficients*:

$$a_0 r^3 + 3a_1 r^2 + 3a_2 r + a_3 = 0, \tag{6.9}$$

where $r \equiv \phi^1/\phi^2$ denotes the *ratio* of the two fundamental components of what in algebraical terms is called a *binary variable* (ϕ^1, ϕ^2). As already stated quite a few times here, we consider this equation as an equation giving eigenvalues of a corresponding 3×3 tensor, specifically a Euclidean tensor, in which case the coefficients a_0, a_1, a_2, a_3 are related to the orthogonal invariants of different degrees constructed with the entries of the corresponding matrix. The most general theory behind the formal procedure of solving this equation has been established by Joseph Sylvester (Burnside & Panton, 1960, Volume II, p. 194), and amounts to putting the cubic from the left-hand side of equation (6.9) in the form of a sum of two perfect cubes, so that the equation takes the form:

$$\beta_1(r-\alpha_1)^3+\beta_2(r-\alpha_2)^3=0. \tag{6.10}$$

In this case it can be easily solved to give

$$r_n=\frac{\alpha_1+\alpha_2 j^{n-1}k}{1+j^{n-1}k},\quad k^3\equiv\frac{\beta_2}{\beta_1},\quad n=1,2,3 \textit{ and } j^3=1. \tag{6.11}$$

The problem of solving the cubic equation is thus translated into that of finding the quantities $\alpha_1,\alpha_2,\beta_1,\beta_2$ from equation (6.10) as functions of the coefficients a_0,a_1,a_2,a_3. This can be done as follows: identifying the equations (6.9) and (6.10) gives the following system:

$$\begin{aligned}\beta_1+\beta_2=a_0,\quad \beta_1\alpha_1^2+\beta_2\alpha_2^2=a_2,\\ \beta_1\alpha_1+\beta_2\alpha_2=-a_1,\quad \beta_1\alpha_1^3+\beta_2\alpha_2^3=-a_3.\end{aligned} \tag{6.12}$$

Now, we are always at liberty to assume that α_1 and α_2 are the roots of a quadratic equation, say of the form:

$$b_0\alpha^2+b_1\alpha+b_2=0$$

which we put in relation with the system (6.12), in order to eliminate α's and β's in favor of a's. The result is a linear system for b_0,b_1,b_2:

$$a_2b_0-a_1b_1+a_0b_2=0,\quad a_3b_0-a_2b_1+a_1b_2=0.$$

This system has its solution defined up to an arbitrary factor, and this solution is given by the coefficients of *Hessian* of the cubic:

$$\frac{b_0}{a_0a_2-a_1^2}=\frac{b_1}{a_0a_3-a_1a_2}=\frac{b_2}{a_1a_3-a_2^2}. \tag{6.13}$$

Therefore that α_1 and α_2 from (6.10) are actually the *roots of a quadratic equation* representing the vanishing of Hessian:

$$(a_0a_2-a_1^2)r^2+(a_0a_3-a_1a_2)r+(a_1a_3-a_2^2)=0. \tag{6.14}$$

Now we must find β_1 and β_2 from (6.10) in terms of the coefficients of cubic. For this we can use any pair from the four equations (6.12); the result is the same up to a factor. Because in solving (6.10) we actually need only the ratio β_2/β_1, we can however use another, more direct method, having the advantage to exhibit directly the algebraic nature of β_1 and β_2. Namely, designating the cubic from equation (6.9) by $f(r)$, and then using (6.10), we find

$$f(\alpha_1) = \beta_2(\alpha_1 - \alpha_2)^3, \quad f(\alpha_2) = -\beta_1(\alpha_1 - \alpha_2)^3$$

whence the ratio β_1 and β_2 is given by equation

$$\frac{\beta_2}{\beta_1} = -\frac{f(\alpha_1)}{f(\alpha_2)}. \tag{6.15}$$

On this equation we read the important conclusion that *β_1 and β_2 have the same algebraic nature as α_1 and α_2*. Specifically, if α_1 and α_2 are real so are β_1 and β_2; if α_1 and α_2 are complex so are β_1 and β_2. This conclusion is important, because, at least in the cases of physical interest, it shows that the nature of the roots of equation (6.9) is indeed decided by the discriminant of the Hessian of cubic:

$$\Delta \equiv (a_0a_3 - a_1a_2)^2 - 4(a_0a_2 - a_1^2)(a_1a_3 - a_2^2). \tag{6.16}$$

This fourth degree quaternary expression, a *quartic* as they say, is also known as *discriminant of the cubic* itself. In cases where $\Delta \neq 0$, for the Hessian two distinct roots, *the ratio* of β_1 and β_2 is uniquely determined by equation (6.15). We then have the general theorem that relates the cubic equation to the equation of the Hessian:

> If the Hessian of a cubic equation has distinct roots then the cubic equation itself has distinct roots. More specifically, there are two cases to consider:
> (a) if the discriminant of the cubic is *positive*, then *the Hessian has two real roots*, and *the cubic itself has one real and two complex roots.*
> (b) if the discriminant of the cubic is *negative*, then *the Hessian has complex roots*, and *the cubic itself has three real roots.*

The proof of this theorem is based on the form (6.11) of the roots of a cubic. Indeed, equation (6.11) shows that if α_1 and α_2 are real, which happens for $\Delta > 0$, k is also real by (6.15), so that from among the solutions of the cubic, r_1 say, is real while the other two, r_2 and r_3 say, are complex, and therefore complex conjugate to each other. On the contrary, if α_1 and α_2 are complex, which only happens when $\Delta < 0$, then by (6.15) k is a complex quantity of unit modulus, and thus all the three r_j from equation (6.11) are real. This completes the proof.

A little more involved reasoning reveals some other properties relating the cubic and its Hessian, that can be proved algebraically exactly like the theorem right above. However, in view of the importance we attach to the physical quantities related to the cubic — the 3×3 matrices or tensors — we suggest an 'interpretative' proof which, despite of not being a genuine proof, has the virtue of better showing how such physical quantities are related to a surface, as in the case of stresses or strains of a material continuum. First we have the theorem already encountered before:

> If a *cubic is a perfect cube*, then it has a *null Hessian*. Reciprocally, if a cubic has a null Hessian then it is a perfect cube.

A perfect cube would mean a matrix proportional to identity matrix, having therefore a sphere as geometrical representative quadratic form. Starting from this, we can go to a more complicate case, *viz.* that of a tensor having only two equal eigenvalues, for the cubic contains a perfect square factor. The geometric image of such a tensor should be a prolate or oblate spheroid, as the case may happen to be. Using equation (6.13), one can easily show that this special quadratic factor is the Hessian of the cubic. In fact, the truth is that the Hessian defines the cubic in a limited way, insofar as in this case we have the theorem:

> If a cubic equation has *a perfect square Hessian*, then such a cubic *contains the Hessian as a factor.*

Therefore, the cubic is referring in this case to a tensor having the representative quadratic form as an ellipsoid of rotation — a prolate or oblate spheroid, as we said — having, for instance, the rotation axis transversal to a given element of surface, in geometrical terms. Such tensors are, in electrodynamics, the Maxwell stress tensors corresponding to a vector field, electric or magnetic, say [see Mazilu *et al.*, 2019, Chapter 4, equation (4.38) ff., and the discussion]. In the case of the planetary model described by us in the beginning of the Chapter 1, such a tensor can be related to a state of electron as described by the *initial velocities* of the Kepler motions of different Hertz material particles from the constitution of electron as a Hertz material point

Up to this stage in our presentation, we concerned ourselves with the determination of a cubic, when we know the roots of the Hessian. If we want to solve the reciprocal problem, *i.e.* to determine the Hessian knowing the roots of the cubic, then we have some alternative options. The first one of these would be given by the Viète relations for our cubic:

$$-3\frac{a_1}{a_0} = \sum r_1, \quad 3\frac{a_2}{a_0} = \sum r_2 r_3, \quad -\frac{a_3}{a_0} = r_1 r_2 r_3,$$

where the sum sign means sum over all cyclic permutations of the indices 1, 2, 3. From these, by performing the appropriate calculations based on equation (6.13), we have the following expressions for the coefficients of the Hessian:

$$\begin{aligned} a_0 a_2 - a_1^2 &= -\frac{a_0^2}{18}\sum (r_2 - r_3)^2, \\ a_0 a_3 - a_1 a_2 &= -\frac{a_0^2}{18}\sum 2r_1(r_2 - r_3)^2, \\ a_1 a_3 - a_2^2 &= -\frac{a_0^2}{18}\sum r_1^2(r_2 - r_3)^2. \end{aligned} \tag{6.17}$$

These are the general relations between the coefficients of the Hessian and the roots of the corresponding cubic equation. We do recognize here the coefficients of equation (4.51): the algebra is therefore quite general. On these expressions, though, we can recognize directly, the results in theorems above as special cases. Another point to emphasize here is that the relations (6.17) allow us to calculate the discriminant of a cubic as a function of the roots. Such an expression is important in that for many cases we can have it in this form, and we must be able to recognize it. To calculate the discriminant we simply use its definition (6.16) and the relations (6.17) above. After a lengthy, but otherwise straightforward calculation, we find

$$\Delta = -\frac{a_0^4}{18}(r_2 - r_3)^2(r_3 - r_1)^2(r_1 - r_2)^2. \tag{6.18}$$

Like (6.17), this expression contains all the previous information. In particular, we can read directly on (6.18) that a given cubic equation cannot have positive discriminant but only in cases where it has two complex roots, as shown before. If it has real roots — which means that the associated tensor is a physical quantity proper, *i.e.* measurable — then the discriminant should be necessarily negative. The term measurable is here understood in the sense of *Novozhilov directional statistics* given in equation [(4.36), Mazilu *et al.*, 2019] which provide r_j according to the formulas from equation [(4.37), *ibid.*].

A second method to deal with the connection between a cubic and its Hessian, is to start directly from the formulas (6.11) and solve them for α_1, α_2 and k. As we are interested mainly in cubics representing measurable physical quantities, we need to settle for the case of negative discriminant, in which α_1 and α_2 are complex conjugate to one another, while k is a complex number of unit modulus. Denoting $\alpha_1 \equiv h$, so that $\alpha_2 \equiv h^*$, we then write (6.11) for this case as

$$r_1 = \frac{h + h^* k}{1 + k}, \quad r_2 = \frac{h + jh^* k}{1 + jk}, \quad r_3 = \frac{h + j^2 h^* k}{1 + j^2 k}. \tag{6.19}$$

In real terms, defined according to $h \equiv u + iv, k \equiv e^{i\phi}$, these are actually the numbers given in equation [(4.37), Mazilu *et al.*, 2019], so they carry a statistical meaning. Solving the system (6.19) for h and k, gives

$$h = -\frac{r_2 r_3 + j^2 r_3 r_1 + j r_1 r_2}{r_1 + j^2 r_2 + j r_3}, \quad k = -\frac{r_1 + j r_2 + j^2 r_3}{r_1 + j^2 r_2 + j r_3}. \tag{6.20}$$

These complex quantities are analytic *in the Pierre Humbert's sense* [see Humbert (1929b); see also Devisme (1933)], *i.e.* they satisfy to *the tricomplex generalization of the Laplace equation* [see also Olariu (2002), for an outstanding presentation, closer to present times]:

$$\Delta_3 h = \Delta_3 k = 0; \quad \Delta_3 \equiv \partial_1^3 + \partial_2^3 + \partial_3^3 - 3\partial_1 \partial_2 \partial_3 \tag{6.21}$$

with $\partial_1 = \partial / \partial r_1$, etc. We will come again to this issue later on, for details of connection between the tricomplex analyticity and physics.

For now we are only interested into noticing that this method of presentation reveals an outstanding physical meaning of the phase ϕ.

As one can appreciate from the above development, the Hessian of a cubic is a key tool in constructing the roots of that cubic. Thus, the previous elaboration shows that, given the roots of the Hessian only, one *cannot* know the corresponding cubic equation without ambiguity. In fact equations (6.10) and (6.11), as well as (6.19) for that matter, show that *to a given Hessian corresponds a one-parameter family of triplets of numbers*; each one of these triplets represents a given cubic equation, up to a factor. The parameter of this family of cubics is the phase ϕ. On occasion, this indeterminacy is independent of the known property of indeterminacy allowed by the relations between roots and coefficients. As a matter of fact, it is even deeper than equation (6.10) shows it, in the sense that the ratio k, which by equations (6.11) and (6.15) depends, apparently, only on quantities related to the cubic equation, may hide in itself an *external phase* completely independent of the cubic equation, and therefore of the tensor from which it originates. This property is entirely analogous to the property of the exponential factor in the case of damped harmonic oscillator: it is not an intrinsic property of the oscillator, but of the time sequences defining the Lagrangian [see equations (2.15) and (2.16)]. In order to reveal the nature of this problem, we use the following identity between the cubic itself (f), its Hessian (H) and its Jacobian (T):

$$4H^3 = \Delta f^2 - T^2 \tag{6.22}$$

with Δ — the discriminant of cubic given in equation (6.18). We follow here the presentation of Dan Barbilian, whereby this identity, is regularly called a *syzygy*, as in the old algebraic theory of quantics, by analogy with the things astronomical (Barbilian, 1971). The expression in right-hand side of equation (6.22) can be decomposed into two factors each of third degree, for the cubic and its Jacobian are prime with respect to each other. On the other hand, the left-hand side of (6.22) is a product of two perfect cubes, for the Hessian is a quadratic polynomial. Therefore, we are at liberty to read the identity (6.22) as showing that each factor of the right-hand side is proportional to a factor of the expression from the left-hand side, and this proportionality can be taken in two ways at will. Indeed, the Hessian can be written as a product of two factors in infinitely many ways, represented by an arbitrary number, m say:

$$H = mU\frac{1}{m}V. \tag{6.23}$$

Here U and V are first degree binomials, and m is any non-zero number. Thus the identity (6.22) can be written, for instance, as the system

$$\sqrt{\Delta}f + T = 2m^3U^3, \quad \sqrt{\Delta}f - T = 2m^{-3}V^3 \tag{6.24}$$

or as a system in which the factors U and V of the Hessian change their places. Let us, nevertheless, settle for the system that we just wrote here: the dual case leads formally to the very same results, so that nothing would change in our general argument. Now, adding together the two expressions from (6.24), gives as result the cubic from the previous equation (6.10), only in a slightly different form, showing clearly *how the external arbitrariness gets into play*: it is a genuinely uncontrollable arbitrariness entering with the factor m from equation (6.23), *i.e.*

$$\sqrt{\Delta}f = m^3U^3 + m^{-3}V^3. \tag{6.25}$$

This is, indeed, exactly the formula of Sylvester used in equation (6.10), as we expected actually, but it shows that the ratio of $\beta's$ is arbitrary, and only its arithmetical nature can be decided: real or complex. One can further decompose the right-hand side here into linear factors, to the effect that (6.25) becomes

$$\sqrt{\Delta}f = (mU + m^{-1}V)(jmU + j^2m^{-1}V)(j^2mU + jm^{-1}V),$$

where j is the cubic root of unity. This factorization allows us to find the roots of the cubic equation $f = 0$ in the form given by equation (6.10), but, this time, with $k \equiv m^{-2}$ thus suggesting that k must be an arbitrary number, to be introduced externally. In case the roots of the cubics are all real, k must be complex of unit modulus as before, and the arbitrariness is transferred into its phase. For the sake of completeness, we can mention that the Jacobian of a cubic can also be obtained from (6.24), but this time as a *difference* of two cubes:

$$\begin{aligned} T &= m^3U^3 - m^{-3}V^3 \\ &= (mU - m^{-1}V)(jmU - j^2m^{-1}V)(j^2mU - jm^{-1}V). \end{aligned}$$

This shows that the roots of Jacobian are of the same arithmetic nature as the roots of the corresponding cubic itself: all real, or one real and two of them complex. In the formula (6.11) — or, in fact, (6.19) — we do not have to change but the sign, in both the denominator and numerator, in order to get the roots of the corresponding Jacobian. This discussion also shows that the form (6.11) of the roots of a cubic equation is valid independently of the nature of the roots.

It is now important to give a hint of *physical interpretation* for the external factor k occurring when one wants to construct the cubic given its Hessian. For this we will consider only the case where the cubic has *real* roots, *i.e.* k is complex of unit modulus. Now take the *vector* of components r_1, r_2, r_3. We are legitimate in using this image, for there is a space reference frame we can construct in every point of space where a symmetric matrix is defined, as given by three special orthogonal vectors — the principal directions of the said symmetric matrix. Thus, like in the Shpilker's coordinates case (see Mazilu *et al.*, 2019), the principal values of such a matrix can be represented by the column matrix, defined by its transposed as

$$\langle r| \equiv (r_1, r_2, r_3), \tag{6.26}$$

which is plainly a dual vector in matrix representation, inasmuch as each principal value can be interpreted as the component of the vector along the corresponding principal direction. We can decompose this vector with respect to the plane cutting the axes of reference frame in the points situated at unit distance from origin — the Olariu's nodal plane. In engineering applications this plane or, in fact, any plane parallel to it, is called an *octahedral plane*, as we already mentioned before, for it represents a face of an octahedron in space. The *normal component* of the vector (6.26) on this plane is given by the 'ket':

$$|r_n\rangle \equiv |n\rangle\langle n|r\rangle = \frac{1}{\sqrt{3}}\begin{pmatrix}1\\1\\1\end{pmatrix}\frac{1}{\sqrt{3}}(1\ 1\ 1)\begin{pmatrix}r_1\\r_2\\r_3\end{pmatrix} = \frac{r_1+r_2+r_3}{3}\begin{pmatrix}1\\1\\1\end{pmatrix}$$

in view of the fact that the normal of the nodal plane is along the tridiagonal direction of components $(1, 1, 1)$. Here we used the consecrated physics' notations, with an obvious use of the 'bra' symbol for transposed vectors. The in-plane (*tangential*) component of (6.26) is then given in the form usually employed in engineering problems, *i.e.*

$$|r_1\rangle \equiv |r\rangle - |r_0\rangle = \frac{1}{3}\begin{pmatrix}2r_1 - r_2 - r_3\\ -r_1 + 2r_2 - r_3\\ -r_1 - r_2 + 2r_3\end{pmatrix}. \tag{6.27}$$

It is this last vector, usually called *octahedral shear vector* in engineering applications, which allows us to interpret the complex number k externally introduced. Namely, the Sylvester form (6.25) of our cubic allows us to identify its binomial coefficients in terms of the quantities h, h^* and k, up to an arbitrary factor, as

$$a_0 = 1 + k^3, \quad a_1 = -(h + h^*k^3), \quad a_2 = h^2 + h^{*2}k^3,$$
$$a_3 = -(h^3 + h^{*3}k^3).$$

From this, or otherwise calculating directly by equations (6.19), we have right away

$$\frac{1}{3}\sum r_1 = \frac{h + h^*k^3}{1 + k^3} \quad \therefore \quad |r_1\rangle = \frac{(h - h^*)k}{1 + k^3}\begin{pmatrix}k - 1\\ j(jk - 1)\\ j^2(j^2k - 1)\end{pmatrix}. \tag{6.28}$$

Now take as reference the value $\phi = 0$, corresponding to the octahedral shear vector with $k = 1$. This is one of the cases when the roots of cubic are exclusively determined by the roots of Hessian, with no ambiguity, so that we have

$$|r_1^0\rangle = \frac{(h-h^*)}{2}\begin{pmatrix} 0 \\ j(j-1) \\ j^2(j^2-1) \end{pmatrix}. \tag{6.29}$$

Then we calculate the angle θ of orientation of the generic shear vector in octahedral plane with respect to this reference vector, by the well-known formula:

$$\cos\theta = \frac{\langle r_1^0|r_1\rangle}{\sqrt{\langle r_1^0|r_1^0\rangle\langle r_1|r_1\rangle}}. \tag{6.30}$$

Using equations (6.28) and (6.29) in calculating the necessary products here, gives

$$\langle r_1^0|r_1\rangle = -\frac{3(h-h^*)^2k(k+1)}{2(1+k^3)},$$

$$\langle r_1|r_1\rangle = -\frac{6(h-h^*)^2k^3}{(1+k^3)^2},$$

$$\langle r_1^0|r_1^0\rangle = -\frac{6(h-h^*)^2}{4}$$

so that equation (6.30) becomes

$$\cos\theta = \frac{1}{2}\left(\sqrt{k}+\frac{1}{\sqrt{k}}\right) \equiv \cos\phi, \tag{6.31}$$

where the phase of k is taken as 2ϕ, in view of the fact that it is a square. We conclude that, indeed, knowing the Hessian does not determine the cubic uniquely. In the case of a cubic with real roots the Hessian actually determines a family of cubics whose roots are defined as a one parameter family. The parameter of this family is basically given by *the angle of orientation of the octahedral shear vector in the octahedral plane.*

The algebraical structure of Barbilian group (6.8) is that of an SL(2, R) type structure, which we take in the standard form given by commutation relations of the vectors (2.43) or (2.45). Because the action (6.8) is simply transitive, its generators — B_k say, suggesting the initials of *Beltrami* or *Barbilian* — can easily be found as the components of the *Cartan frame* [see Fels and Olver (1998, 1999) for a relatively recent work on the notion] from the formula

$$\begin{aligned} d(f) &= \sum (\partial_k f) dr_k \\ &= \left[\omega^1 \left(h^2 \frac{\partial}{\partial h} + h^{*2} \frac{\partial}{\partial h^*} + (h - h^*) k \frac{\partial}{\partial x_k}\right)\right. \\ &\quad \left. + 2\omega^2 \left(h \frac{\partial}{\partial h} + h^* \frac{\partial}{\partial h^*}\right) + \omega^3 \left(\frac{\partial}{\partial h} + \frac{\partial}{\partial h^*}\right)\right](f), \end{aligned} \tag{6.32}$$

where ω^k are the components of the *Cartan coframe* to be found from the system of differentials

$$\begin{aligned} dh &= \omega^1 h^2 + 2\omega^2 h + \omega^3, \\ dh^* &= \omega^1 h^{*2} + 2\omega^2 h^* + \omega^3, \\ dk &= \omega^1 (h - h^*) k. \end{aligned}$$

Thus we have immediately both the infinitesimal generators and the corresponding coframe, by identifying the right-hand side of equation (6.32) with the standard dot product of $\mathbf{sl}(2, \mathrm{R})$ algebraical structure:

$$\omega^1 B_3 - 2\omega^2 B_2 + \omega^3 B_1.$$

To wit, we have for the frame:

$$\begin{aligned} B_1 &= \frac{\partial}{\partial h} + \frac{\partial}{\partial h^*}, \quad B_2 = h \frac{\partial}{\partial h} + h^* \frac{\partial}{\partial h^*}, \\ B_3 &= h^2 \frac{\partial}{\partial h} + h^* \frac{\partial}{\partial h^*} + (h - h^*) k \frac{\partial}{\partial k} \end{aligned} \tag{6.33}$$

while for the coframe we have

$$\begin{aligned} \omega^1 &= \frac{dk}{(h - h^*) k}, \quad 2\omega^2 = \frac{dh - dh^*}{h - h^*} - \frac{h + h^*}{h - h^*} \frac{dk}{k}; \\ \omega^3 &= \frac{h dh^* - h^* dh}{h - h^*} + \frac{h h^*}{h - h^*} \frac{dk}{k}. \end{aligned} \tag{6.34}$$

In real terms, defined, as usual, by $h = u + iv$, and $k = e^{i2\phi}$, the frame and coframe can be written as

$$
\begin{gathered}
B_1 = \frac{\partial}{\partial u}, \quad B_2 = u\frac{\partial}{\partial u} + v\frac{\partial}{\partial v}, \\
B_3 = (u^2 - v^2)\frac{\partial}{\partial u} + 2uv\frac{\partial}{\partial v} + 2v\frac{\partial}{\partial \phi}, \\
\omega^1 = \frac{d\phi}{2v}, \quad \omega^2 = \frac{dv}{v} - \frac{u}{v}d\phi, \\
\omega^3 = \frac{u^2 + v^2}{2v}d\phi + \frac{vdu - udv}{v}.
\end{gathered}
\tag{6.35}
$$

Nevertheless, in his original work from 1938, Dan Barbilian does not work with these one-sided invariant differential forms — they are only left-invariant — but with the following differentials, *absolute invariant* in his terms — they are both left- as well as right-invariant — with respect to action (6.8):

$$
\begin{gathered}
\omega^1 = \frac{dh}{(h - h^*)k}, \\
\omega^2 = -i\left(\frac{dk}{k} - \frac{dh + dh^*}{h - h^*}\right), \\
\omega^3 = -\frac{kdh^*}{h - h^*}.
\end{gathered}
\tag{6.36}
$$

The invariance can be easily verified by direct calculation. However, these differential forms do have an apparent shortcoming: they are not real like their counterparts from equation (6.34). The only real one among them is ω^2, while ω^1 and ω^3 are complex conjugate to each other. The purely real counterparts Ω^1, Ω^2 and Ω^3, say:

$$
\begin{gathered}
\Omega^1 = d\phi + \frac{du}{v}, \quad \Omega^2 = \cos\phi\frac{du}{v} + \sin\phi\frac{dv}{v}, \\
\Omega^3 = -\sin\phi\frac{du}{v} + \cos\phi\frac{dv}{v}
\end{gathered}
\tag{6.37}
$$

are typical $\mathbf{sl}(2,\mathrm{R})$ differential 1-forms, generating a three-dimensional Lorentz metric:

$$\omega^1\omega^3 - \frac{1}{4}(\omega^2)^2 \equiv (\Omega^1)^2 - (\Omega^2)^2 - (\Omega^3)^2$$
$$= (d\phi)^2 + \frac{2}{v}(d\phi)(du) - \frac{1}{v^2}(dv)^2. \qquad (6.38)$$

This circumstance, as Barbilian himself noticed, defines the variable ϕ as the 'angle of parallelism' of the hyperbolic plane (the connection). In fact, recalling that in modern terms (du/v) represents the connection form of the hyperbolic plane (Flanders, 1989), equations (6.37) then represent *a general Bäcklund transformation* in that plane (Sasaki, 1979; Rogers and Schief, 2002; Reyes, 2003). We will come back to this geometry after another insight into the problem of time from the point of view of its sequencing differentia.

6.2 The Connection between Phase and Time: The Physics of a Compass

As we know that at this juncture the time average of Lagrangian should be classically involved (in the form of action), we need to assume that between the phase, ϕ say, and average Lagrangian, t say — suggesting that it represents *an average content of time* along a certain time sequence — there is a homographic relation, which we take in the form extensively used by Elie Cartan as an essential example in building his geometrical theory of moving frame (Cartan, 1931):

$$\phi = \phi_0 - \frac{t}{\alpha t + \beta}. \qquad (6.39)$$

This correlation has the advantage of presenting the average Lagrangian as an independent, albeit quite particular, statistics, whereby the parameters (ϕ, α, β) have notable interpretations.

For once, if the average Lagrangian is zero, we have unconditionally

$$\phi = \phi_0, \qquad (6.40)$$

which gives interpretation to the parameter ϕ. This means free particle phase: *no content of time*. On the other hand, if the average Lagrangian goes to infinity, we have, again, a known interpretation of the situation:

$$\phi = \phi_0 - \alpha^{-1}. \tag{6.41}$$

The parameter α should therefore be the *reciprocal of a known phase*, which is also a phase, to be obtained in the cases when the content of time goes to large values. If this reciprocal phase is zero, we have a linear relationship between phase and the content of time, which is the classical case of identification of phase with the action:

$$\phi = \phi_0 - \beta^{-1}t. \tag{6.42}$$

Therefore β is non-dimensional, but should have the physical meaning of a time.

Now, regarding the statistics related to t, assume that the phase is constant, and we want to find the circumstances in which this requirement is satisfied. In order to do this we have for the average Lagrangian a Riccati differential equation, representing a generalized Planck equation like (5.41):

$$dt + \omega^1 t^2 + \omega^2 t + \omega^3 = 0. \tag{6.43}$$

This expresses the fact that, referred to an appropriate parameter of continuity, θ say, whereby $t'(\theta)$ is the variance of the exponential distribution of mean t, this distribution is from a family with quadratic variance functions. Here $\omega^{1,2,3}$ are the following differential 1-forms:

$$\begin{aligned}
\omega^1 &= -(\alpha^2/\beta)d\phi_0 - d\alpha/\beta, \\
\omega^2 &= -2\alpha d\phi_0 - d\beta/\beta, \\
\omega^3 &= -\beta d\phi_0.
\end{aligned} \tag{6.44}$$

We can say something about this family of ensembles if we can associate a continuous parameter in order to transform the equation (6.43) into an ordinary differential equation. In other words, the circumstances in which the phase is constant are given by an ensemble of average Lagrangians which are solutions of equation (6.43), in case it can be reduced to an ordinary differential equation of Riccati type. A procedure of such a reduction presents itself right away: the differential forms (6.44) are representing a basis of $\mathbf{sl}(2,\mathrm{R})$ Lie algebra, which can be organized as a metric space in the parameters (ϕ_0, α, β). Along the geodesics of this metric, the basis differential forms are exact differentials in the arclength, $d\theta$ say, which thus induces, via equation (6.43), an ordinary differential equation of Riccati type for t itself. Let us write this story in equations.

First, along the geodesics of the $\mathbf{sl}(2,\mathrm{R})$ metric space we can write:

$$\begin{aligned}(\alpha^2/\beta)d\phi_0 + d\alpha/\beta &= a^1(d\theta),\\ 2\alpha d\phi_0 + d\beta/\beta &= 2a^2(d\theta),\\ \beta d\phi_0 &= a^3(d\theta)\end{aligned} \tag{6.45}$$

where $a^{1,2,3}$ are constants, and θ is the arclength, as described above. In this case, (6.43) becomes

$$t'(\theta) = a^1 t^2 + 2a^2 t + a^3, \tag{6.46}$$

which is an ordinary Riccati equation. There are three possibilities for the solutions of this equation, according to the sign of the discriminant of the quadratic in right-hand side of this equation. We choose the first of these, whereby the discriminant is negative, which leads to the solution:

$$a^1 t + a^2 = \sqrt{\Delta}\tan[\sqrt{\Delta}(\theta - \theta_0)], \quad \Delta \equiv a^1 a^3 - (a^2)^2 \tag{6.47}$$

with θ_0 a constant; this constant is, however, not arbitrary with respect to the parameters of our problem.

In the case given by (6.47), we shall also have a certain solution for *the differential system* (6.45), which is easy to integrate, and will show how equation (6.39) needs to be read in general. We have:

$$\begin{aligned}\phi_0(\theta) &= t_0 + \theta_0 \tan[\sqrt{\Delta}(\theta - \theta_0)],\\ \beta(\theta) &= \beta_0 \cos^2[\sqrt{\Delta}(\theta - \theta_0)],\\ \alpha(\theta) &= \alpha_0 \cos[\sqrt{\Delta}(\theta - \theta_0)]\\ &\quad \times \{a^2 \cos[\sqrt{\Delta}(\theta - \theta_0)] + \sqrt{\Delta} \sin[\sqrt{\Delta}(\theta - \theta_0)]\},\end{aligned} \tag{6.48}$$

where t_0, α_0, β_0 and θ_0 are constants of integration. The situation has this statistical interpretation: we have an ensemble of average Lagrangians of mean t given in equation (6.47), having a distribution function with quadratic variance given by

$$a^1 t^2 + 2a^2 t + a^3 = \frac{\Delta}{a^1 \cos^2[\sqrt{\Delta}(\theta - \theta_0)]}. \tag{6.49}$$

This is actually a family of such distributions, indexed by the parameter θ. $[\beta(\theta)]^{-1}$ measures the variance of this ensemble.

The parameters α, β from equation (6.48) can be transcribed independently of the parameter ϕ_0, for instance as in the system:

$$\begin{aligned}\frac{\alpha}{\beta} &= \frac{\alpha_0}{\beta_0}[a^2 + \sqrt{\Delta} \tan \sqrt{\Delta}(\theta - \theta_0)],\\ \frac{1}{\beta} &= \frac{1}{\beta_0}[1 + \tan^2 \sqrt{\Delta}(\theta - \theta_0)].\end{aligned} \tag{6.50}$$

These can be interpreted as *Appell coordinates* in a Kepler problem of reduction of field of forces (see Mazilu *et al.*, 2019), for which $\alpha/\alpha_0 \equiv u$ and $\beta/\beta_0 \equiv v$. Parametrically, equation (6.50) represents a conic. Indeed:

$$\begin{aligned}r &\equiv \frac{u}{v} = a^2 + \sqrt{\Delta} \tan \sqrt{\Delta}(\theta - \theta_0),\\ z &\equiv \frac{1}{v} = 1 + \tan^2 \sqrt{\Delta}(\theta - \theta_0).\end{aligned} \tag{6.51}$$

In the implicit form, this conic is a parabola:

$$(r - a^2)^2 = \Delta(z - 1) \tag{6.52}$$

or, in an expanded form:

$$\Delta z = r^2 - 2a^2 r + a^1 a^3. \tag{6.53}$$

By restoring the coordinates (u, v) we get an ellipse:

$$(u - a^2 v)^2 + \Delta\left(v - \frac{1}{2}\right)^2 = \frac{1}{4}$$

or in the parameters α and β:

$$(\beta_0 \alpha - a^2 \alpha_0 \beta)^2 + \Delta\alpha_0^2\left(\beta - \frac{\beta_0}{2}\right)^2 = \frac{\alpha_0^2 \beta_0^2}{4}. \tag{6.54}$$

We read the following in these results: the equal probability time sequences are described by equation (6.46). They correspond to the same value of the phase. The parameters of the time transformation from phase describe time cycles when the phase is 'locked'. These parameters are in connection with the inertial properties of matter.

6.3 The Relativity as Absolute Geometry

There is no way of representing a physical point in a space but either by some physical quantities, or by some kinematical or dynamical quantities describing that point. Once this representation is accomplished, the remaining problem is purely theoretical: *the connection of these quantities with positions in the space of our intuition.* To start with, the points in the four-dimensional manifold of events — the spacetime — have as coordinates the positions of uniform motions starting from origin, and the radius of the sphere of propagation with the center in origin. The condition of limitation of possible velocities of inertial matter, epitomizes, within framework of space kinematics, the mathematical idea of confinement as presented in Mazilu *et al.* (2019, Chapter 4). From the very same point of view, the matter in a planetary model shaped by a Kepler motion, can be thought of as properly confined (see Mazilu *et al.*, 2019), provided we assume spatial extension of the material components of this model. Then the difference between the two models of mathematical confinement — the relativistic and the Keplerian — rests upon the sole fact that while relativistically the limit velocity is a constant, in the case of Kepler motion analysis the limit velocity is variable with the Hertz material particle from the matter in revolution, as well as with the material particle from the matter thought as creating the central field. Both cases, however — the relativity as well as the Kepler problem — are *about a velocity space.* This calls for a general presentation of the mathematical confinement of the velocities, whereby the limit velocity is variable, in order to be able to describe the most general case.

The equation of a Kepler ellipse in polar coordinates:

$$\frac{\dot{a}}{x} = \frac{k}{\dot{a}} + v_1 \cos\phi + v_2 \sin\phi \tag{6.55}$$

is indeed an equation for velocities. To wit, it connects the velocity $(\dot{a}/x)$, to the magnitude of actual position vector $\mathbf{x}$ of the mobile material point along its orbit. This material point is identified by its initial velocity (v_1, v_2), chosen to describe that orbit. One can say that a Kepler orbit is a trajectory corresponding to the given initial velocity, which is a closed curve — an ellipse, as Kepler inferred for the first time from the analysis of the data of Tycho Brahe for the planet Mars — if and only if this initial velocity satisfies the inequality:

$$\left(\frac{k}{\dot{a}}\right)^2 - v_1^2 - v_2^2 > 0. \tag{6.56}$$

As we already have noticed, this condition describes a particular relativistic ensemble of particles, out of all possible Hertz material particles from the ensemble giving the interpretation of the kernel, or nucleus, of an electron in the planetary model. Indeed, if the electron has a closed orbit — assuming the condition an ideal one, in the sense that the shape is the same after traversing the orbit — then, from a classical point of view, all of the constituent material particles from the interpretative ensemble of the nucleus of that electron should have closed orbits in the interior of a torus in space. The ideal condition also assumes the Hertz material particles have their trajectories all along the Kepler orbit of the electron.

Now, if we denote by X a point in this space of velocities, then a coordinate representation is given by a quadruple of (until further notice, real) numbers:

$$X \equiv (c, \mathbf{v}). \tag{6.57}$$

The norm of these points is then given by the quadratic form:

$$(X, X) \equiv c^2 - \mathbf{v}^2. \tag{6.58}$$

Thus, the points which satisfy the condition $(X, X) = 0$ are geometrically shaping an absolute for this geometry. Among other things, physically they represent the propagation of light if the limit velocity c is a constant. Specifically, the points with positive norm represent *inertial motions*, while the points with negative norm represent, for instance, *de Broglie waves* in the regular case of special relativity, or some other ensembles of Hertz material particles in the case of the Kepler motion.

The norm (6.58) induces an internal multiplication of points by the known 'polarization' procedure (see, for instance, Pierpont, 1928):

$$(X_1, X_2) \equiv c_1 c_2 - \mathbf{v}_1 \cdot \mathbf{v}_2. \tag{6.59}$$

Here an obvious correspondence is understood, between indices of points and indices of coordinates. This internal product helps in describing a straight line in space, which is the essential concept necessary in constructing a metric, at least from the differential geometric point of view. The straight line joining two points X_1 and X_2 is the locus of the points

$$X = \lambda X_1 + \mu X_2 \tag{6.60}$$

with λ and μ variable numbers, representing homogeneous parameters of points along the line. This straight line intersects the absolute in two points, having the homogeneous parameters partially established by the quadratic equation:

$$(X, X) \equiv \lambda^2 (X_1, X_1) + 2\lambda\mu (X_1, X_2) + \mu^2 (X_2, X_2) = 0. \tag{6.61}$$

To wit, from this equation we can determine only the ratios of these two parameters — the non-homogeneous coordinates along the line — as its two roots, *viz.*

$$t \equiv \frac{\lambda}{\mu} = -\frac{(X_1, X_2)}{(X_1, X_1)} \pm \frac{\sqrt{(X_1, X_2)^2 - (X_1, X_1)(X_2, X_2)}}{(X_1, X_1)}. \tag{6.62}$$

As it turns out, though, these ratios are just enough for our purpose of building a metric of the space.

Indeed, a metric represents the distance between two infinitesimally close points, therefore what we need is to give a distance between points. The quantity that reduces to distance between two positions in the Euclidean case turns out to be the *cross ratio* of four points on a straight line, two of which are fixed and used as a reference frame of that straight line. More precisely the distance is, up to a numerical factor, *the logarithm of this cross ratio*, the so-called *Laguerre formula* [for a brief pertinent presentation see, for instance, Duval and Guieu (2000)]. Given two points $X_{1,2}$, the straight line joining them contains all points of the form $X = tX_1 + X_2$, where t is a non-homogeneous parameter. In order to define the distance between the two points, we choose arbitrarily a reference frame formed by two other points on their straight line, $X_{3,4}$ say, and *define the cross ratio of the four points.* As the cross ratio of points is simply the cross ratio of the corresponding non-homogeneous parameters t, we have, for instance:

$$(X_1.X_2; X_3, X_4) = \frac{t_1 - t_3}{t_1 - t_4}\frac{t_2 - t_3}{t_2 - t_4}. \tag{6.63}$$

By Laguerre's formula, the distance is simply proportional to the logarithm of this quantity. It depends, of course, on the *reference frame* — the second pair of points on the straight line — but this ambiguity can be substantially reduced if we refer the construction to the absolute of space, and take the reference frame on the line as given by the two points of the straight line located on the absolute. First, notice that the parameter t has the values $t_2 = 0$ for the point X_2 and $t_1 = \infty$ for the point X_1. In view of this, the cross ratio (6.63) takes the simple form of a usual ratio:

$$(X_1.X_2; X_3, X_4) = \frac{t_4}{t_3}. \tag{6.64}$$

Incidentally, we need to take notice of the fact that the choice of the reference frame on the line as given by two points on the absolute, has the advantage of allowing a standardization of the construction of metric, inasmuch as every pair of points in space has a corresponding pair of points on the absolute: the points where the corresponding straight line intersects the absolute, as we said. The parameters $t_{3,4}$ are then given by the two ratios from equation (6.62), so that equation (6.64) becomes

$$(X_1.X_2; X_3, X_4) = \frac{(X_1, X_2) + \sqrt{(X_1, X_2)^2 - (X_1, X_1)(X_2, X_2)}}{(X_1, X_2) - \sqrt{(X_1, X_2)^2 - (X_1, X_1)(X_2, X_2)}}. \tag{6.65}$$

With this we can construct a differential version of the distance, which is the metric of space. Assuming that the two points X_1 and X_2 are infinitesimally close, therefore $X_1 = X, X_2 = X + dX$, we can calculate the necessary quantities in equation (6.65) as the mixed — finite and differential — forms:

$$\begin{aligned} (X_1, X_2) &= (X, X) + (X, dX), \\ (X_2, X_2) &= (X, X) + 2(X, dX) + (dX, dX), \\ (X_1, X_2)^2 - (X_1, X_1)(X_2, X_2) &= (X, dX)^2 - (X, X)(dX, dX). \end{aligned} \tag{6.66}$$

Now, at least in the real range we can accept that the quantity $(X, dX)/(X, X)$ is *infrafinite* of the first order in the sense of Georgescu-Roegen, while also forcing the condition that in the arithmetic of this first order, the square root of $(dX, dX)/(X, X)$ is such an infrafinite. Thus the cross ratio (6.65) can be expanded in a series and, to the first infinitesimal order it is

$$(X_1.X_2; X_3, X_4) = 1 + 2\sqrt{\left(\frac{(X, dX)}{(X, X)}\right)^2 - \frac{(dX, dX)}{(X, X)}}. \tag{6.67}$$

Further on, the logarithm of this quantity is, to the same infinitesimal order, the second term from the right-hand side of this equation, so that the absolute metric can be written, up to an arbitrary constant factor, in the form:

$$(ds)^2 = \left(\frac{(X, dX)}{(X, X)}\right)^2 - \frac{(dX, dX)}{(X, X)}. \tag{6.68}$$

It turns out that this expression is also valid in larger conditions of space definition: complex points, the definition of the absolute as a general quadric in this space, or even as a general quantic, etc. Thus, we have here a general form of the absolute metric valid for any type of quadric, independent of dimension of space and the nature of its elements, and even for any other kind of quantic, as in the example given by Dan Barbilian (Barbilian, 1937), and reproduced by us (Mazilu *et al.*, 2019). Nevertheless, our momentary interest is actually more limited, insofar as we follow the objective wave-particle duality — the duality in the sense of Louis de Broglie — and by this a proper generalization of the special relativity, necessary in order to construct a scale relativistic physics.

So, let us go back to our specific case, as given by equation (6.58). If the coordinates of points are real then the absolute quadric is a two-sheeted hyperboloid, as we know it from analytic geometry. This description can also be used for the case of complex coordinates. However in that instance the geometrical image is not quite as simple as in the real case: there, we have to do with the intersection of two real quadrics, therefore with a real conic in space. Regardless of these details, momentarily we shall refer to a direct calculation of the metric given by equation (6.68). This will be based on the identification $dX \equiv (dc, d\mathbf{v})$, which gives the following quadratic form as a metric based on this representation:

$$(ds)^2 = \frac{c^2(d\mathbf{v})^2 - (\mathbf{v} \times d\mathbf{v})^2}{(c^2 - \mathbf{v}^2)^2} + \frac{\mathbf{v}^2(dc)^2 - 2c(\mathbf{v} \cdot d\mathbf{v})(dc)}{(c^2 - \mathbf{v}^2)^2}. \tag{6.69}$$

This metric can be related to some known, and physically important situations. Let us, therefore, consider the equation just obtained in a greater detail.

We wrote here the absolute metric as a sum of two terms for a particular reason: in the framework of special relativity the first term from equation (6.69) is the metric of velocity space, and is a direct consequence of the law of composition of relativistic velocities [(Fock, 1959, §§16,17); see especially equation (17.01) of this reference]. In this taking, one can say that the whole metric (6.69), itself, generalizes the relativistic metric of the velocity space, which can be obtained from (6.69), for instance in cases where the first component of the point X is a constant: the differential of a constant c is always zero, and the second term in (6.69) vanishes. In other words, *there is no infrafinite four-vector velocity in the relativistic physics* as we inherited it from Maxwell electrodynamics. However, as we have shown (Mazilu *et al.*, 2019, Chapter 4) such a metric also characterizes a classical Kepler orbit, as in equation (6.55), a description which can be obtained by identifying the vector $\mathbf{v}$ with the *initial condition of motion*, and then taking $c \equiv k/\dot{a}$. Every closed Kepler orbit is thus described by some initial velocity having a magnitude smaller than this value, as in equation (6.56) above. The metric (6.69) could therefore be applied for describing the structure of *matter contained in the nucleus of revolving body of the planetary model* of physics, but there is a drawback. Indeed, in this interpretation, the matter of the revolving material point can be taken as a 'swarm' of Hertz material particles, as described by Joseph Larmor [see Larmor (1900); see also Chapter 2 here, equation (4.34) ff.]. Then the value of c is *variable with each particle* of the material point in revolution. And, as each particle is uniquely labeled by its initial conditions, there should be a correlation between the limit velocity c and the *label of the particle.* The nature of such a correlation can be inferred directly from (6.69), because this formula discloses still another condition in which the metric reduces to the usual 'relativistic' one, besides maintaining c a constant. This condition is given by the differential equation

$$vdc - 2cdv = 0 \quad \therefore \quad v^2/c = const. \tag{6.70}$$

which guarantees the vanishing of the second term in (6.69). In other words, in our present generalization of special relativity, *there are two cases in which the metric becomes purely relativistic.* One of these is the usual case of constancy of the value of limit velocity, which precludes the infrafinite differential measure for the time component in the realm of velocities taken as four-vectors: there are only regular finite four-vectors and infrafinite three-vectors there. On the other hand, there is a case of differential measure in this range, given by equation (6.70), involving the connection between the time component and space components of the four-vectors. These last velocities are 'reciprocal' with respect to the 'sphere' given by the initial conditions of the Kepler problem. Such a reciprocity is akin to a 'de Broglie duality', as it were, between the phase and wave-group velocities. However, this time *it is the inertial matter which imposes it upon light.* In general, therefore, one can assume that c is not a constant, regardless of this last situation, and describe a universal special relativity starting from a 'double' relativistic metric, to be obtained from (6.69) in two special cases: one for considering *the light in ether*, the other considering *the matter in ether*. For the first case, we cannot decide anything yet, but the last case can always be described, indeed, by two absolute metrics corresponding to two limit velocities.

In order to get the gist of the method, let us assume the case of constant c: only after analyzing this case we can properly improve on it. The absolute metric (6.69) turns out to be:

$$(ds)^2 = \left[1 - \left(\frac{\mathbf{v}}{c}\right)^2\right]^{-2} \left[\left(\frac{d\mathbf{v}}{c}\right)^2 - \left(\frac{\mathbf{v}}{c} \times \frac{d\mathbf{v}}{c}\right)^2\right]. \tag{6.71}$$

In the three-dimensional velocity space of relativity, this metric can be written in the form

$$(ds)^2 = \left(\frac{d\mathbf{q}}{1-\mathbf{q}^2}\right)^2 - \left(\frac{\mathbf{q} \times d\mathbf{q}}{1-\mathbf{q}^2}\right)^2, \quad \mathbf{q} \equiv \frac{\mathbf{v}}{c}, \tag{6.72}$$

while in the two-dimensional case (see Mazilu *et al.*, 2019) it can be written in the form

$$(ds)^2 = \left(\frac{d\mathbf{e}}{1-\mathbf{e}^2}\right)^2 - \left(\frac{\mathbf{e} \times d\mathbf{e}}{1-\mathbf{e}^2}\right)^2, \quad \mathbf{e} \equiv \frac{\mathbf{v}}{c}, \quad c \equiv \frac{k}{a}, \tag{6.73}$$

where, this time, $\mathbf{e}$ is precisely the eccentricity vector of the orbit. There is no formal difference between the two formulas, (6.72) and (6.73), except the dimension of the velocity space, and the fact that the last one of them is controlled by gravitation according to Newtonian precepts. Keeping this last condition in reserve, (6.73) can be obtained from (6.72) just by choosing to work in one of the planes of coordinates of the velocity three-dimensional space.

In order to settle our ideas, we choose to work on the formula (6.73), with the Cartesian components of the eccentricity vector, for then it is more obvious how the space extension of matter can enter our reasoning. Indeed, the eccentricity vector $\mathbf{e}$ represents the relative position of the center of orbit, in the dynamical Kepler problem, with respect to the center of force. Therefore, it can be taken as a rough measure of the *space extension of the matter generating the force field.* One can surely assume that the geometry of the space containing the matter which generates the force field in the classical Kepler problem is not an Euclidean geometry, but a Lobachevsky geometry. In other words, we have a geometry of the *space region containing forces*: it is the hyperbolic geometry. To make this statement even more obvious, the absolute metric (6.73) can be cast into the Beltrami–Poincaré form

$$(ds)^2 = -4\frac{dh\,dh^*}{(h-h^*)^2} \equiv \frac{(du)^2+(dv)^2}{v^2} \tag{6.74}$$

by the transformation

$$h = u + iv \equiv \frac{e_2 + i\sqrt{1-e^2}}{1-e_1}; \quad h^* \equiv u - iv. \tag{6.75}$$

There is an element of arbitrariness left with this representation, leading to the necessity of our reverse interpretation, which makes the objective of the present chapter of this work. Indeed, the metric is conformal to an Euclidean metric, and this last one is known to be invariant with respect to rotations in plane. We need to point out how this arbitrariness shows up in the hyperbolic geometry.

It turns out that in (u, v) coordinates this geometry exhibits a special invariance connected with the *Bäcklund transformation.*

Following a known routine, we take a point P in the hyperbolic plane as being described in the coordinates (u, v), so that its vector variation can be written as (Flanders, 1989, p. 134)

$$\mathbf{d}P = \begin{pmatrix} du \\ dv \end{pmatrix} \equiv \sigma^1 \mathbf{e}_1 + \sigma^2 \mathbf{e}_2, \quad \mathbf{e}_1 = \begin{pmatrix} v \\ 0 \end{pmatrix}, \quad \mathbf{e}_2 = \begin{pmatrix} 0 \\ v \end{pmatrix}. \tag{6.76}$$

This formula makes the fact obvious, that the hyperbolic plane is conformal to the Euclidean one. If we take the two differential forms

$$\sigma^1 = \frac{du}{v}, \quad \sigma^2 = \frac{dv}{v}$$

as a coframe, the Beltrami–Poincaré (6.74) is the sum of their squares. Thus, the hyperbolic plane can be treated as a *constant curvature surface*, whereby the parallel transport of a vector, $\mathbf{V}$ say, is given by the equations

$$dv^\alpha + v^\nu \Omega_\nu^\alpha = 0, \tag{6.77}$$

where v^α are the components of the vector $\mathbf{V}$ in a local reference frame, and $\boldsymbol{\Omega}$ is a Frenet–Serret matrix of differential elements. Now, along the paths on which the vector $\mathbf{V}$ is transported by parallelism there is only normal component of the variation of the vector, which can be written as

$$d\mathbf{V} = (v^\alpha \boldsymbol{\Omega}_\alpha^3)\hat{\mathbf{e}}_3 \quad \therefore \quad \hat{\mathbf{e}}_3 d\mathbf{V} = v^\alpha \boldsymbol{\Omega}_\alpha^3. \tag{6.78}$$

Here the normal vector to hyperbolic plane is taken as being of unit length in order to simplify the argument. Applying this theory to the displacement vector from equation (6.76), we have that, if

$$d\sigma^\alpha + \sigma^\nu \Omega_\nu^\alpha = 0,$$

then the vector $\mathbf{d}P$ is parallel-transported in the hyperbolic plane. The paths along which this transport is realized are then geodesics of the plane (Struik, 1988). Their equations can therefore be written in a purely differential form, independent of any continuity parameter:

$$d\frac{du}{v} - \frac{du}{v}\frac{dv}{v} = 0, \quad d\frac{dv}{v} + \frac{du}{v}\frac{du}{v} = 0. \tag{6.79}$$

This form of equations involves just symmetric differentials, *i.e.* differentials in the sense of Leibniz — not exterior differentials — and we use it here in order to draw attention on the fact that the equations can even be fractal: there is no need to refer them to a continuity parameter. Once this reference is done, however, the differentials become 'fluxions' in the plain sense of Newton's initial theory, representing the convention of definition of 'finite' with the associated 'infrafinite' and 'transfinite', in the acceptance of Nicholas Georgescu-Roegen. Now, by a proper identification, equation (6.79) makes the connection differential form (du/v) conspicuous. Expanding now the equation by performing the due differentiations, we have

$$\frac{d^2u}{v} - 2\frac{du\,dv}{v^2} = 0, \quad \frac{d^2v}{v} + \frac{(du)^2 - (dv)^2}{v^2} = 0$$

which are equations of the geodesics one can obtain by using the metric (6.74) as a Lagrangian, and then solving the appropriate associated variational problem. One can calculate then the components of $\mathbf{d}^2P$ along the geodesics, and with them the second fundamental form of the surface locally representing the hyperbolic plane. According to equation (6.78), the second fundamental form of this surface coincides with its first fundamental form, *i.e.* with the Beltrami–Poincaré metric itself:

$$\hat{\mathbf{e}}_3 \cdot \mathbf{d}^2P = -\{(\sigma^1)^2 + (\sigma^2)^2\}. \tag{6.80}$$

The unit vector of the Beltrami–Poincaré metric (Shirokov, 1988) is therefore given by a conformity, starting from the unit vector of the Euclidean plane:

$$\mathbf{v} = (\cos\phi)\mathbf{e}_1 + (\sin\phi)\mathbf{e}_2, \quad v^1 = \cos\phi, \quad v^2 = \sin\phi. \tag{6.81}$$

According to (6.77) and (6.79), this vector is parallel-transported in the hyperbolic plane if and only if

$$d\phi + \frac{du}{v} = 0, \quad \omega_2^1 = -\frac{du}{v}.$$

Now, since the Beltrami–Poincaré metric is itself conform-Euclidean, it does not change in a rotation of the coframe, which comes to a

Bäcklund transformation of the differentials

$$\sigma^1 = \cos\phi\frac{du}{v} + \sin\phi\frac{dv}{v}, \quad \sigma^2 = -\sin\phi\frac{du}{v} + \cos\phi\frac{dv}{v}. \tag{6.82}$$

However, something still changes, and this is *the connection form.* Indeed, by differentiation here, we have

$$d\sigma^1 = (d\phi)\sigma^2 + \cos\phi\, d\frac{du}{v} + \sin\phi\, d\frac{dv}{v},$$
$$d\sigma^2 = -(d\phi)\sigma^1 - \sin\phi\, d\frac{du}{v} + \cos\phi\, d\frac{dv}{v}$$

so that along the geodesics given by equation (6.79), we must have

$$d\sigma^1 = \left(d\phi + \frac{du}{v}\right)\sigma^2; \quad d\sigma^2 = -\left(d\phi + \frac{du}{v}\right)\sigma^1. \tag{6.83}$$

Thus, the newly rotated coframe (6.82) is still parallel-transported along the geodesics, only with a new connection form, which represents 'an update', as it were, of the original one by the differential of the rotation angle. This is the meaning of the Bäcklund transformation (6.82).

But there is more to it: as it turns out, the system of the three differential forms (6.37), formed, as one can see, with the new connection form and the new coframe (6.82), is an algebraically closed system in the sense that they constitute a coframe for an $\mathbf{sl}(2,\mathrm{R})$ type algebra. This fact is not quite so obvious, at 'the first sight', as it were, but it can be made obvious by an exterior differentiation of the indicated 1-forms, for the Maurer–Cartan relations here turn out to be of the characteristic form:

$$d\wedge\Omega^1 = \Omega^2\wedge\Omega^3, \quad d\wedge\Omega^2 = -\Omega^3\wedge\Omega^1, \quad d\wedge\Omega^3 = -\Omega^1\wedge\Omega^2. \tag{6.84}$$

In fact, the Killing–Cartan quadratic form that describes this algebra as a Riemannian space is that from equation (6.38), a typical three-dimensional quadratic Lorentz–Minkowski form. It describes a Riemannian space which reduces to the hyperbolic plane in cases where ϕ is itself the 'angle of parallelism', as Barbilian used to say, *i.e.* a connection. This space is the central feature of our physics here, as the representative of matter *per se*. Let us elaborate on this aspect of the problem, for it is destined to illuminate us upon the kind of forces which are described by the Riemannian geometry, and what is their relationship with the manifold of events, *i.e.* with the spacetime.

The transformation (6.75) makes clear what physical kind of geometrical beings are represented by the complex variables h and h^*: they are *gauging lengths* proper, or parameters related to these lengths, like the semiaxes of the Kepler orbits and their eccentricity or, in the case of light, the semiaxes of the 'ellipsoid of elasticities' of ether. If we take this last case as typical for gauging a continuum, then the concepts are algebraically settled by the theory of cubic equations, describing the semiaxes of an ellipsoid, which are always expressed by the roots of a cubic equation as eigenvalues of a tensor. In this case the geometry is no more an Euclidean one, but should be submitted to some rules of gauging. These rules are pretty obvious, if we express them in terms of the two basic tools of the theoretical physics of universe: *the Wien–Lummer hohlraum* and *the Einstein elevator*. By definition, these tools ask that the expressions of laws governing the universe must be the same *no matter of the size of this universe*, and invariant for the transition between different sizes. In a word, they should be *scale invariant*.

Chapter 7

Gravitation: The Case of Anisotropic Relativity

It is, we think, the proper moment to concern ourselves with the issue raised by Henri Poincaré, and detailed by us in the Chapter 1 of this work: the problem of gravitation. As it was conceived by Poincaré [see equation (1.20), and discussion], if it is to draw a parallel between gravitation and light as two forms of action at a distance, the problem of gravitation is connected to that of propagation. In Chapter 1 we deepened this connection, considering the classical property of gravitation of acting radially — a property valid as such for both the classical physics and general relativity — which leads to a groupal description involving the SL(2, R) action in two variables with three parameters [see equation (1.21) and the discussion following it]. The 'proper moment' here, to treat this important issue, comes with the observation that the absolute geometry may be able to provide a proper way of treatment, by a property of matter, 'dual', as it were, to that of light [see equations (6.70)–(6.71), and discussion]: formally, the matter can also be described by a limit velocity, like the light itself.

Marcel Brillouin was the first to draw attention to the fact that, as long as we consider the light as a material phenomenon — obviously along with the ether, which physically supports it as such — the quantum manifestation at the level of planetary atom asks for the existence of *two rapidities*, or *celerities*, in propagation, in order to have a certain proper classical explanation. Quoting:

> — ***Besides the speed of light, the universal medium*** (*ether*) ***possesses a celerity of propagation*** much smaller [*of the order of a few tens of kilometers per second* (?)]. *The quantal phenomena occur when the*

> *electrons move with a speed superior to this celerity, along quasi-periodic orbits, such that the electron is at every moment in the field of a finite number of its earlier positions.* [(Brillouin, 1919); *our translation; original Italics and question mark*]

Therefore, in this acceptance, the light emission would appear in the manner in which the sonic boom does, when a supersonic accelerating aircraft surpasses the speed of sound in air. A routine phenomenon for today's life on Earth, but just making its entrance in the common knowledge by the time when the work from which we excerpted the above fragment was issued. In fact it was this very phenomenon that inspired Marcel Brillouin, leading him to the idea of a double rapidity of an ether perturbation [for details one can also consult Brillouin (1920, 1922)].

One could say that the modern wave mechanics itself is established on the principle of this duality of speeds having, however, the particular aspect of the inversion of velocities, as in equation (6.71). Based on this, de Broglie's argument follows a more convincing line, *namely that of associating a propagation to motion*, completely covered by the concept of energy (de Broglie, 1923). This mathematical exactness can be logically explained if we leave aside the classical theory, and consider the light — physically, the quintessential representative of the phenomenon of propagation! — from the point of view of the quantization of individualities, specifically from the point of view of what we call today a *photon*, which comes only naturally with the concept of interpretation. For this we have to turn back in time though, to the idea that introduced this kind of quantization in world's consciousness (Einstein, 1965). Quoting:

> Monochromatic radiation of low density (within the range of validity of *Wien's radiation formula*) behaves thermodynamically *as though it consisted of a number of independent energy quanta* of magnitude $R\beta\nu/N$. [(Einstein, 1965); *our Italics*]

Without any further explanation of other symbols, insofar as this work of Einstein is now famous, we must mention however that the monomial $(R\beta/N) \equiv h$, is known today as the *Planck's constant.* Also, we need to add, related to the idea of quantization related to the classical statistical theory of the thermal radiation, that the name 'photon' was coined even a little later than the de Broglie's own idea of associating a frequency to a massive particle (Lewis, 1926).

One must recall though, that this well-known work of Einstein's is actually a kind of 'indictment', if we may say so, addressed to *the classical concept of ideal gas*, from the point of view of the probabilistic thermodynamics, whereby the assessment of probabilities is done *by constructing them with the assistance of thermodynamic volume*. By the beginning of the last century, the general idea took shape, mostly through Einstein, that in order to use the ideal gas as a standard concept in thermodynamics, the very theory of this gas has to be 'completed', so to speak, in the first place. Therefore, Einstein 'completes' it, and thus, based on this accomplishment, he extracts both the quoted above conclusion, and a few others just as important, but which do not interest us momentarily. Intimations about using the theory as such for the real gases and for solids, foreshadow his later attempts to elucidate issues related to these physical systems (Einstein, 1907; Einstein and Hopf, 1910; Einstein and Stern, 1913).

One of these essential issues, related to Einstein's conclusion above, led to the foundation of the wave mechanics. Louis de Broglie took that conclusion face value, as it were. Thus, he noticed that it brings up a hidden problem, which can easily pass unnoticed, but which we consider here *from the very point of view of Einstein himself*. Namely, *in order that the theory may be applied even to the ideal gas* — not to mention molecular gases in general, or solids for that matter — *we must associate a frequency to a point mass*, there is no way around! Therefore, the de Broglie's theory of wave-particle duality, would appear as a quantal alternative to the classical one of Marcel Brillouin, to which in fact Louis de Broglie even makes a brief reference in a footnote of his fundamental work from 1923. Therefore, if the quantum theory of Einstein is true, then the rest energy of a material point, of mass m_0 say, can be expressed in two different equivalent manners, based on the theory of relativity. In equations, this equivalence amounts to

$$h\nu_0 = m_0c^2 \quad \therefore \quad \nu_0 = \frac{m_0c^2}{h}, \tag{7.1}$$

where h is Planck's constant and c the speed of light in vacuum. Equation (7.1) can be read in the manner in which de Broglie himself did it: *we can associate a frequency* ν_0 *to a particle of rest mass* m_0 *— the de Broglie frequency.* In this form it is not hard to see that the process of association is in fact a two-way process: first we have here an interpretation proper, whereby *one* particle is associated to *one* wave; secondly we have an 'inverse-interpretation', if we may, whereby *one* wave is associated to *one* particle. Indeed, the essential feature of the wave is that of *a continuum*, while the corresponding feature of particle — at least from the classical point of view! — is that of *a position in space.*

Louis de Broglie noticed that, as it turns out, this association comes in with certain contradictions though, which meet immediately the eye if we ask ourselves what frequency we should assign to a *moving material point.* In other words, inasmuch as the frequency refers to a *wave*, which, by its very nature, *is a propagating continuum*, the problem can also be formulated like this: can we associate a propagation to a motion, *which is a 'discrete act' of moving*? The solution given by de Broglie to this problem reveals a fundamental contradiction of our knowledge, which in turn brings out the necessity of the very wave mechanics. Indeed, denoting $\nu \equiv \beta c$, the *mass* of the moving particle appears to a rest observer as given by Einstein's relativistic formula:

$$m = \frac{m_0}{\sqrt{1-\beta^2}}.$$

This shows that an observer at rest can associate to this particle a *clock frequency* defined by the equation

$$\nu = \frac{\nu_0}{\sqrt{1-\beta^2}}. \tag{7.2}$$

On the other hand, if we take this definition for good, and apply it to the *time period* of a clock which would have the rest frequency ν_0, then according to Lorentz's transformation laws the frequency in motion would be, in fact, smaller, as given by the equation:

$$\nu = \nu_0\sqrt{1-\beta^2}. \tag{7.3}$$

Obviously, the frequencies given by equations (7.2) and (7.3) do not coincide for one and the same material point, but only in cases where the particle is at rest, which leads indeed to a contradiction: the attempt to associate a frequency according to the energetic idea from equation (7.1) has, generally, no physical meaning, *since the associated frequency cannot be unique*. However, de Broglie shows that this contradiction is only apparent, the way of association being entirely natural from the very classical point of view. More specifically, we have the following theorem of a classical extraction:

> *If, at the beginning, the internal phenomenon of the moving body is in phase with the wave, this harmony of phase will always persist.* [(De Broglie, 1924); *original Italics*]

The proof of this theorem of de Broglie is, as a matter of principle, a little more involved than it would appear at the first sight. First of all, if we associate a wave to a particle in a reference frame where the particle is at rest, then this wave should be described by a trigonometric function, say

$$\Psi(x_0, t_0) \equiv a_0 \sin(2\pi\nu_0 t_0)$$

with ν_0 given by equation (7.1). On the other hand, in a reference frame moving with the speed βc, de Broglie finds just natural to admit that the associated wave is form-invariant with respect to Lorentz transformation and, moreover, that *the amplitude of the wave is invariant* too. Although this last condition may be proved by the Lorentz's theorem of 'corresponding states' (Lorentz, 1916, §§162 ff.), in the framework of special relativity it can be secured 'by law', as it were, *if the wave in question is either electromagnetic*, or transversal, in a broader sense, because *as a vector* — and at least within a proper gauging — the amplitude of the wave is then perpendicular to the direction of propagation. What we want to say here, is that, in general, we have to assume that the electromagnetism is only a special case of a more general condition. But, no matter what the case may be, the associated wave will be given by the previous expression, whereby the time is replaced by its Lorentz transform (the Lorentz's 'local time'):

$$\Psi(x_0.t) = a_0 \sin[2\pi\nu(t_0 - x_0/v_{ph})]$$

with ν given by equation (7.2) and a *phase velocity*, higher than the speed of light, and defined by the obvious relation $v_{ph} \equiv c^2/v$. Now, if the particle moves infinitesimally with (vdt), the phase of the wave associated by propagation, will change either by $\nu_0\sqrt{(1-\beta^2)}dt$, or by $\nu_0(1 - v/v_{ph})dt/\sqrt{(1-\beta^2)}$, depending on the observer's point of view. However, if the periodic phenomenon remains in phase with the wave, the two variations must be equal, so that we must have

$$\sqrt{1-\beta^2} = \frac{1}{\sqrt{1-\beta^2}}\left(1 - \frac{v}{v_{ph}}\right).$$

By the definition of phase velocity this equation is in fact an identity in view of the definition of the phase velocity, which proves *the de Broglie's theorem* as enunciated in the excerpt above.

The weak point of this association particle-wave — therefore of the association of a motion with a propagation — is that we do not know anything about the *physical relation* of its two terms. They cannot be identical, and we can even call down a positive principle in order to prove this: from relativistic point of view the phase velocity is not real for a particle in the free space. On the other hand, though, a monochromatic wave is, in fact, just as unrealistic from the point of view of the experimental implementation. Indeed, what we can always accomplish from an experimental point of view is only a *group of waves* of a certain width in frequency. This fact allowed Louis de Broglie to take notice that it would be only the association *between particle and wave group* the one that makes sense anyway we look at it: either relativistically or experimentally. And in this case the things go normal.

Indeed, let us assume that all of the frequencies of a group of waves are located in the neighborhood of a certain frequency ν, given by equation (7.2). The phase of one of these waves can then be represented by the equation

$$\Phi \equiv 2\pi(\nu+\varepsilon)\left(t - \frac{\beta(\nu+\varepsilon)}{c}x\right), \quad |\varepsilon| \leq \Delta v,$$

where $\beta(\nu)$ is a function of frequency that can be written explicitly based on equation (7.2), and ε is a small variation of the frequency in the vicinity of ν. To a displacement dx in the time interval dt, two of the waves from the group, corresponding to the frequencies $\nu + \varepsilon_1$ and $\nu + \varepsilon_2$ say, will get phase variations, which to the first order in the variations of frequency are:

$$\Delta\Phi_1 = 2\pi(\nu + \varepsilon_1)\left\{dt - \tfrac{1}{c}[\beta(\nu) + \varepsilon_1\beta'(\nu)]dx\right\},$$
$$\Delta\Phi_2 = 2\pi(\nu + \varepsilon_2)\left\{dt - \tfrac{1}{c}[\beta(\nu) + \varepsilon_2\beta'(\nu)]dx\right\}.$$

Here the accent denotes the derivative with respect to frequency, and we have used a Taylor expansion for the function $\beta(\nu + \varepsilon)$, retaining the first two terms. These variations are equal in the chosen expansion order, if and only if

$$dt - [\beta(\nu) + \nu\beta'(\nu)]\frac{dx}{c} = 0 \tag{7.4}$$

because, the waves are distinct, and thus $\varepsilon_1 \neq \varepsilon_2$. We thus define a velocity:

$$v \equiv \frac{dx}{dt} = \frac{c}{[\nu\beta(\nu)]'} \tag{7.5}$$

corresponding to a position from the space covered by the group of waves, *where all of them are in phase*. Using equation (7.2) we have right away

$$\nu\beta(\nu) = \sqrt{\nu^2 - \nu_0^2}, \quad [\nu\beta(\nu)]' = \beta^{-1}$$

so that

$$v = \beta c$$

whence the conclusion: *it is only the point of phase concordance of a wave group that moves with the velocity of a particle.* Now, because from the point of view of classical mechanics the energy is carried by a particle, de Broglie draws the natural conclusion:

> *The velocity of the moving body is the energy velocity of a group of waves having frequencies* $\nu = (1/h)m_0c^2/\sqrt{(1-\beta^2)}$ *and velocities* (c/β) *corresponding to very slightly different values of* β. [(De Broglie, 1924); *original Italics*]

From this point on, the wave mechanics started unfolding, naturally we might say, based on the general idea that *to a given motion there corresponds a multiple propagation*. Regrettably though, it grew almost exclusively based on de Broglie's own works, who on a lonely path, as it were, followed the objective line of defining the duality particle-wave. We witnessed (see Mazilu *et al.*, 2019) quite a few details of de Broglie's contributions, that can be taken to illustrate this statement. For now, we are only interested in the law of duality *per se*, which can be best articulated by a suggestive sentence: *the relation between phase velocity and group velocity can be expressed by a homography between their magnitudes*, of the type shown in the last equality from equation (6.71). Such a homography is therefore very special, and we write it in the form connecting the magnitudes of the two vectors:

$$v_{\text{Group}}\, v_{\text{Phase}} = c^2. \tag{7.6}$$

In this relation, the celerity of propagation of light is a standard in inversion, just as in equation (6.71) the speed of motion of a 'body' is such a standard. Not quite the law itself, as much as its mathematical basis interest us momentarily, for through this mathematics the law itself can thus be turned into *a relation between motion and propagation* as follows in the rest of the presentation of this chapter.

7.1 The Group of Waves as an Ensemble

If we consider equation (7.5) as an ordinary differential equation for the group velocity, then the relativistic formula (7.2) of transformation for frequency is just a particular solution. One can get this solution assuming that the group velocity cannot be greater than the speed of light in vacuum. Then the initial frequency is that associated to the particle at rest. From this (purely mathematical!) point of view, the alleged physical condition of limitation of velocities to values smaller than the speed of light in vacuum, is what one can see sometimes termed in the literature as a 'theoretical folklore'. Indeed, such a condition is *physically explained* as a consequence of the relativistic mass transformation formula, whereby if the velocity of body approaches the speed of light, the mass associated by an observer at rest goes necessarily to infinity. It is then only natural to think that something strange, unphysical, may happen there, so we have to limit the considerations to speeds smaller than that of light in order to do physics properly. It is this way that the limitation of velocities came up as a physical condition. Useless to say, the modern theoretical physics found mathematical methods to deal with superluminal velocities, from the very point of view of the concept of *interpretation* as we understand it here: the *tachyons* were invented [see Bilaniuk, Deshpande and Sudarshan (1962), Bilaniuk and Sudarshan (1969), and the literature cited there]. This fact compels us to somehow formally place the initial statement which was the basis of de Broglie's association between group of waves and particle, in a larger framework, in order to delineate it from unsecured additions. One way to do this, worthy of consideration inasmuch as it suggests a fruitful approach of the problem in its utmost generality, is indeed to formally accept as physical all the velocities, smaller as well as higher than that of light, and analyze the consequences. This fact can easily be accomplished in case we seek a three-dimensional extension of the theory.

Notice that the core of association wave-particle is the idea of frequency as coming directly from Planck's quantization through the concept of quantum. While the original Planck's considerations were referring to blackbody radiation for which the notion of frequency makes perfect sense, Louis de Broglie made an unsecured extension of the area of application of that notion. That should have made him uneasy, to the extent that he felt necessary to further elaborate on the subject. However, as his later works prove, de Broglie was exclusively dealing with the *frequency per se*: no relation with the phase, or with things other than the energy; he struggled to clarify the nature of the frequency he introduced. As the history proves though, the nature of that frequency thus introduced in physics is still unclear even today.

It is just natural then, to explore alternatives under the broader idea of the association wave-particle. The first reaction in this direction seems to present itself naturally by the fact that the frequency has always a mathematical definition related to phase (Mandel, 1974; see Mazilu *et al.*, 2019, Chapter 2): *try to avoid the notion of frequency.*

This can be done easily, by exploiting a relation between velocities as suggested when using the frequency itself, but on the interpretation side of the issue. More to the point, the relation between phase and group velocities as given in equation (7.6) is the core of the whole construction: mathematically it represents an *inversion* which, while made possible by using the frequency, can be described, in general, by dispensing with this concept. In case one seeks an extension of this relation for the three-dimensional vectors, the most natural and direct one is the *inversion of velocity vectors with respect to the sphere of speeds of light in vacuum.* In order to do this it is necessary to look at equation (7.6) as representing, for instance, a relation between *magnitudes* of two vectors representing the group and phase velocities. Then the relation between the two vectors can be written as

$$\mathbf{u} \cdot \mathbf{v} = c^2 \quad \therefore \quad \mathbf{u} = c^2 \frac{\mathbf{v}}{v^2}, \tag{7.7}$$

which is, indeed, the expression of a geometrical inversion of the velocity vectors with respect to the sphere

$$\mathbf{v} \cdot \mathbf{v} = c^2 \tag{7.8}$$

representing the possible speeds of light in vacuum. Thus if $\mathbf{v}$ is a vector inside the sphere (7.8), $\mathbf{u}$ is necessarily a vector outside that sphere, and vice versa. This way we have a clear distinction between the vector velocities smaller than that of light and those greater than that of light. If we deny physical reality to these last ones, then *they do not represent physical entities*: in de Broglie's language they represent just *phase waves*. In this formalism, a superluminal theory cannot therefore refer but to phase waves, some kind of mathematical beings that have no physical reality.

By equation (7.7) the correspondence between *phase waves* and *particles* is one-to-one: there is only one phase wave associated to a particle, and vice versa. And there is no contradiction in this. The idea of group of waves enters the stage here as related to *the indecision of the reference frame.* Indeed, equation (7.7) should be valid in a reference frame at rest. In a reference frame moving with a constant velocity, $\mathbf{w}$ say, the associated wave will have the speed $\mathbf{u}'$ where

$$\mathbf{u}' \equiv \mathbf{u} + \mathbf{w} = c^2 \frac{\mathbf{v}}{v^2} + \mathbf{w}. \tag{7.9}$$

Assume that this is the velocity of a phase wave. Applying the inversion formula to it, gives the *particle* velocity

$$\mathbf{v}' = \frac{\mathbf{v} + \mathbf{s}v^2}{1 + 2\mathbf{s} \cdot \mathbf{v} + s^2\mathbf{v}^2}, \quad \mathbf{s} \equiv \frac{\mathbf{w}}{c^2}. \tag{7.10}$$

The very same goes for phase waves: one can write

$$\mathbf{u}' = \frac{\mathbf{u} + \mathbf{s}u^2}{1 + 2\mathbf{s} \cdot \mathbf{u} + s^2\mathbf{u}^2}, \quad \mathbf{s} \equiv \frac{\mathbf{w}}{c^2} \tag{7.11}$$

thereby associating an ensemble of waves to a phase wave, in one-to-one correspondence with the ensemble of reference frames.

A little digression seems now in order. First of all, the correspondence between particles and waves is here one-to-one and, as already noticed, there is no contradiction in this association, inasmuch as the interior of the light sphere in velocity space is exclusively reserved for 'physical' particles, while the exterior is exclusively reserved for 'unphysical' phase waves. There is, however, a subtle issue here. Notice, indeed, that the idea of group of waves, which is the key in an energetic description of particles, may become itself unphysical if we do not accept that the classical vector composition applies to velocities greater than that of light. Even at a superficial consideration, one can see the trouble boiling here, because we need to answer at least to a question like: has this purely geometrical speculation any bearing on some physical facts? In other words, if mathematically does not matter, the physics, as always, asks for an observational support. Now, if one forgets about the relativity, there certainly exists such a support. Indeed, let us take one of equations (7.10) or (7.11), and calculate the magnitude of the vector from its left-hand side. We have, for instance,

$$\mathbf{v}'^2 = \frac{v^2}{1 + 2\mathbf{s}\cdot\mathbf{v} + s^2v^2} \quad \therefore \quad v' = v\Big(1 - \frac{vw}{c^2}\cos\theta\Big), \tag{7.12}$$

where θ is the angle between the direction of the motion of reference frame and that of the velocity $\mathbf{v}$. When v and v' represent speeds of light in a *transparent medium*, which is plainly a physical situation, equation (7.12) gives the *Fresnel dragging* as a purely kinematic phenomenon. The second expression in equation (7.12) has been written here in order to compare this inversion formula with the formula used by Lorentz in deducing the equation of the Fresnel dragging from his theorem of 'corresponding states' (Lorentz, 1916, §§161–163).

However, a question still remains, in view of the fact that the observational support generated the special relativistic views of the world: is this kind of association between particle and wave the right one from the physical point of view? In order to ask this question we need an excursion into some relativity views which lay special emphasis on the parameter associated with the *velocity of propagation*. This will help us in distinguishing what is the true meaning of the *velocity of motion*.

7.2 A One-Dimensional Anisotropic Relativity

In his argument above [see equations (7.1) ff.], Louis de Broglie referred to the *spacetime* continuity, mixing, so to speak, two essentially different points of view: the 'global' continuity of the events, and the continuity representing the motion, which must be, at least in a classical taking, a 'differential' continuity. For instance, (x, t) represents, from a relativistic point of view, an event in a two-dimensional manifold supporting the idea of 'globality'. The global continuity is here represented by *transformation between events*: Lorentz transformations in the case of de Broglie. When talking of a motion, 'x' must be a continuous function of 't', and in order to associate a velocity to the motion, this continuity should even be differentiable. In general, the differential spacetime continuity is represented by a connection between the *differentials* of the coordinate and of time. In a global representation, this connection can be a kind of Lorentz transformation only in the case it is linear and homogeneous — in order to be compatible with the differential continuity of the motion — so that it can be expressed by a matrix.

By the reason of what was just signaled here, one should make, therefore, a distinction between *the Lorentz transformation proper*, and the *transformation between differentials.* This means that if the Lorentz transformation is usually taken as a transformation of moments and positions, as points, and therefore takes the time as a sequence, the transformation between differentials must be regarded as somewhat independent, not always obtainable simply by a differentiation. In other words, it is only the transformation between differentials that takes *the time and the coordinates in a 'measured' continuity*, so to speak, necessary for the description of motion by a speed. In order to grasp the right connection between the two kinds of continuity, and, moreover, even to justify it from all the proper points of view, we shall have to reproduce here, first, the mathematics presented by Victor Lalan a long time ago, and based exclusively on the theory of continuous parametric groups (Lalan, 1937). Then, once we discern what the group theory asks for in this problem in implementing the idea of continuity, we shall have to use this point of view for the transformations of differentials themselves. For the general theory of the families of parametric transformations, in the form used by Victor Lalan himself, one can also consult (Vrânceanu, 1962).

Victor Lalan takes notice of the fact that in the case of a one-dimensional physical theory, whereby the events are located by the values of a space coordinate decided in an external reference frame and of a time moment decided by an external clock, the *invariance with respect to the origin of events* imposes that the possible transformations be of the form

$$x' = A(x - vt), \quad t' = Bx + Ct, \tag{7.13}$$

where the coefficients A, B, C depend on the relative velocity of the reference frame in which the coordinates are (x', t') with respect to the reference frame where the coordinates of events are (x, t). The condition that this one-parameter family of transformations should contain the identity transformation for $v = 0$, comes down to assuming that A, B, C depend on the relative velocity in such a way that, within the first order in the value of relative velocity, we have

$$A(v) = 1 + \alpha v, \quad B(v) = \beta v, \quad C(v) = 1 + \gamma v$$

with α, β, γ are constants. The evolution differential equations then become

$$\frac{dx'}{d\varphi} = \alpha x' - t', \quad \frac{dt'}{d\varphi} = \beta x' + \gamma t', \tag{7.14}$$

where φ is a 'canonical parameter' of the transformation: *not the velocity itself*! The requirement of continuity imposes that for $\varphi = 0$ we should have the 'initial' coordinates x and t. The system (7.14) can be integrated if we write it in the form

$$\frac{dx'}{\alpha x' - t'} = \frac{dt'}{\beta x' + \gamma t'} = d\varphi$$

and then construct, based on it, exact differentials of the form, as we proceeded in Chapter 4 [see equations (4.17) ff.; see also Vrânceanu (1962)]:

$$\frac{m dx' + n dt'}{(m\alpha + n\beta)x' - (m - n\gamma)t'} = d\varphi,$$

which can be integrated directly when the coefficients of the denominator are proportional with those of the numerator:

$$m\alpha + n\beta = \lambda m, \quad m - n\gamma = -\lambda n. \tag{7.15}$$

In this case we have, with the chosen initial condition,

$$mx' + nt' = (mx + nt)e^{\lambda\varphi}$$

and, in order to adequately write the transformation, everything comes down to finding the constants m, n and λ. Using, for instance, the first of equations (7.15), the transformation (7.13) can be written in the form:

$$x' - ct' = (x - ct)e^{\lambda\varphi}, \quad c \equiv \frac{\beta}{\alpha - \lambda}. \tag{7.16}$$

Now, m and n are solution of the linear system (7.15), generally compatible for only two values of parameter λ, the roots of quadratic equation:

$$\lambda^2 - (\alpha + \gamma)\lambda + \alpha\gamma + \beta = 0.$$

By (7.16), this means two possibilities of the same algebraical nature for the parameter c, given by the quadratic equation:

$$c^2 - (\alpha - \gamma)c + \beta = 0. \tag{7.17}$$

Assume now that the roots of this equation are real and different. The finite transformation that interests us can be drawn from the two specimens of equation (7.16), corresponding to the two values of c from (7.17):

$$x' - c_1 t' = (x - c_1 t)e^{\lambda_1\varphi}, \quad x' - c_2 t' = (x - c_2 t)e^{\lambda_2\varphi}. \tag{7.18}$$

One handles better this transformation if given in the form of a vector equation:

$$\begin{pmatrix} x' \\ t' \end{pmatrix} = \begin{pmatrix} 1 & -c_1 \\ 1 & -c_2 \end{pmatrix}^{-1} \begin{pmatrix} e^{\lambda_1\varphi} & 0 \\ 0 & e^{\lambda_2\varphi} \end{pmatrix} \begin{pmatrix} 1 & -c_1 \\ 1 & -c_2 \end{pmatrix} \begin{pmatrix} x \\ t \end{pmatrix}$$

whose matrix is the 2×2 table

$$\frac{1}{c_1 - c_2} \begin{pmatrix} c_1 e^{\lambda_2\varphi} - c_2 e^{\lambda_1\varphi} & c_1 c_2 (e^{\lambda_1\varphi} - e^{\lambda_2\varphi}) \\ e^{\lambda_2\varphi} - e^{\lambda_1\varphi} & c_1 e^{\lambda_1\varphi} - c_2 e^{\lambda_2\varphi} \end{pmatrix}. \tag{7.19}$$

From equation (19.18) we have, obviously,

$$(x' - c_1 t')^{\lambda_2} (x' - c_2 t')^{-\lambda_1} = (x - c_1 t)^{\lambda_2} (x - c_2 t)^{-\lambda_1}, \tag{7.20}$$

which means that this algebraic expression is an invariant of the family of transformations. In order to grasp something of a meaning of this expression, notice that if $\lambda_1 + \lambda_2 = 0$, it becomes the *usual quadratic Lorentz invariant*, $x^2 - c^2 t^2$. Therefore a possible meaning should be connected to a metric form, upon which we will come again later.

A quick comparison between the matrix of transformation (7.13) and the matrix (7.19), gives the connection between the *velocity parameter* of the family of transformation and the *canonical parameter* φ, in the form:

$$v = \frac{c_1 c_2 (e^{\lambda_1 \varphi} - e^{\lambda_2 \varphi})}{c_2 e^{\lambda_1 \varphi} - c_1 e^{\lambda_2 \varphi}} \quad \therefore \quad e^{(\lambda_1 - \lambda_2)\varphi} = \frac{c_1 v - c_2}{c_2 v - c_1}. \tag{7.21}$$

Using the last relation from this equation, we can transcribe the additivity property of the canonical parameter $\varphi : \varphi_3 = \varphi_1 + \varphi_2$, as a composition property of the corresponding velocities, *i.e.*:

$$\frac{v_3 - c_2}{v_3 - c_1} = \frac{c_1}{c_2} \frac{v_1 - c_2}{v_1 - c_1} \frac{v_2 - c_2}{v_2 - c_1}.$$

After due calculations, we have the *Lalan's relation of composition of velocities* [*loc. cit*, equation (10')]:

$$v_3 = \frac{v_1 + v_2 - (c_1 + c_2)\dfrac{v_1}{c_1}\dfrac{v_2}{c_2}}{1 - \dfrac{v_1}{c_1}\dfrac{v_2}{c_2}} \tag{7.22}$$

which reduces to the usual relativistic rule for $c_1 + c_2 = 0$ (Einstein, 1905). Consequently, either this equation or the homonymic one resulting from it — by the transformation given in equation (7.16) — for the eigenvalues $\lambda_{1,2}$ of the transformation matrix, must be of a special physical importance, which can be revealed via the theory of groups. In order to conclude this mathematical line, we can say that there are, indeed, *by law*, as it were, two limit speeds for the description of the kinematics of the classical material points, so long as this kinematics has to respect relativistic precepts. In cases where these speeds are equal in magnitude but of different algebraical signs, we have the special relativity with its Lorentz transformations and, consequently, with everything that follows logically from this observation.

In order to realize where, from a physical point of view, the importance of the groupal transformation resides, let us first notice the form of the transformation matrix (7.19) in this last special case. It is:

$$\begin{pmatrix} \cosh(\lambda\varphi) & c\sinh(\lambda\varphi) \\ (1/c)\sinh(\lambda\varphi) & \cosh(\lambda\varphi) \end{pmatrix}, \quad \begin{matrix} c_1 = -c_2 \equiv c, \\ \lambda_1 = -\lambda_2 \equiv \lambda. \end{matrix} \tag{7.23}$$

This is clearly an usual Lorentz transformation which, however, admits a physical interpretation based on the homography (7.6), *involved in the objective duality wave-corpuscle* from the initial theory of de Broglie. In order to make this fact more obvious and, in fact, more explicit, let us take notice that such a duality is expressed in (7.6) by the linear-rational action of the 2×2 matrix

$$\mathbf{Q} \equiv \begin{pmatrix} 0 & c^2 \\ 1 & 0 \end{pmatrix}, \tag{7.24}$$

which is an involution: its square is proportional to identity matrix. The set of *Cayley transforms* generated by this involution is a two-parameter family of matrices, which in fact represents a two-parameter family of Lorentz transformations. Indeed, we have

$$a\mathbf{1} + b\mathbf{Q} \equiv \begin{pmatrix} a & bc^2 \\ b & a \end{pmatrix} \tag{7.25}$$

so that the Cayley transform is the matrix

$$\begin{aligned} \mathbf{L} &\equiv (a\mathbf{1} + b\mathbf{Q}) \cdot (a\mathbf{1} - b\mathbf{Q})^{-1} \\ &= \frac{1}{a^2 - b^2c^2} \begin{pmatrix} a^2 + b^2c^2 & 2abc^2 \\ 2ab & a^2 + b^2c^2 \end{pmatrix} \end{aligned} \tag{7.26}$$

which represents a Lorentz transformation. This fact becomes more obvious if we use the classical notation

$$\frac{v}{c} \equiv \tanh\xi = \frac{2abc}{a^2 + b^2c^2}. \tag{7.27}$$

Taking this into equation (7.26) gives the Lorentz transformation in the form (7.23):

$$L = \begin{pmatrix} \cosh\xi & c\sinh\xi \\ (1/c)\sinh\xi & \cosh\xi \end{pmatrix}. \tag{7.28}$$

Indeed, we have here the transformation (7.23), in which, however, the parameter φ acquires a definition that on occasion can be related to physics. We can assign an invariant velocity, for instance, using the 'electromagnetic intensities', by the ratio between the 'Poynting vector' and the 'energy of electromagnetic field', in a 'Poincaré interpretation', as it were (see Chapter 1):

$$\boldsymbol{\beta} = \frac{\sqrt{\lambda\mu}(\mathbf{e} \times \mathbf{b})}{(1/2)(\lambda\mathbf{e}^2 + \mu\mathbf{b}^2)}. \tag{7.29}$$

This means that we assign a length to the vector (7.29) by the formula

$$\beta \equiv \frac{v}{c} = \frac{2ebc}{e^2 + c^2b^2} \tag{7.30}$$

then the direction of this vector is the general normal to the plane containing the vectors $\mathbf{e}$ and $\mathbf{h}$. For the rest, these vectors can be anything having some 'electromagnetic' properties, like, for instance the gauge vectors of Zenaida Uy [see Chapter 4, equations (4.53) ff.]. We can even identify this vector, for an important instance, with a vector having as magnitude the limit 'initial velocity' from the Kepler problem: $v \equiv k/\dot{a}$. In this case the components of the limit *initial velocity* from the classical Kepler problem, taken *as a three-dimensional vector*, are oriented along the normal direction to the Kepler orbit determined by this initial velocity, *i.e.* a vector of the form

$$\beta_1 = \frac{k}{c\dot{a}} \sin\theta \cos\varphi, \quad \beta_2 = \frac{k}{c\dot{a}} \sin\theta \sin\varphi, \quad \beta_3 = \frac{k}{c\dot{a}} \cos\theta, \tag{7.31}$$

where φ is the direction of the perihelion of the revolving material point. However, the implications of this association, conditioned by the objective duality wave-particle — and therefore by the representation of the motion by propagation or reciprocally, of the propagation by motion — are way deeper, even fundamental we should say, and will be revealed along our discussion, in a pertinent moment for its suitable understanding of its message.

7.3 Vladimir Boltyanskii: A Way of Describing the Gravitation

It is now time to draw some 'methodological' conclusions, if we may say so, from the display above, in order to apply them in the case of differentials. In the groupal presentation of Victor Lalan, the Lorentz geometry, as expressed, specifically, by a quadratic metric form, is obtained only as a particular case where the two speeds that describe the light — therefore, from the most general perspective of de Broglie association wave-particle, a special propagation! — are equal in magnitude and opposite as sign. The problem occurs, however, if in the cases where the two propagation speeds are different, the geometry can still be Lorentzian, and under what condition this geometry preserves the quadratic character of the metric. Vladimir Grigorevich Boltyanskii gives an answer to this question, using, indeed, the linear transformations on the differentials of coordinates and time — which is the essential feature of the Lorentz transformation — not the coordinates themselves (Boltyanskii, 1974). The importance of the works of Boltyanskii on this subject rests upon the fact that he makes out of the problem of gravitation a *problem of control* (Boltyanskii, 1981), whereby the control is exerted through the intermediary of a wind, as it were, specifically, a *Zermelo wind* if it is to take into consideration the physical jargon used in such instances for a long while [see Gibbons and Warnick, 2011) for details on the concept]. Besides, the most important feature of Boltyanskii's approach, is that the motion to which this control is referring is generated by a 'gravitational level' formally calculated by the sum of the two limit speeds from the Lalan's case. Let us, therefore, briefly render the development of essentials of the idea of Vladimir Boltyanskii.

To start with, we write the transformation of the differentials in the form of a linear transformation

$$dx' = Adx + Bdt, \quad dt' = Cdx + Ddt \tag{7.32}$$

and this can be taken as a homographic transformation between velocities:

$$\frac{dx'}{dt'} = \frac{A(dx/dt) + B}{C(dx/dt) + D}. \tag{7.33}$$

In this instance of the transformation, we have to deal with a *homography between the magnitudes of velocities*, which generalizes the de Broglie's homography between phase and group velocities. Boltyanskii then takes notice of the fact that such a homography is well defined by the condition of existence of two speeds invariant by homography and the condition that a classical material point maintains its identity at rest, *i.e.* at zero velocity, and at any arbitrary speed, v say. Indeed, in these conditions we can write the system of linear algebraical equations

$$c_1 = \frac{Ac_1 + B}{Cc_1 + D}, \quad c_2 = \frac{Ac_2 + B}{Cc_2 + D}, \quad v = \frac{B}{D}$$

where c_1 and c_2 are the two invariant speeds. For these, Boltyanskii uses in fact, the symbols ρ and σ; however we continue to further use the Lalan's notations, with the hope that the invariant speeds, like those of light itself, are properties transcending the scale: they are the same in the transfinite, finite and infrafinite scale. The system above completely determines the homography (7.33), and thus, up to an arbitrary factor, the linear transformation (7.32), is written by Boltyanskii in the form:

$$\begin{aligned} dt' &= D\left(dt - \frac{v}{c_1 c_2} dx\right), \\ dx' &= D\left\{vdt + \left[1 - v\left(\frac{1}{c_1} + \frac{1}{c_2}\right)\right] dx\right\} \end{aligned} \tag{7.34}$$

where D is, this time, that arbitrary factor. By a direct calculation we get the invariance of a *quadratic metric* — not of the *general form* metric (7.20)! — under condition that the transformation has a unit determinant. Therefore the metric is quadratic from the very beginning, and it can be written as:

$$(dx)^2 - (c_1 + c_2)(dx)(dt) + c_1 c_2 (dt)^2 \equiv (dx - f\,dt)^2 - c^2 (dt)^2, \tag{7.35}$$

where, this time in Boltyanskii's notations, we put: $f \equiv (c_1 + c_2)/2$, and $c \equiv (c_1 - c_2)/2$. Among others, Boltyanskii obtains the general rule of composition of velocities given by Victor Lalan, and reproduced by us here in equation (7.22), but with important observation that any two reference frames have relative velocities satisfying the condition

$$v_1^{-1} + v_2^{-1} = c_1^{-1} + c_2^{-1}. \tag{7.36}$$

Therefore, if c_1 and c_2 are absolute constants, the harmonic mean of the relative velocities must be a constant, well defined by these velocities. Seems just normal: if the invariant velocities obtained based on the transformation of the differentials are those from Lalan's theory — which refers to the coordinates themselves, not to their differentials — then these velocities should nevertheless satisfy to some further restrictions. Noticing that $v_1 + v_2 = 0$ is the usual case of the material point in vacuum, equation (7.36) can be read as saying that even the invariant velocities refer to such points. In this case it tells us much more about the cause which might determine the breaking of this classical vectorial symmetry.

Indeed, noticing that for $f = 0$ the metric (7.35) reduces to usual Lorentz metric, Boltyanskii has then an idea which seems quite natural, namely that the condition $c_1 + c_2 \neq 0$ *would be due to the gravitational field*, so that this should be the correct way of introducing this field starting directly from the special theory of relativity. Thus, as we said, Vladimir Boltyanskii translates the regular problem of motion of a classical material point into a problem of control. Quoting:

> The velocity of any motion is restricted in every reference frame X_α by the inequalities like $\rho \leq v_\alpha \leq \sigma$ (the limit values correspond to light motions, the intermediary ones — to ≪ gravitational motions ≫), *i.e.* $-c \leq v_\alpha - f \leq c$. If we introduce the parameter $u = v_\alpha - f = \dot{x}_\alpha - f$, we get the relations
>
> $$\dot{x}_\alpha = f + u, \quad -c \leq |u| \leq c. \tag{5}$$
>
> Thus, the consideration of all possible motions in the frame X_α leads to the *controllable object* (5). The light motions are optimal (by the rapidity of action) trajectories of this object. Notice that passing from a frame X_α to another frame X_β the number $f = (\sigma + \rho)/2$ does not change, *i.e.* equation (5) is *invariant* with

respect to passing from a frame to another, so that this invariance appears as a consequence of the relativistic postulate (ρ and σ are the same in all reference frames) [Boltyanskii, 1979; *our translation, original Italics and captions*]

The theory can be extended to the case of three-dimensional velocity vectors (Boltyanskii, 1995), which allows for a generalization of the metric (7.35) to a stationary metric of the spacetime, in the so-called "$3 + 1$ form", that proves necessary to both the theory of the Ernst's complex potential and the membrane model of the black holes (Mazilu and Porumbreanu, 2018). This fact shows that as long as the motions are represented, for instance, in the manner of Wolfgang Rindler, by lines on a one-sheeted hyperboloid (Rindler, 1960), and the matter can be represented by matrices offering transformations between 'light motions' [see equation (7.30) and the following ones], the logical proof of the existence of two such motions must be sought for in the theory of general relativity, and in all its physical consequences. The great merit of the theory of Boltyanskii rests, therefore, with the accomplishment of this possibility, which, from this perspective, offers us the following explanation.

Now, since we have involved here the name of Wolfgang Rindler, let us take this opportunity for another important notice, related to his name. The 'Schrödinger's hyperboloid', to which Rindler makes reference in the work just cited right above, and which describes the cosmological universe of de Sitter (see Chapter 1), represents *the topology of radial directions*. Only referred to such a direction the two speeds of the light must be understood in the anisotropic theory of Boltyanskii. For a proper understanding, to which we shall have to come back later, let us take the concrete case of the Earth as a spatially extended particle: the radial motion implies a center of this particle, from which the radial direction extends outward. Along this direction, the matter of the central nucleus of the Earth transforms itself by physical rearrangement; further on, according to the relativistic cosmology the universe expands. Therefore, there should be, indeed, two fundamental radial speeds: one of them describes the physical arrangement of the central matter, therefore the dimension adjustment of the central nucleus of the

Earth, the other describing the external expansion of the universe. The Boltyanskii's case refers to an extreme possibility, whereby the matter is vacuum, and disappears with the speed with which the vacuum extends, *i.e.* the speed of light. This is, in fact, *the whole drive of the special relativity*: the matter of vacuum transforms into physical structure at the rate of light propagation in space — the Huygens principle. Generally however, the situation should be that defined by the opinion of Marcel Brillouin, theoretically validated, as shown before, by Victor Lalan, according to which there are indeed two speeds characterizing the ether, therefore the matter in general. According to Boltyanskii the sum of these represents the gravitation, and this would be, in fact, *the whole drive of the general relativity*! We shall came back to this conclusion in the closure of the present work.

It would be here the case to add two important things related to the ideas of Boltyanskii. First, comes the fact that it is not absolutely true that the universe is in expansion: occasionally it was found that it is contracting! We would say that Boltyanskii's theory, and especially its mathematical product and basis, *i.e.* the equation (7.36), shows us the manner in which the universe depends on the central matter in the vicinity of which the observer that considers the universe is located. As far as we are concerned we profess the idea, which we even tried to document (Mazilu and Porumbreanu, 2011), that *only the quantization is, in fact, a law in the world we inhabit.* Reckoned from this point of view, the red shifting, which validates that expansion of the universe, might just as well be indication that the atoms producing the light are contracting (Sambursky, 1937, 1938). This too is a conclusion in perfect accordance with observations, but especially with the idea of adiabatic invariance lying at the foundation of the modern theoretical physics. The chapters to follow will deepen and justify theoretically this conclusion.

Secondly, mention should be first made here, of an important approach, involving directly some *Finsler structure of the spacetime* (Bogoslovsky, 2013). Long time ago, George Yu. Bogoslovsky took notice of the fact that the invariance property from equation (7.20) calls for what we deemed here as an explicit recognizance of the fact that Einstein's construction of special relativity leaves something to be desired (Bogoslovsky, 1973). As we have shown here, that 'something' has been undertaken by Louis de Broglie in the objective duality wave-corpuscle, and this is the main point of Bogoslovsky's intervention. Quoting:

> The basis of the proposed special relativistic theory of *anisotropic space-time* is given by the idea that in nature exists another law, more general than the Lorentz transformations connecting coordinates of e- vents in different inertial reference frames, *whose noticeable difference from the Lorentz transformations occurs at speeds extremely close to the speed of light.* The question arises, how to approach the construction of a theory, generalizing a special Einstein's theory of relativity along this line. First of all, it is clear that the new law of transformation of the coordinates of the event must obey a group property and be linear if we want to remain within a homogeneous space–time. Next, we need to require that the new transformations leave Maxwell equations invariant, and *according to Michelson's experiments would assure for events related with photon propagation, the vanishing of invariant quantity* $(\Delta x_0)^2 - (\Delta x)^2$. It is well known that all these requirements are satisfied by ordinary transformations from the inhomogeneous Lorentz group, which with mathematical points of view are movements in the event space endowed with a pseudo-Euclidean metric; but, as we show now, *there is another group* of event coordinate transformations that satisfies the above requirements and realizes movements (in the sense of preserving the metric) in another metric space of events. [(Bogoslovsky, 1973), *our emphasis*]

Notice the careful observation that Michelson's experiment is not referring to electrodynamics, but to the space of events — the space–time — and that, in completing the Lorentz group, we should have always in mind if the propagation is referring to photons. It is this property that determined, to Bogoslovsky, the choice of metric, still as a quadratic form, however, affected by a factor deriving from the general invariant (7.20), and representing the properties of anisotropy of the space of events (the space–time). The metric thus chosen can be regarded as a general *Finsler metric*, of the type occurring with the 'Zermelo wind', and covering also the problem of control of Boltyanskii's approach. However, we shall follow here a middle line, as it were, imposed by the astrophysical observations, whereby the forms of the matter can be perceived by its physical properties. Among these, of course, it is the important case of the electrical charge of the matter.

7.4 A View to Celestial Matter

In the earliest attempts to describe the spiral galaxies classically, based on the data of their internal motions, the use of Newtonian theory led to the result that the spiral arms as paths of motion of stars would not be an option firmly sustained by the experimental data at hand (Jeans, 1923). This could be counted, we suppose, as a first sign of divorce between the Newtonian theory and the cosmic data. Nevertheless, the data could sustain the idea that spiral arms are some envelopes of Kepler orbits of the stars around the center of the galaxy (Brown, 1925). Useless to say, none of these ideas had the chance to be clearly assessed by experimental data. In view of this, the spiral arms are most commonly described today as *density waves* in the spirit of the works of Wilczynski cited above (Toomre, 1977).

However, speaking of Kepler orbits and their envelopes, Newton himself had an important point on this issue, related to the so-called problem of 'revolving orbits'. In his system of natural philosophy, Newton was able to describe the spiral as an orbit determined by a central force with magnitude depending exclusively on distance, and being inversely proportional with the third power of distance (*Principia*, Book I, Proposition IX). Then he proved that this force has another important property — that of being a *force of transition* between two identical Kepler orbits rotated with respect to each other (*Principia*, Book I, Propositions XLIII and XLIV). Thus he succeeded in describing the revolving orbits by a central force with magnitude depending exclusively on distance. The magnitude of force is here a linear combination of the gravitation force proper and the transition force. It is well known that this description is not quite accurate, having to be corrected by the general relativity, which turned the attention of physicists and astronomers in an entirely different theoretical direction. However, the Newtonian problem of revolving orbits has apparently something of eternal ingenuity, so that it started attracting again the attention of some astrophysicists (Lynden-Bell, 2006).

The ingenuity of the Newtonian description of the problem of revolving orbits might come to light from another fact, this time related to quantum mechanics: it can classically explain the quantum jumps. Indeed, the process of motion of a particle along a revolving orbit can be decomposed in an infinitesimal motion along an ellipse, followed by an infinitesimal motion along a logarithmic spiral, and then by an infinitesimal motion along the next ellipse, etc. In other words the jump between Keplerian orbits is realized by a motion along the logarithmic spiral. In this case, the whole process of quantization can be described in terms of the *Hannay angle* related to the family of rotating ellipses (Mazilu *et al.*, 2014).

Fact is that, in case we accept the idea of "enveloping", this problem of revolving orbits is entirely analogous to the problem of structure of spiral arms of galaxies, at least from geometrical point of view. One, as well as the other, is referring to *a family of Kepler orbits.* As the motions they indicate seem to be flat, at least by their tendency, one is led to the conclusion that, if they are described by forces in the Newtonian manner, then these forces *have to be central.* However, the Keplerian motion means more than the flatness of the motion: *it also requires the area law to be respected.* This issue, however, cannot be properly assessed by the astronomical observations of galaxies, mostly because we do not have the chance to see much of their internal motion during our lifetime.

Nevertheless, with the advent of the spectroscopic technologies, the possibilities were open for measurements of internal velocities of the nebulae, at least for the cases where they are not perpendicular to the line of sight. Such results by van Maanen were used by Jeans and Brown in their theoretical speculations mentioned above. They seem to contradict the Newtonian wisdom. In time, Newton's hypothesis regarding the relation between mass and gravitation, which is quite arbitrary by itself (Poincaré, 1897), came to be directly challenged. It was thus realized that there is a huge discrepancy between the "shining matter" and "opaque matter", as Newton used to call the kinds of matter we perceive in space (in the letters to Bishop Bentley). The specific issue is that, while the "shining matter" reveals certain speeds of revolution in the spirals, these do not seem to agree with the speeds calculated by the Newtonian dynamics' recipes, thus indicating that there may be more "opaque matter" located in the cosmic formations than the "shining matter" lets us see (Sofue and Rubin, 2001). Thus, the whole Newtonian theory came under attack from this angle. Of course, there are reactions to such conclusions (Bekenstein and Milgrom, 1984), but none of them seem to properly address the problem of mass hypothesis of Newton, which in our opinion is essential in case we ever need to build solidly upon previous achievements of science. Moreover, the mentioned reactions do not even seem addressed to defending or criticizing the Newtonian natural philosophy *per se*, but mainly to show that the idea of dark matter is not really necessary: there are a lot of other possibilities opened to the physical speculation. So the dark mass, like the mass itself to Newton, is out of critical question. It seems to be taken merely a kind of option due to the infinite possibilities open to our spirit.

In our opinion, a big problem here is if the Newtonian spirit is indeed respected in inferences from spectroscopic measurements of the internal velocities of spiral galaxies. Only after this issue was decided can one go to deeper problems, depending, of course, on the result of decision. Having this in mind as a general program, we aim to first decide if the flat matter formations, which the heavens present to us, do have something in common. In view of the universality of Newtonian gravitation they should, indeed, have. Specifically, the task of the present work will be twofold: (1) to decide if the flat cosmic shapes *we see* in the sky are all of the same nature. The "same nature" will be specified in due time, as we proceed. The partial answer, involving only the geometrical point of view, is affirmative. Then comes the second task: (2) to determine the position of *the area law* in connection with this kind of data of our knowledge. It will be shown that, with respect to the area law, the shapes we see in heavens are divided in two groups: ones are obeying the area law, the others are not obeying it. The first group is related to evolving Kepler orbits. However, the evolution leads by no means to spirals, but to formations like the Saturn rings or the asteroid belt. The second group, not respecting the area law, comprises the spirals, general parabolic motions (like the ballistic motions), and the comets' structure.

We apply here the Newtonian philosophy in its initial spirit. First, recall that Newton had at his disposal Kepler's synthesis, with clear (*we don't say precise!*) quantitative data. Therefore a synthesis is the first thing to do. We have precise data on the very same problem, that seem to baffle us on occasions. Then, we have clear (*again, we don't say precise!*) data on the shape of our planetary system, galaxies, and structure of incoherent cosmic matter. A unitary geometrical synthesis of these is missing from the current view of astrophysicists, and this seems to be reflected in the whole science today. Thus, for instance, if from astrometric measurements we can infer a geometrical shape, this one does not come directly with its physical explanation. However, it never came with that physical explanation, even to Newton: the synthesis of geometrical shape, like that of the orbits of Kepler, preceded the physical explanation, in order that a physical theory may be assessed. Specifically, Newton had first the geometry well described, and then, on that very base, he proceeded at the description of the forces. Let's therefore describe the geometry of flat heavenly formations of matter. This presentation follows one of our previous works (Mazilu *et al.*, 2014).

Just about the same time with the works of Marcel Brillouin, two other works of Edwin Bidwell Wilson stand again witness to the fact that Newtonian spirit was not completely lost into the avalanche of the new facts of positive knowledge from the beginning of the previous century (Wilson, 1919, 1924). Obviously, there are many more such works on that side of our spiritual movement, but we chose Wilson's works because they have a direct message for the triumphant quantization of today, a message that might explain the difference between the 'shining matter' and the 'dark matter', *i.e.* between what we see in the sky, and what we don't see. In order to make our point, we follow closely the second of Wilson's works cited above. This work is explicitly aligned to the Newtonian natural philosophy, not only in that Wilson lists the facts that should lead to the expression of force responsible for them, but he is even critical about what should be considered as fact, and what should be taken as a product of imagination. In other words, Wilson is explicitly aligned to Newton's celebrated adage *hypotheses non fingo*, with the proviso that, if we cannot stop imagining, we have, at least, to be critical.

Now, the Wilson's list of facts to be considered in a classical treatment of the planetary atom is

$$mr\omega^2 = -F = dV/dr, \tag{i}$$

$$mr^2\omega = n\,\hbar, \tag{ii}$$

$$E_2 - E_1 = 2\pi\hbar\nu, \tag{iii}$$

$$\nu = N\left(\frac{1}{n_1^2} - \frac{1}{n_2^2}\right), \tag{iv}$$

$$E = V + mr^2\omega^2/2. \tag{v}$$

The captions here are in Wilson's original order. The first condition is what he calls the "force condition", expressing the fact that, *for circular orbits*, the force is just the centripetal force deriving from a potential V, function only of the distance of the electron from the center of the orbit. The electron has mass m, angular velocity ω and revolves on a circular orbit of radius r. The second condition represents "the quantum condition", whereby the *kinetic moment* is quantized. The third condition is Bohr's "frequency condition", where ν is the frequency of light and E the level of total energy of the orbit. The fourth condition is Rydberg's formula, or the "spectral law", where n_1 and n_2 are numbers, and N is a constant (today's Rydberg's constant). The fifth condition is the expression of the *total energy* assignable to a certain orbit — the "energy equation".

Wilson recognizes that from among the five conditions listed above, only the Rydberg's formula (iv) is an experimental fact. Two of them, namely (ii) and (iii), are simply hypotheses, while the other two, *i.e.* (i) and (v), are classical definitions. We are now in position to accept these definitions, as they are part of a theory which is scale invariant in the sense of Berry–Klein theory: there should be nothing wrong with them *at any space scale*, including the microscopic world. The assumptions, however, add constraints to the selection of possible forces responsible for the structure of the planetary atom. Wilson inserts, as a last observation, the comment that the theory might not mean too much, but: "There is some advantage in replacing the hypothetical Coulomb law by an experimental fact". Truth be told, all the advantage should be there! As it turns out, the way in which the only experimental fact — *Rydberg's formula* — enters our considerations about the force is not quite so clean as it should be, but it helps us, nevertheless, in discerning what the 'shining matter' means — a condition, as we have seen, inexplicable to Newton. In other words, if the fact itself does not enter the theory quite as pure as it should — as those which Newton had at his disposal, for instance — permeated by Bohr's hypotheses, it adds positively to the classical natural philosophy.

Wilson started his analysis by trying to obtain an equation for *the potential.* First he noticed that, when combining the hypothesis (iii) with the experimental condition (iv), a certain "conservation law" emerges, in the form

$$E_1 + \frac{2\pi\hbar N}{n_1^2} = E_2 + \frac{2\pi\hbar N}{n_2^2}.$$

When combining this further with the definition (v), and using the fact that the potential energy is defined up to an additive constant, we end up with the equation of the form:

$$V + \frac{mr^2\omega^2}{2} + \frac{2\pi\hbar N}{n^2} = 0. \tag{7.37}$$

This equation is transformed by Wilson into a Clairaut-type equation, using the definition (i) and the quantization hypothesis (ii). It is

$$V = uV' + \frac{A}{V'}, \quad ur^2 = 1, \quad A \equiv \frac{\pi N\hbar^3}{m}. \tag{7.38}$$

The general solution of this equation is function of a constant representing V:

$$V = Cu + \frac{A}{C}. \tag{7.39}$$

Using the second of equations (7.38), this gives a force going inversely with the third power of distance:

$$F = \frac{2C}{r^3}. \tag{7.40}$$

The constant C can be calculated using, in the reverse now — on the belief that we are allowed to do it — the definition (i) and the assumption (ii), so that the final result is

$$F(r) = \frac{\hbar^2}{m}\frac{n^2}{r^3}. \tag{7.41}$$

We therefore get the important result that the experimental constraint (iv) combined with classical dynamics and the two Bohr hypotheses, *lead to the quantization of the magnitude of force.* The force is therefore quantized — not the orbit! However, this quantized force is not the Coulomb force, that dictates the transcendence of the planetary model, but a force inversely proportional with the cube of the distance between the moving material point and the force center.

Nevertheless, the force from equation (7.40) is not the only solution of the Clairaut's equation (7.38); there is also the so-called *singular solution*, obtained by eliminating the constant C between solution (7.39) and its derivative with respect to C, when this derivative is zero:

$$V = cU + \frac{A}{C} \quad \textit{and} \quad 0 = u - \frac{A}{C^2} \quad \Rightarrow \quad V(r) = 2\frac{\sqrt{A}}{r}. \tag{7.42}$$

Obviously, this is the potential giving Newtonian inverse-square forces. As a matter of fact, it was to be expected, due to the scale invariance of the classical equations of motion for the corresponding

Kepler problem. This force will therefore include the Planck's constant in its final expression:

$$F(r) = \frac{2\sqrt{A}}{r^2}, \tag{7.43}$$

where A is the constant defined in equation (7.38). However, its value will not be quantized, like the value of force having magnitude inversely proportional with the third power of distance.

The force from equation (7.41) is the one assumed by Irving Langmuir in his theory of *static atoms.* According to the static theory:

> If *in addition to the Coulomb forces* between charged particles we assume the existence of another force (quantum force) equal to... (7.41)... acting between an electron and a nucleus, we find that a stationary electron is in a stable equilibrium when its distance r from a nucleus is the same as the radius of a circular orbit corresponding to a stationary state in Bohr's theory. [(Langmuir, 1921); *our emphasis*]

At a slightly later stage of development of the old quantum theory, G. C. Evans has shown that the two systems — Bohr's and Langmuir's — cannot be obtained from one another by a global transformation (Evans, 1923), therefore they cannot be equivalent. They are, however, 'locally equivalent', as it were, in that we have to limit our considerations to a *Bohr atom with circular orbits.* This observation brings us to the core of our argument: once we have used the definition (i) and the assumption (ii) for getting the solution (7.40) we may not be allowed to use them again back, in order to 'quantify' the force.

Like any theory trying to settle the classical ideas within the new streak of facts indicating the necessity of concept of quantization, Wilson's argument was apparently not taken seriously, and was, in fact, soon to be forgotten by the mainstream physics. This case, like Marcel Brillouin's also discussed in this chapter — and a host of others, in fact — is one of those historical cases that obligate us to uphold the general thought that the ideas have their own 'objective' dynamics, to wit, a dynamics independent of personalities. For, if anything, Wilson's idea, as we see it, is not destined to oblivion: we have here in equation (7.40), if properly worked out, of course, the reason for that background 'inverse cubic' acceleration imposed by the Berry–Klein theory *via* Ermakov–Pinney equation. In hindsight, again, one can rightfully say that the concept of field is not complete when derived by disregarding its true origin, which origin also involves the concept of quantization, this time of the force. According to this observation, the description of connection between force and acceleration is mediated by *two* theoretical facts: the first one is *quantization*, referring, however, to the force; it is just natural, we should say, in a classical Newtonian framework, to think that if the orbit is quantized, so must be the force that comes with it. The second theoretical fact is the idea of *fluctuations*: the idea of Carlton Frederick has to be taken as seriously as it gets (Frederick, 1976). However, on account of the scale transition, we have some genuine paths to follow. First, there are fluctuations in the velocity, with respect to the 'exact' velocity provided by the vanishing of the Lagrangian (2.35). This 'exact' velocity, provides, *via* a proper transport theory, of course, the intensity of the field necessary to a continuous theory, which is kinematical by its nature. Then, secondly, there are fluctuations in the acceleration proper, with respect to the background provided by the 'inverse cubic acceleration'. These fluctuations are dynamical by their nature and, if properly referred to a force, this one should be quantized. Such observations must be applied, for instance, in case one needs to obtain the Schrödinger equation *via* a stochastic approach. So much the more, they should be applied when establishing a fractal calculus, insofar as this one is, in all fairness, always connected to a scale transition.

In this respect, we got the idea that the human knowledge indicates a universal *holographic universe* connected with the idea of *asymptotic freedom*, as the true kind of freedom one can speak of, in theoretical physics (Mazilu, Agop, and Mercheş, 2019). The story goes simply as follows: any complex function, having the modulus and the phase as functions of position *in space* and the moment in time, is a solution of the Schrödinger equation *for a free particle* of a classical fluid. The amplitude, though, defines a potential in the fluid, that goes with the wave, and this is, as we see it, a genuine holographic property required by the concept of interpretation in the wave mechanics. One can thus construct, for instance, a genuine theory of the brain as a universe, based on cosmological principles (Mazilu, 2019).

Chapter 8

Classical Theory of Affine Surfaces in Space

We uphold the idea that from the point of view of physics, the best approach of the theory of surfaces is the one in which we use the Cartan's ideas, exploiting the properties of differential forms (Mazilu and Porumbreanu, 2018) [see also Mazilu *et al.* (2019, Chapters 7 and 8)]. This approach of the theory in question is, in our opinion, the closest to any physical point of view that might be related to the existence of surfaces. The present chapter is an illustration of this statement, taken to the 'next level' so to speak, of the affine theory of surfaces. We reserved this place for such an illustration, expressly in order to have at our disposal the Barbilian theory of the Riemann spaces associated to families of cubic equations, previously elaborated here [see equation (6.7) in Chapter 6]. As we see it, this item of knowledge is the main piece in the puzzle of interpretation. It is, in fact, its pivot, allowing mainly for what we have called in this work the reverse interpretation: the interpretation of an ensemble of 'detached structures', as a continuum. This reverse interpretation involves, as we said, a further step in the theory of surfaces, namely the *affine theory.* Its clearest concept, expressed in differential forms of course — otherwise a physical theory of surfaces could not exist! — was provided, according to our views, by Cheng and Yau (1986). The following version is a reproduction of that material [see also the classical work of Shiing-Shen Chern and Chuu-Lian Terng, which inspired, in fact, many other valuable works in the spirit of the one just cited (Chern and Terng, 1980)].

In this version of surface geometry, we take three non-collinear vectors, $\mathbf{e}_1, \mathbf{e}_2$, and $\mathbf{e}_3$ defined at the same space location, as a reference frame. Theoretically, these vectors are in a general position with respect to each other; specifically, there is no homogeneous linear relation among them, as from affine point of view we do not know what is a relative direction in space. Thus, the non-collinearity is simply defined here by the property of linear independence. After all, the direction, as described by an angle or two, is a metric concept, and we do not have a metric at our disposal just yet. Assume that $(\mathbf{e}_1, \mathbf{e}_2)$ is a frame in general position on the surface. Looked upon from the environment, the surface here appear as embedded in space, in the very same way it is embedded in the Euclidean space. Thus, each displacement vector of the *generic* point of the surface can still be defined in the form

$$d\mathbf{m} = s^k \mathbf{e}_k \equiv s^\alpha \mathbf{e}_\alpha + s^3 \mathbf{e}_3 \quad \therefore \quad d\mathbf{m} = s^\alpha \mathbf{e}_\alpha; \quad s^3 = 0, \tag{8.1}$$

but this time the two vectors in surface are not necessarily orthonormal. Neither may be the third vector: the most to be asked now is that it should point out of surface, *i.e.* should be *transversal* to surface. Again, the last equation here defines a connection between the position in space and the surface. It is sometimes convenient to write the displacement in the form

$$d\mathbf{m} = s^k \mathbf{e}_k \equiv \langle s|\mathbf{e}\rangle \tag{8.2}$$

with $|\mathbf{e}\rangle$ denoting here the reference frame in space. Then, the equations of compatibility between frame and position, come out from the fact that $d\mathbf{m}$ is an exact differential, to which we add an assumed Frenet–Serret frame evolution:

$$|d\mathbf{e}\rangle = \boldsymbol{\Omega} \cdot |\mathbf{e}\rangle, \tag{8.3}$$

where $\boldsymbol{\Omega}$ is a 3×3 matrix having as elements some differential forms in space coordinates. As we do not have yet a metric here, and the reference frame is arbitrary in this respect, the matrix $\boldsymbol{\Omega}$ is also arbitrary: it has no particular symmetry, like the skew-symmetry imposed in the usual theory of surfaces by orthonormality of the frame. However, *if the volume of reference frame* is constant during evolution — the so-called *equiaffine* case of frame evolution — then the matrix should satisfy the condition of null trace (Flanders, 1965):

$$\mathrm{tr}\,\mathbf{\Omega} \equiv \Omega_k^k = 0. \tag{8.4}$$

The conditions of integrability $d \wedge d\mathbf{m} = \mathbf{0}$, for the displacements given in equation (8.2), lead to

$$d \wedge s^k + \Omega_j^k \wedge s^j = 0, \tag{8.5}$$

while those for the evolution equations (8.3) lead to

$$d \wedge \Omega_j^k + \Omega_i^k \wedge \Omega_j^i = 0. \tag{8.6}$$

Again, we remind here to our reader the summation rule over repeated indices, which we are using in this work. In detail, however, the equations deriving from (8.5) are as follows:

$$\begin{aligned} d \wedge s^1 + \Omega_1^1 \wedge s^1 + \Omega_2^1 \wedge s^2 &= 0, \\ d \wedge s^1 + \Omega_1^2 \wedge s^1 + \Omega_2^2 \wedge s^2 &= 0, \\ \Omega_1^3 \wedge s^1 + \Omega_2^3 \wedge s^2 &= 0, \end{aligned} \tag{8.7}$$

where the condition $s^3 = 0$ was considered in the last one of these equalities, representing the fact that displacement is accomplished in surface. This condition leads to an usual relation, as a consequence of Cartan's Lemma 1 (see Mazilu *et al.*, 2019, Chapter 7):

$$\Omega_\alpha^3 = h_{\alpha\beta} s^\beta, \quad h_{\alpha\beta} = h_{\beta\alpha}. \tag{8.8}$$

Now, assuming that our surface is convex, this equation can be considered as defining a metric, after the manner of definition of second fundamental form in the regular theory of surfaces. This metric is *affinely invariant* if, using (8.8), we write it in the form of what is sometimes termed as *Blaschke metric* of the affine surface (Yau, 1989):

$$(ds)^2 \equiv (1/\sqrt[4]{h})\Omega_\alpha^3 s^\alpha = (1/\sqrt[4]{h}) h_{\alpha\beta} s^\alpha s^\beta, \tag{8.9}$$

where h is the determinant of $\mathbf{h}$. As we said, by comparison with the usual theory of surfaces in Euclidean space, according to equation (8.8) the quadratic form just defined as metric here is actually the equivalent of the second fundamental form of a regular surface. This is why it is usually designated by geometers with II [see also (Chern and Terng, 1980).

The problem now remains, to deal with the other side of the matrix $\mathbf{\Omega}$, namely — the line Ω_3^α — because in the case of affine theory this matrix *is no more skew symmetric* (the frame is not orthonormal). One approach — the regular one in geometry — is to choose, in a first step, Ω_3^3 as an *exact differential, namely.* $d\wedge\Omega_3^3 = 0$, in which case from the corresponding equation (8.6) we must have

$$\Omega_3^1 \wedge \Omega_1^3 + \Omega_3^2 \wedge \Omega_2^3 = 0,$$

$$\begin{pmatrix} \Omega_3^1 \\ \Omega_3^2 \end{pmatrix} = \begin{pmatrix} b^{11} & b^{12} \\ b^{12} & b^{22} \end{pmatrix} \begin{pmatrix} \Omega_1^3 \\ \Omega_2^3 \end{pmatrix}, \tag{8.10}$$

where Cartan's Lemma 1, guaranteeing the existence of a convenient symmetric matrix $\mathbf{b}$, was used. There is not a correspondent of this relation in the regular theory of surfaces, but it is usually taken as the equivalent definition of what we have designated as the curvature vector (Mazilu *et al.*, 2019, Chapter 7). In a word, in the differential affine geometry of surfaces, the curvature vector components are bilinear forms in the entries of *two* conveniently introduced 2×2 matrices. This means that our possibilities to introduce physics here are 'doubled', as it were. Using equation (8.8), we can get from (8.10) a quadratic form as follows:

$$\Omega_3^\alpha = b^{\alpha\nu} h_{\nu\beta} s^\beta \equiv b_\beta^\alpha s^\beta \quad \therefore \quad III \equiv \Omega_3^\alpha \Omega_\alpha^3 = b_{\alpha\beta} s^\alpha s^\beta \tag{8.11}$$

where the notation III seems to be, again, geometers' preference: this is the *third fundamental form* of affine differential theory of surfaces. Recall that in the Euclidean differential geometry of surfaces, the third fundamental form is usually the square of what we designated as the curvature vector.

Now, that we have a metric, we can define a direction via relative angles, and therefore a frame in the tangent plane to the surface, $(\hat{\mathbf{e}}_1, \hat{\mathbf{e}}_2)$ say, orthonormal with respect to this metric, and attach a space vector $\mathbf{n}$ such that the volume $(\hat{\mathbf{e}}_1, \hat{\mathbf{e}}_2, \mathbf{n})$ remains constant. Then, the following theorem due to Harland Flanders can be proved

(Yau, 1989): there is a *unique* space vector $\mathbf{n}$ satisfying the following structural equations

$$d\mathbf{m} = \phi^1\hat{\mathbf{e}}_1 + \phi^2\hat{\mathbf{e}}_2; \quad \begin{aligned} d\hat{\mathbf{e}}_1 &= \Omega_1^1\hat{\mathbf{e}}_1 + \Omega_1^2\hat{\mathbf{e}}_2 + \phi^1\mathbf{n}, \\ d\hat{\mathbf{e}}_2 &= \Omega_2^1\hat{\mathbf{e}}_1 + \Omega_2^2\hat{\mathbf{e}}_2 + \phi^2\mathbf{n}, \\ d\mathbf{n} &= \psi^1\hat{\mathbf{e}}_1 + \psi^2\hat{\mathbf{e}}_2; \quad \Omega_1^1 + \Omega_2^2 = 0. \end{aligned} \tag{8.12}$$

One can see from this system that the vector $\mathbf{n}$ is, formally at least, as close as possible to the normal vector from regular differential theory of surfaces in Euclidean space [cf. Mazilu *et al.*, 2019, equation (7.6)], and is indeed geometrically known as the *affine normal to surface.* Given any *affine* reference frame in space, $(\mathbf{e}_{10}, \mathbf{e}_{20}, \mathbf{e}_{30})$ say, the affine normal to a surface can be written in the following general form:

$$\mathbf{n} = a^1\mathbf{e}_{10} + a^2\mathbf{e}_{20} + h^{1/4}\mathbf{e}_{30}, \tag{8.13}$$

where the auxiliary vector $|a\rangle$, which belongs to surface — more to the point, it describes a physical property defining specifically the surface — and serves for this definition of the normal to surface, is established by the following differential equation (Yau, 1989, Proposition 1.1):

$$\langle a|\mathbf{h}|\phi\rangle + d(h^{1/4}) + h^{1/4}\Omega_3^3 = 0. \tag{8.14}$$

The proof of this theorem, which is only quoted as such in the work of Chi-Ming Yau just cited above, can be found in Flanders (1965, §7). One can see here that, having the metric at our disposal, there are alternatives to choose from: we can either define $|a\rangle$ by choosing Ω_3^3, or define Ω_3^3 by choosing $|a\rangle$. Again, the geometers' preference seems to be the first procedure, and for a good reason at that (Cheng and Yau, 1986). This reason becomes, indeed, quite obvious if we notice that the definition from (8.12) of the affine normal is pending on constraints represented by equation (8.14) which can be rewritten in the form

$$\langle a|\mathbf{h}|\phi\rangle + h^{1/4}\left\{\frac{1}{4}d\ln(h) + \Omega_3^3\right\} = 0$$

and if one chooses Ω_3^3 to be an exact differential — as we did before in a first step — defined by relation

$$\frac{1}{4}d\ln(h) + \Omega_3^3 = 0, \tag{8.15}$$

then, with one further special choice, namely $|a\rangle = |0\rangle$, the affine normal vector can be defined as follows:

$$\mathbf{n} = h^{1/4}\mathbf{e}_{30}. \tag{8.16}$$

Such a choice is therefore particularly attractive in the geometrical exploits of this mathematical theory, for in the cases where $h = const.$ we can write

$$\Omega_3^3 = 0; \quad \mathbf{n} = h^{1/4}\mathbf{e}_{30}. \tag{8.17}$$

and $\mathbf{n}$ can be taken as a *unit vector* normal to surface, as in the classical differential theory of surfaces. This way, the surface itself is equiaffine, and it is concomitantly described in an equiaffine space reference frame.

Notice, however, for the incidental benefit of physics — and, in fact, even for the benefit of affine geometry of surfaces altogether, as we shall see shortly — that while still under the spell of affine invariance condition, but possibly even with $h \neq constant$ — therefore in a non-equiaffine theory of the surface itself — equation (8.15) can offer Ω_3^3 as the differential of a function depending exclusively on external parameters already introduced through the metric. Therefore, the choice of Ω_3^3 as an exact differential can be subjected to the very same physical considerations to which the matrix $\mathbf{h}$ itself is subjected.

The bottom line is that, taking equation (8.14) for guidance of our logic, we still have a few other possibilities of defining the vector $|a\rangle$, depending on the location in the tangent plane of the affine surface. Again, a nonzero vector $|a\rangle$ should be especially attractive from physical point of view, as a way to introduce *physically specific considerations*. Such possibilities are offered, for instance, by the nontrivial choices:

$$\langle a|\mathbf{h}|\phi\rangle = 0 \quad \therefore \quad \frac{a^1}{h_{12}\phi^1 + h_{22}\phi^2} = \frac{-a^2}{h_{11}\phi^1 + h_{12}\phi^2},$$

$$\langle a|\mathbf{h}|\phi\rangle + \frac{1}{4}d\ln h = 0 \quad \therefore \quad \Omega_3^3 = 0. \tag{8.18}$$

In the first case here, the condition (8.15) is also secured, and with it the fact that Ω_3^3 is related to the determinant of the metric tensor, so that it is manifestly an exact differential. The surface itself is equiaffinely described from an intrinsic point of view. However, the reference frame is not equiaffine, insofar as the Ω_3^3 does not vanish. It is only in the second case (8.18) that the theory can be made equiaffine, for both surface and ambient reference frame, and this means that $|a\rangle$ must be so chosen that the bilinear form $\langle a|\mathbf{h}|\phi\rangle$ is an exact differential. We shall return soon to this discussion.

The bottom line here is that the affine theory of surfaces provides an exquisite theoretical tool for physics. If for nothing else, but only for the idea of the volume of the space reference frame in describing the embedding of a surface, this approach would still be extremely valuable. Think of the fact that in the physics of any interpretable theory, the physical systems are described by confining them to given volumes, in order to settle the boundary conditions. Then again, the very physical significant reference frame is defined by a volume, before being defined by directions. The Wien–Lummer cavity serving for the study of the blackbody radiation is such a reference frame, which was extended to the whole universe, based on the background radiation observations. The theoretical basis of this extension from the *finite scale to transfinite scale* is the invariance of the blackbody radiation spectrum to space extension of volume, whose expression is the Wien displacement law [see Chapter 5, equations (5.31)]. The same should be happening with the Einstein elevator, which is modern incarnation of the classical Cartesian reference frame, but unfortunately it is not the current case in physics. Realizing that, and the fact naturally we always need a *physical surface* in order to build our experience — recall the Einstein's 'Earth's crust plays such a dominant role in our daily life' — will, hopefully solve the problem of inertia, in a physically sound manner. Let us, therefore, show what other physical incentives has the affine theory of surfaces in store for us.

8.1 Conditions of Physical Nature on a Surface: The Fubini–Pick Cubic

The 'physically specific considerations' we have in mind here, are algebraically brought to light by the theory of binary cubic forms, as presented before in Chapter 6. The physical situation they characterize concerns, for instance, the interpretation of continuous matter from within the nucleus of the revolving particle of the planetary model — in fact even of the continuous matter within the nucleus of the particle generating the field — by ensembles of Hertz material particles. We have here a surface delimiting the matter *per se* in its incessant change, which can be locally described by the exclusive variation of the curvature parameters. This is a condition which guarantees (Mazilu *et al.*, 2019, Chapter 6) that the variation of a quadratic binary form is due strictly to the variation of some physical parameters, embodied in its coefficients. Such parameters can be related — for instance by a formula like that in [Mazilu *et al.* (2019, equation (6.28)), *loc. cit.*] — to a local field of velocities labeling the Kepler orbits in the flux of Hertz material particles inside the toroidal de Broglie tube containing their instantaneous orbits. Perhaps useless to utter it again but, nevertheless, we still emphasize it, the description of this situation is of primary importance for the theoretical physics.

In order to carry out a theoretical implementation according to previous lines, we need to bring here another concept from the affine differential theory of surfaces. Indeed, the definitions in (8.18) can be tied up to an important algebraical concept that we are set out to exploit here — again, in the interest of physics and this is the concept of *algebraic apolarity*. The affine differential theory of surfaces makes the fact obvious that the apolarity is not quite as direct as presented in the classical theory of surfaces, but involves also a cubic invariant — the so-called *Fubini–Pick cubic form* [for a clear but extended account of the Fubini–Pick tensor and related concepts, one can consult the work of Shirokov and Shirokov (1959)]. This form can be introduced in the manner that follows, involving Cartan's calculus (Chern and Terng, 1980). By exterior differentiating equation (8.8) — but still within the choice from equation (8.18) — and using the corresponding equation from (8.6), we get

$$Dh_{\alpha\beta} \wedge \phi^{\beta} = 0; \quad Dh_{\alpha\beta} \equiv dh_{\alpha\beta} - h_{\alpha\mu}\Omega^{\mu}_{\beta} - \Omega^{\nu}_{\alpha}h_{\nu\beta}. \tag{8.19}$$

Using again Cartan's Lemma 1, we further get

$$Dh_{\alpha\beta} = \Phi_{\alpha\beta\gamma}\phi^{\gamma}. \tag{8.20}$$

The newly introduced third-order tensor $\mathbf{\Phi}$ — *the Fubini–Pick tensor* — is symmetric in all its three indices, as required by its very definition. It satisfies some special conditions imposed by the idea of metric volume. These conditions come easier to light if we work formally. Namely, with (8.19), we can write equation (8.20) in the form:

$$d\mathbf{h} = \mathbf{h} \cdot \mathbf{\Omega} + \mathbf{\Omega} \cdot \mathbf{h} + \mathbf{F}, \quad F_{\alpha\beta} \equiv \Phi_{\alpha\beta\gamma}\phi^{\gamma}.$$

Left multiplying here by $\mathbf{h}^{-1}$ and then taking the trace of resulting matrix leads us to

$$\mathrm{tr}(\mathbf{h}^{-1} \cdot d\mathbf{h}) = \mathrm{tr}(\mathbf{h}^{-1} \cdot \mathbf{F}). \tag{8.21}$$

Here we have used the last property in (8.12), *i.e.* the equiaffine surface: $\Omega^{\alpha}_{\alpha} = 0$, and some routine properties of the operation of trace. Equation (8.21) is the condition of apolarity sought for, and we now explain its meaning in some detail, in order to make it more palatable, for instance, by comparison with our previous developments (Mazilu *et al.*, 2019). The space frame evolution is not equiaffine, but the departure from this condition is due to the physics of surface embodied in the Fubini–Pick tensor. Indeed, using (8.21), the condition (8.15) becomes

$$\Omega_3^3 + \frac{1}{4}\,\mathrm{tr}(\mathbf{h}^{-1}\cdot\mathbf{F}) = 0 \tag{8.22}$$

and, therefore, in this case everything depends on the symmetric tensor **F**, obtained by contraction of the Fubini–Pick third-order tensor with the fundamental displacements. For instance, the condition (8.22) is satisfied if there is an affine normal to the surface — according to the definition from (8.12) — which, expressed by (8.13) in a certain affine reference frame, imposes the conditions (8.14) on this frame. Now, if we choose the ancillary vector $|a\rangle$ *according to the first one of conditions from* (8.18), then (8.22) represents a mandatory condition of the theory. Let us write this last condition in detail. It looks like:

$$\Omega_3^3 + \frac{1}{4}F_\alpha\phi^\alpha = 0, \qquad \begin{aligned} F_1 &\equiv h^{11}\Phi_{111} + 2h^{12}\Phi_{121} + h^{22}\Phi_{221}, \\ F_2 &\equiv h^{11}\Phi_{112} + 2h^{12}\Phi_{122} + h^{22}\Phi_{222}. \end{aligned} \tag{8.23}$$

Therefore, the classical geometers' choice $\Omega_3^3 = 0$, securing the definition of the equiaffine evolution of the space reference frame connected to surface as in (8.12), can be satisfied in more general terms, even without such a condition, if between the metric tensor **h** and the Fubini–Pick tensor $\mathbf{\Phi}$ there is some physically accountable connection.

In order to uncover that connection, and how specifically is it physically accountable, let us consider the Fubini–Pick cubic form associated with the tensor $\mathbf{\Phi}$. It is a binary cubic in the binary variable (ϕ^1, ϕ^2), which, expressed in the general form, looks like

$$\Phi \equiv \Phi_{\alpha\nu\sigma}\phi^\alpha\phi^\nu\phi^\sigma. \tag{8.24}$$

An important property of this binary cubic form is that if it vanishes for a certain affine surface, then that surface is locally a quadric (Cheng and Yau, 1986). We are inclined to consider this property as coming extremely handy, not only in experimental, but mainly in highly theoretical physical problems. Suffice it to recall only that the first modern physical theory of light in Fresnel's take, is intimately connected with the theory of quadrics in two significant ways: on one hand, from the purely geometrical point of view of representing a physical continuum, and, on the other hand, from a physical point of view, whereby the quadric is defined by the elastic properties of that continuum: the ether sustaining light (Fresnel, 1827). However, there is more to it, from a modern physical point of view: the most general shape of an enclosure allowing for the conclusions of Berry–Klein scale transition theory [see Mazilu *et al.*, 2019, Chapter 9, equation (9.34)] is a *quadric* (Klein, 1984). Groupal reasons allow us to say even more: as the theory of quadrics stays at the foundations of the so-called *theory of superquadrics* (Kindlmann, 2004), and as these last geometrical things are able to closely visualize any kinds of physical magnitudes *in a general theory of reverse interpretation*, as it were, one can say that the concept of affine surface is the right path in constructing any scale transient mathematical theory in physics. The Fubini–Pick tensor is here the driving power of the whole physics.

Fact is that, if we take into consideration the symmetry properties of the tensor $\mathbf{\Phi}$, then it has just four independent components, as follows:

$$\begin{aligned}\Phi_{111} &\equiv a_0, \quad \Phi_{112} = \Phi_{121} = \Phi_{211} \equiv a_1,\\ \Phi_{221} &= \Phi_{122} = \Phi_{212} \equiv a_2, \quad \Phi_{333} \equiv a_3.\end{aligned} \tag{8.25}$$

With these notations, the Fubini–Pick cubic can be written in the typical binomial form (see Burnside and Panton (1960) for the theory of binary quantics, and the apolarity of quantics in general)

$$\Phi \equiv a_0(\phi^1)^3 + 3a_1(\phi^1)^2\phi^2 + 3a_2\phi^1(\phi^2)^2 + a_3(\phi^2)^3. \tag{8.26}$$

If we assume that this cubic is known, which means that its coefficients are known, then the two components of the covector defined in equation (8.23), can be written as

$$\begin{aligned}hF_1 &\equiv h_{22}a_0 - 2h_{12}a_1 + h_{11}a_2,\\ hF_2 &\equiv h_{22}a_1 - 2h_{12}a_2 + h_{11}a_3.\end{aligned} \tag{8.27}$$

If this is a *null covector*, the binary quadratic form representing the affine metric is simultaneously apolar to the binary quadratic forms having the coefficients $(a_0, 2a_1, a_2)$ and $(a_1, 2a_2, a_3)$, respectively.

Practically, in this case, *if we know the Fubini–Pick tensor*, then we have in (8.27), a system of two linear equations with three unknowns — the entries of the affine metric tensor of the surface — which is compatible, and has a simple infinity of solutions that can be expressed by the system

$$\frac{h_{11}}{a_0 a_2 - a_1^2} = \frac{2h_{12}}{a_0 a_3 - a_1 a_2} = \frac{h_{22}}{a_1 a_3 - a_2^2}. \tag{8.28}$$

This purely algebraic result gives the entries of affine metric tensor of surface as being proportional to *the coefficients of the Hessian of Fubini–Pick cubic* [see Chapter 6, equation (6.17)]. For what is physically worth then, this shows that not quite any affine metric tensor should qualify for the condition of constant determinant, necessary to an equiaffine theory of a surface. Summarizing, in the case of affine surfaces we have the following statement.

If we define *the affine normal to surface in an arbitrary reference frame according to an ancillary vector orthogonal to* $|\phi\rangle$ in the metric $\mathbf{h}$:

$$\langle a|\mathbf{h}|\phi\rangle = 0 \quad \therefore \quad \frac{a^1}{h_{12}\phi^1 + h_{22}\phi^2} = \frac{-a^2}{h_{11}\phi^1 + h_{12}\phi^2}$$

the affine metric form of surface is the Hessian of the Fubini–Pick cubic [geometrically this condition is emphasized especially in Shirokov and Shirokov (1959)]. *The space reference frame adapted to such a surface has a constant volume when* $\Omega_3^3 = 0$. If we imagine a space where the parameters (a_0, a_1, a_2, a_3) are taken as coordinates, the condition $\Omega_3^3 = 0$ is then only satisfied on a surface of fourth degree, represented by the determinant of metric tensor of surface written in terms of the entries of the Fubini–Pick tensor [see equation (8.16)]. We shall return in due time to this kind of physical theory.

Up to this point the story is told in a pure mathematical way: one can only say *where* the physics gets really involved, as we actually did a few times. However, there is a very simple observation that firmly establishes the general place of entrance of physics into this mathematical theory. Namely, up to its interpretation defined by

C. G. Darwin, a continuum — a fluid for instance, of the kind we need for developing SRT — needs to be described physically as a continuum, of course. Ever since the times of Cauchy, such a continuum was mathematically described based on matrices or, more specifically, tensors. The values of a matrix quantity are three — the eigenvalues — in three different space directions, and they are the roots of a *cubic equation*, the well-known characteristic equation of the matrix. This characteristic equation can always be taken as a Fubini–Pick cubic, and used *to update the geometry of continua* by an affine differential process, based on the previous line of ideas. This certainly gives a well-established place where the physics can enter our mathematical considerations. Nevertheless, the physics needs a little more.

Fact is, that having thus secured the place of access of the physics into our mathematical theory, we obviously need to see where the geometry stops and where the physics begins, as it were. To this end, notice that if the tensor $\mathbf{F} \equiv \mathbf{\Phi}|\phi\rangle$ defined above vanishes, the surface characterized by such a condition can be described with the so-called *Levi-Civita connection*, for which, by definition, $D\mathbf{h} = \mathbf{0}$. This would indicate a geometry of the *space with no matter*, therefore a geometry of a surface *separating the space from matter*, or *the matter from space*, as the case may occur. This categorization is only made based on the simple fact that a Levi-Civita connection belongs to the classical geometry, which is known as referring exclusively to space. In our context, however, the observation gains a few different nuances depending on *how*, specifically, the tensor $\mathbf{F}$ vanishes. Its vanishing *per se* means a system of three linear equations:

$$a_0\phi^1 + a_1\phi^2 = 0; \quad a_1\phi^1 + a_2\phi^2 = 0; \quad a_2\phi^1 + a_3\phi^2 = 0. \quad (8.29)$$

In writing these equations we have used the identifications from equation (8.25). This system must have a nontrivial solution for the binary variable (ϕ^1, ϕ^2), otherwise the Levi-Civita geometry itself would have no object. What we can algebraically decide from the system (8.29) is, nevertheless, only the ratio of the two components of the binary variable. And even for this much, the essential components of the Fubini–Pick tensor cannot be independent. They must satisfy the conditions

$$\frac{a_1}{a_0} = \frac{a_2}{a_1} = \frac{a_3}{a_2}\left(\equiv -\frac{\phi^1}{\phi^2}\right) \tag{8.30}$$

which have an exquisite meaning from physical point of view. In order to uncover that meaning we have to use the Fubini–Pick *binary cubic.* As we mentioned, if this cubic is vanishing, the surface it describes is a quadric. Now, the binary cubic corresponding to the conditions (8.30) is a perfect cube:

$$a_0(\phi^1 + \lambda\phi^2)^3 \tag{8.31}$$

which means that the surface it describes is, in fact, a sphere. From (8.30), we get the quadratic relations

$$a_0a_2 - a_1^2 = a_0a_3 - a_1a_2 = a_1a_3 - a_2^2 = 0 \tag{8.32}$$

which mean a null Hessian for this Fubini–Pick cubic. Certainly, therefore, this would not mean a proper metric geometry of a surface — in fact, it would not even mean a surface — at least from a metric point of view, according to the above developments. We are therefore compelled to admitting that the Hessian of the Fubini–Pick cubic determines actually a *deformation of the metric of surface,* rather than the whole metric itself.

8.2 General Theory of Equiaffine Reference Frames in Space

The above theory can only be used in the construction of a reference affine surface, based on physical considerations. Using this reference surface, a family of equiaffine frame evolutions connected to surface can be constructed, by extending those physical considerations as follows. If in equation (8.14) we define the ancillary vector $|a\rangle$ such that

$$\langle a|\mathbf{h}|\phi\rangle + d(h^{1/4}) = 0, \tag{8.33}$$

then $\Omega_3^3 = 0$ automatically. In view of this, and the last condition from (8.12), the space affine reference frame is equiaffine itself.

In this case, using (8.21) we have

$$\langle a|\mathbf{h}|\phi\rangle + \frac{1}{4}h^{1/4}\,\mathrm{tr}(\mathbf{h}^{-1}\cdot\mathbf{F}) = 0 \tag{8.34}$$

and with (8.27) this becomes

$$\langle a|\mathbf{h}|\phi\rangle + \frac{1}{4}h^{-3/4}\langle F|\phi\rangle = 0. \tag{8.35}$$

One important situation where this condition is satisfied occurs for the case

$$\langle a|\mathbf{h} + \frac{1}{4}h^{-3/4}\langle F| = \langle 0| \quad \therefore \quad |a\rangle = -\frac{1}{4}h^{-3/4}(\mathbf{h}^{-1}\cdot|F\rangle). \tag{8.36}$$

The last condition here shows that in the case of sheer apolarity — *i.e.* $|F\rangle = |0\rangle$ — the only possible auxiliary vector is the null vector, which, as we have seen, is the geometric preference. This shows that in order to have a possible non-trivial choice, the physical theory, based on geometry but, nevertheless, adding something to that geometry, one needs to choose an auxiliary vector that describes a deformation, to be included in the vector $|F\rangle$. In other words, the vector (8.27) needs to be 'updated', as it were, to

$$\begin{aligned} hF_1 &\equiv h_{22}(a_0 + \delta a_0) - 2h_{12}(a_1 + \delta a_1) + h_{11}(a_2 + \delta a_2), \\ hF_2 &\equiv h_{22}(a_1 + \delta a_1) - 2h_{12}(a_2 + \delta a_2) + h_{11}(a_3 + \delta a_3), \end{aligned} \tag{8.37}$$

where the symbol δ means a variation of the symbol following it. In this case, if the metric tensor of the reference surface is the one given by apolarity conditions, the Fubini–Pick vector (8.37) reduces to

$$\begin{aligned} hF_1 &\equiv h_{22}(\delta a_0) - 2h_{12}(\delta a_1) + h_{11}(\delta a_2), \\ hF_2 &\equiv h_{22}(\delta a_1) - 2h_{12}(\delta a_2) + h_{11}(\delta a_3). \end{aligned} \tag{8.38}$$

To close with a methodological note, these last formulas allow us a theoretical description of one of the most desirable, in physics, connections ever, *i.e.* the connection between a coordinate system and a reference frame which serves in the physical interpretation of these coordinates. To start with, the regularization procedure via three-dimensional analyticity conditions [Chapter 4, equation (4.29)], needs three solutions of a Poisson equations, in order to define the three complex functions that play the part of what we have called here the Appell coordinate system. Such a coordinate system can aptly replace any other coordinate system in physics. To wit, it is, first of all, a coordinate system of classical extraction, inasmuch as it can aptly be connected with a classical dynamics through the Appell equations of motion (4.31). Secondly, it can appropriately generalize the *harmonic coordinate system* — *i.e.* three solutions of the Laplace equations — in the three-dimensional space. Indeed, they satisfy the analyticity conditions in the form given in equation (4.33), which are valid in a material continuum, that might be even the vacuum. And last, but by no means the least of all, this coordinate system can be rationally associated with a reference frame connected to that material continuum, by the theory just presented right above. More importantly, that theory can be realized as a confining theory, via an absolute geometry of the Cayleyan type and, even more importantly, from the topological point of view it copes with the classical planetary model as we presented it in Chapter 1 here.

The mathematical theory was presented by Dan Barbilian almost a century ago (Barbilian, 1938), and is based on the apolarity of the Fubini–Pick cubics presented above. Specifically, we have the following observation: in the coordinates given by the coefficients of a binary cubic as in equation (6.9) here, the set of all cubics having the Hessians apolar with the Hessian of a given cubic forms a locus described by a quadratic form — a quadric. If the given cubic has real roots, then the corresponding quadric thus constructed is a one-sheeted hyperboloid, *topologically* isomorphous with a *torus*, or with a *canal surface*, which can be imagined as a de Broglie capillary tube representing the electron in a planetary model as a ray of Hertz material particles, like the ones described by Joseph Larmor in 1900, as a 'cloud of meteors'. Now, a cubic with real roots can represent a continuum: it can serve in a 'reverse interpretation' of a certain flux, as a second-order tensor, and thus as a reference frame. In view of the affine theory of surfaces, in a cubic with real roots we always have a reference frame that can be modeled as a portion of an affine surface, as described above, in the present chapter. Any Hessian apolar with the Hessian of a cubic with real roots must have real roots too. Therefore, cubic corresponding to such a Hessian must have one real and two complex roots, that *can be taken as a system of general Appell coordinates*.

Therefore, we have the following general conclusion regarding the connection between reference frame and coordinate systems: for any physical reference frame, there is a continuous set of coordinate systems, allowing the 'Appell dynamics' described by equation (4.31), which may be taken as defining the time connected with coordinates. This continuous set of coordinate systems can be geometrically modeled as a Riemannian space, with the metric of Cayleyan type, which can be constructed with respect to the 'absolute of coordinate systems', as it were. This metric is indefinite, of Lorentzian form, given in equation (6.38) (Barbilian, 1938). This seems to us as being the most general connection between the reference frames and the coordinate systems, that can serve for any physical theoretical purpose.

8.3 The Existing Physical Incentives

Perhaps it is time to insist a little more on the subject of cubic equations, which, in our opinion, has an overwhelming importance for the physics of matter. Specifically, if one can talk of an interpretation for a continuum like the matter filling all of the available space — the matter *per se*, as we called it every now and then — this interpretation cannot be mathematically achieved but only in terms of matrices or, more restrictively, in terms of tensors. The vectors come into play only in case we have to do with an interpretation proper, *i.e.* interpretation of matter by ensembles of Hertz material particles having one of the powers of continua. Along this line of reasoning, a comparison between the two 'cases' of interpretation can be conceived, which is facilitated by it a special comparison between a vector and a matrix, be it even a tensor or, in fact, *especially* in the case it is a tensor.

Let us say, just for the sake of argument momentarily, that the tensors we are talking about represent tensions or deformations.

According to current human experience and knowledge, such tensors represent the universal property of matter to turn the result of action of a certain force upon it, *away from the direction of that action.* Indeed, our experience shows that whenever a force acts upon a piece of matter, sideways displacements of matter are to be expected. This effect is always described mathematically by a certain theory of continuum, on the results of which one then comes with some 'tweaking' hypotheses, in order to approach the real situations as best as possible. One of these hypotheses is that, in the theory of continua, we have to work with the eigenvalues of matrices or tensors. These are the three extreme values of the quantity represented by such mathematical objects, occurring along three different directions in the continuum. In cases where the matrix or tensor is symmetric, these directions are reciprocally orthogonal, thus qualifying for a Euclidean reference frame in the very classical sense word. Physically, this reference frame is therefore given by those directions along which the magnitude represented by the tensor in question has extremum values. Let us retain this case for our argument, the extension to more complicated cases being then straightforward.

As we are not interested here in anything else but physics to the extent it is related to natural philosophy, we are bound to keep the case as close as possible to natural philosophical issues. And as a surface in a continuum characterized by the Fubini–Pick cubic cannot be affinely or, in fact, otherwise described but by an interpretation according to the necessities of wave mechanics, such issues often occurred in history. In the case of classical ether — as described by Samuel Earnshaw, for instance, in the fragment we excerpted earlier in this work — they were not recognized as such. However, these issues popped up as soon as the man has recognized that the ether can be the support of light. Recently (see Mazilu and Porumbreanu, 2018) we made a special point from the fact that they ought to be accounted for in the formalism of the first modern physical theory of light (*loc. cit. ante*, Chapter 3; see also Poincaré, 1889). The manner of this account revealed the importance of geometrical theory of surfaces. Again, this concept was necessary to Lorentz for building his model of electric matter, as shown previously in our work.

But there is more to it: in our recent work just cited, we also advanced the idea that the two essential auxiliaries of human philosophical natural reasoning — the *Wien-Lummer hohlraum* on the experimental side, and the *Einstein elevator* on the theoretical speculative side — must be considered as two differentiae of the very same concept, namely that of *physical reference frame.* The archetype of a reference frame is then the very structure of the elementary particle of modern physics, described along the lines initiated by Enrico Fermi in the year 1950 (see Mazilu and Porumbreanu, 2018, Section 17; see also Mazilu *et al.*, 2019, Chapter 6]. Here we have the occasion to extend this very concept of reference frame by recounting two original ideas of Alphonsus Fennelly regarding the interpretation as related to a specific physical structure of elementary particles. Quoting:

> I note also that the *parton model* of hadrons containing an *almost infinite number of almost light-like particles* lends itself easily to *a statistical-fluid description of the hadron* in terms of its constituents. So I conclude that it is useful to examine a fluid description of the hadrons in which the cosmological constant also may have meaning. *One possibility is Gödel's universe...* [(Fennelly, 1974); *our Italics*]

Of course, in view of our general task here, the emphasis in this excerpt should be placed, first and foremost, upon that 'statistical-fluid description' of the constituent particles of nuclear matter. However, the approach to the problem of scale physics urges us to emphasize two very important ideas of Fennelly, that can easily pass unnoticed, as actually they already did. First comes the idea of *interpretation of the nuclear continuum:* it already exists in theoretical physics, in the form of the *parton model*, by an ensemble of partons 'containing an almost infinite of almost light-like particles', as Fennelly would say. It seems to us, indeed, that Richard Feynman's partons (Feynman, 1969) are the modern reincarnation of the old idea of Hertz material particles. Thus, what Fennelly has here is simply an interpretation of an extended physical particle — the hadron — whereby the Hertz material particles are to be taken in the sense of

Louis de Broglie (Mazilu *et al.*, 2019). Secondly, Alphonsus Fennelly has the first ever reference to the idea that *an extended physical particle is actually a universe*, in the sense we take it in this work: based on the concept of interpretation, not necessarily a physical structure. This, according to Fennelly, cannot be just any universe, but that universe in which, as we have shown previously, the inertia is born in a moment that makes it different from gravity, *i.e.* the Gödel universe. The main point of the development just following is to show that, from the point of view of the very concept of interpretation, Alphonsus Fennelly was right: the space of residence of the matter *per se* is, indeed, *the space of a Gödel universe.*

Now, while a model universe can be just cursorily built, up to the point of interpretation, a Gödel model of universe should ask for a lot more. Of course we did not elaborate on the constitution and rationale of such a universe yet, but the idea of a universe applied to the structure of what are known today as elementary particles can bring a certain methodology in understanding such a constitution and rationale. And we think this is the right place to put the issue on the table, so to speak. Quoting from the pioneering work of Enrico Fermi, the one that triggered a whole current of ideas in the physics of elementary particles and inspired the Alphonsus Fennelly's ideas presented above:

> When two nucleons collide with very great energy in their center of mass system this energy will be *suddenly* released in a *small volume* surrounding the two nucleons. We may think pictorially of the event as of a *collision* in which the *nucleons with their surrounding retinue of pions* hit against each other so that *all the portion of space occupied by the nucleons and by their surrounding pion field* will be *suddenly* loaded with a very great amount of energy. Since the interactions of the pion field are strong we may expect that rapidly this energy will be distributed among the *various degrees of freedom present in this volume* according to statistical laws. One can then compute statistically the probability that in this *tiny volume* a certain number of pions will be created with a given energy distribution. It is then assumed that the *concentration of energy will rapidly dissolve* and that the particles into which the

energy has been converted *will fly out in all directions.* [(Fermi, 1950); *our Italics*]

Two essential physical aspects of the problem of close encounters are touched here, one involving *the thermodynamics*, the other involving *the theory of regularization.* Both of them were presented by us before from a perspective shown in this excerpt as being imperiously necessary: *the scale transition*, made by Fermi the ruler of physics in the close encounters of elementary particles.

The thermodynamics is here devoid of the essential condition of existence of its classical epitome, the ideal gas: the existence of an enclosure confining the gas. The 'tiny volume' of encounter has an ephemeral existence, and is accordingly decided by the rate of that 'distribution of energy' over the available degrees of freedom. This fact takes us in the realm of sufficiency, opened by works like that of Arthur Compton presented by us in Chapter 2 above. Which brings us to the second problem deriving from the very notion of close encounters, namely the theory of regularization: it brings along the oscillators, which should be followed in the structure of a Gödel universe, if the ideas of Fennelly are correct. In reality, this is indeed the case!

Fact is that, insofar as the pure statistics are involved, the things can be related to the idea of matrix and actually to a statistics for the eigenvalues, at least to the extent they are of physical interest, are always explicitly related to some statistics. Indeed, as we have shown (Mazilu *et al.*, 2019), the measurement of a tensorial magnitude always results in two values, *which can be described with respect to a plane in space.* In ideal conditions, the averages of these magnitudes involve therefore two statistics, that we associated with the name of Valentin Novozhilov [equations (4.35–4.36), *loc. cit. ante*]. These two statistics can then be used to construct a one-parameter family of real triplets, representing the roots of a one-B parameter family of cubics, which can be taken as Fubini–Pick cubics. As we have shown in (Mazilu *et al.*, Chapter 4) the Maxwell stress tensor of electromagnetic light is one of the important illustrations of such a procedure.

8.4 Conclusions: Historical Line for a Theoretical Physics

According to Berry–Klein invariance theory of forces, the central forces of Newtonian type play an essential part in the understanding, through scale transitions, of the construction of the universe we inhabit. The first example of scale transition is implicit to Newton himself, and is referring to the time scales: the validation of central forces, acting permanently, by collision forces, acting instantaneously. Here the action of force is time scaled: permanent action means action in finite or transfinite time intervals, while instantaneous action means action in infrafinite time intervals. The notorious case of transition from finite to infrafinite is represented in mathematics by differential calculus, invented by Newton specifically to serve the purpose of physics. Such a transition is based on continuity (see especially Newton, 1974, Volume I, Book II, Lemma II), even though the calculus may not depend on this condition, as in the contemporary fractal calculus.

If it is to build a physics based on the condition of scale transition invariance, we need to take notice of the fact that this invariance places a special importance on the dynamical description of the Kepler problem. This makes out of the planetary model, the structural model of all space and time scales, as shown by us in Chapter 1. Not only this, but considering the forces *per se*, the only forces entitled to existence at any space scale are the central forces of Newtonian character, *i.e.* having a magnitude depending only on distance, namely inversely proportional with the square of this distance. This allows the construction of a differential geometry of physical magnitudes at any scale, based on the mathematical principle of confinement: a Cayleyan or absolute geometry, as we did in Chapter 1. There are three Newtonian forces, whose action is liable to describe an ensemble of Hertz material particles in equilibrium: the gravitational mass and two charges, electric and magnetic. This ensemble is the physical image of the mathematical absolute necessary in constructing the geometry of physical properties of matter. We have to stress once again that, using Poincaré's inspired expression, this ensemble is 'fictitious', more to the point 'logically conjured' if we may. The true, physically verifiable relations, involve nonequilibrium of the Newtonian forces, and characterize the space scale by the domination of forces: at the cosmological scale the gravitation is dominant, while at the microscopic scale the electromagnetic force is dominant. No matter of scale though, the geometry of physical magnitudes turns out to be the same: the hyperbolic geometry. It allows the description of matter by a density, with physical formations represented by solitons. But more importantly this geometry allows us to assert that the scale transitions are described by properties connected with the $SL(2, R)$ group.

In connection with the fictitious equilibrium ensembles of Hertz material particles, there is another connotation of the Berry–Klein theory of transition invariance of the Newtonian forces analyzed by us in Chapter 2. Notice that, from a classical point of view, an ensemble of material points with central forces in equilibrium between them, can be treated as an ensemble of free particles, like the classical ideal gas. Then the radial motion of such a material point is liable to offer us the gauge length of a Berry–Klein gauging, just as the mean path of the gas molecules in the classical case. Only, here, the gauge length describes an acceleration whose fluctuations are, according to the Berry–Klein theory, proportional with the distance between material points. We see this as a generalization of the known Hubble law of contemporary cosmology. A special theorem of H. J. Wagner, given at the end of Chapter 2, then assures us that the motion of material particles of the gauged ensemble is a harmonic oscillator motion. This makes out of the harmonic oscillator an important structure in describing the physical interactions. It turns out to have been just as important in extending the classical dynamical theory of Kepler problem in the infrafinite space range, by the theory of regularization presented by us in Chapter 3. This theory revealed the importance of the analyticity condition in the definition of forces. A three-dimensional analyticity condition is instrumental in the definition of universal space coordinates, as we did in Chapter 4, and related to such coordinates, physical reference frames in the sense of Bartolomé Coll can be defined (Coll, 1985) as in Chapter 6.

Speaking of equilibrium ensembles we need to touch a special subject connected with the definition of physical surfaces, as given in Chapter 8. This is imposed by what we would like to call the *de Broglie interpretation* of the wave mechanics, whose main instrument is the classical light ray. This interpretation allows us to see the electron in the planetary model as a ray of toroidal shape, and the nucleus in that model as a ray of spherical shape. The theory of Chapter 8, allows one to define physically a wave surface for such a ray by the condition of apolarity between the reference frame and coordinate system. The point we want to make here is that such a construction is part of a historical line followed closely by Louis de Broglie all along his work. And everything commenced with Schrödinger's approach of the description of classical planetary atom.

Schrödinger started his program with a logarithmic transformation of the classical variable of action, which helped transform the Hamilton–Jacobi equation for the planetary hydrogen atom into a functional of Dirichlet type (Schrödinger, 1928, p. 12). Then, by a classical variational principle applied to the energy thus calculated, he was able to obtain a Helmholtz-type partial differential equation as the eigenvalue equation of energy, thus providing the simplest quantization condition for the hydrogen atom. Now, as an eigenvalue problem, the quantization thus performed can be carried out in any conceivable detail, just by placing the physical system in some specific environments, via particular boundary conditions. However, classically speaking, the idea of 'physical system' involves an identification of a system by its physical properties. Therefore, the description of a system like the hydrogen atom should be more of a cosmological type, aiming mostly for a universal description of the physical structure of that system, independently of any specific environment. In the words of Charles Galton Darwin, the details connected to a specific environment, belong to 'coordinate space', while the classical rigor always asks for the description of system 'in ordinary space' (see the end of Chapter 3). In hindsight, one can say that this is actually a general aspect of synthetic knowledge — the only kind of knowledge involving human creativity — implanted in human's mind by the necessity of describing the fundamental 'bricks' in a 'construction' of a universe. Such a universal description was apparently quite handy for the method of approach of physics, which Schrödinger devised, so he wasted no time in adopting, and effectively using it: one simply needs to *get rid of the eigenvalues in the very problem of description of the hydrogen atom*, for the eigenvalues carry the particular details.

Indeed, the problem of energy eigenvalues of the hydrogen atom carry the specific of this physical system by two features. First, we have the potential describing the physical system as such, namely only in 'cosmological' boundary conditions: the potential has to vanish *at infinity*. Secondly, we have particular aspects of our experience with the hydrogen atom — behavior in an electric or magnetic field, in a molecule, in an ensemble of atoms forming a thermodynamical system, etc. — all these involving ultimately only the eigenvalues. Thus, getting rid of eigenvalues in the description of a planetary system may be taken as having the connotation of a description which is independent of at least some environments, if not all of them. This last kind of description is, indeed, a 'cosmological' description of the system, satisfying the classical rigor, so we can very well conjecture an equivalence of this kind of rigor with the 'elimination of the eigenvalues'. This can be taken as a clear advantage of the new method over the classical description of a system: it provides the means to extract the essentials about a system from the very particular situations in which it makes itself observable to the mankind.

Fact is that Schrödinger proceeded this way, by using a special periodical time dependence of the wave function he introduced *via* his initial logarithmic transformation. And such a dependence, in the case of a complex function, means an imaginary exponential with the phase linear in time. As Darwin expresses it, 'a stationary state is merely a solution of the wave equation that happens to be harmonic in time' (Darwin, 1927, p. 260), so that such a time dependence seems to be the hallmark of the eigenvalue problem altogether. The method of elimination is just as obvious as its well-known and epoch-making result: the partial differential equation resulting from the variational principle is time independent. However, the solution of that equation is defined up to an arbitrary time-dependent factor. If this factor is complex and periodic in time the eigenvalue can be eliminated leading to the conclusion that the universal description of the hydrogen atom stays in a non-stationary partial differential equation, which is the Schrödinger equation as we know it today. In so doing, however, Schrödinger apparently started being embarrassed. Duly reflected in the work cited above is his insistence on the fact that, with consideration of the time exponential factor, the eigenvalue can be eliminated by a time derivative of *any* order. And issuing from this possibility, there are at least a couple of sources of embarrassment to be considered.

In the first place, this raises questions on the very concept of energy as formally represented by a Hamiltonian. So, in order to maintain the 'first derivative elimination', as it were, one should ask for the identity between Hamiltonian and energy as a separate hypothesis of wave mechanics, a condition that came later under special scrutiny (Kennedy and Kerner, 1965). One might suspect that the complex nature of the wave function is the real problem: after all, the classical optics works in real domain with trigonometric functions, and everything goes just fine. Secondly, assuming just real wave functions, in trying to trigonometrically eliminate the eigenvalue, one is forced to multiple time derivatives specifically two of them in view of the duality of the trigonometric functions satisfying the conditions of periodicity. At the time when Schrödinger elaborated his fundamental work, the idea of two-component wave function was a far cry from what it became later through the works of Dirac and Pauli. So it seemed that the complex wave function with the first time derivative still remains the best way to eliminate the eigenvalue, in order to describe the system from a classical point of view.

Be it as it may, fact is that the complex character of the wave function became so important for Schrödinger, that he took it as a persisting 'flaw' in the theory, which would irrevocably exclude any idea of 'complete' classical rigor, as long as this rigor is to be satisfied in real mathematical terms. Namely, while the stationary problem of energy eigenvalues could be so conducted as to be accomplished with a function ψ in real terms, when assuming that it has to satisfy the new 'universal' nonstationary Schrödinger equation, *that function had to be essentially complex.* Today this fact would hardly be a problem, and one would virtually accept the condition with not so much concern about it. However, a century ago the things were entirely different when talking about the classical rigor. Thus, in asking why, and how could this happen, Schrödinger found no satisfactory answer in just the obvious mathematical procedure imposed by the particular choice of the time dependence of the wave function: he needed a more profound reason, perhaps even a physical reason. Which he could not find at the time, so he served the eternity only with a footnote joke regarding the human condition in general. This joke carries the unmistaken signature of Wolfgang Pauli, notorious for many such incidents in the epoch we are talking about, incidents that even earned him the nickname *Mephisto* among the fellow physicists. Quoting, then:

> The words 'essentially complex' try to cover here a great difficulty. In his desire to visualize, at any rate, the propagation of the waves ψ as a real phenomenon in the classical acceptance of this term, the author has refused for himself to recognize frankly that any development of the theory accentuated more and more clearly the essentially complex nature of the wave function. Yet, this function is determined by an equation (*the nonstationary Schrödinger equation, n/a*) whose coefficients are essentially complex!
>
> But how could $\sqrt{(-1)}$ bring itself into this equation? One answer, which I don't dare indicate here but the general idea, was given to this question by a physicist who once left Austria, but who, despite the long years spent abroad, has not lost completely his biting Viennese humor, and who, besides, is known for his ability to find the right word, all the more fair as it is cruder. Here is that answer: '$\sqrt{(-1)}$ slipped into that equation as something we let accidentally escape, providing us however with an invaluable relief, even though we produced it involuntarily.' [(Schrödinger, 1933), footnote on pp. 166–167; *our translation, n/a*]

Joke aside, the fact remained unsettled until recently: the complex nature of the wave function seemed just an accidental blunder of the common kind suggested by the joke in this quotation, that we have to accept inasmuch as it provides 'invaluable relief', in this particular instance when using the existing mathematical rules of handling, of course. It took almost a century and a simple observation made quite recently (Schleich, Greenberger, Kobe and Scully, 2013), in order to realize that the wave function, such as Schrödinger wanted it, *must be complex at any rate*, for there is no way around this condition. Besides, the consequences of this condition are staggering, first from historical point of view, but mostly from a gnoseological point of view. To mention just the most important of these consequences, *this condition would mean that a wave-mechanical universe is necessarily holographic*!

It is on this note that the present work was conceived. We had to start, therefore, with an obvious methodological question: what would an *interpretable* complex wave function involve? This means, first and foremost, that everything in elaborating an answer to this question should revolve around the concept of interpretation, and we take it according to Darwin's definition, presented by us in a few important historical details in Chapter 1, but in a simpler, more general form: *there should be an ensemble of particles describing the waves* in general. At the time when Schrödinger published his ideas there were already two wave- mechanical objective interpretations of the wave function, on which, by and large, we have elaborated recently (Mazilu, Agop, and Mercheş, 2019). When we say 'objective' here, we understand interpretations in the very sense of Schrödinger, more to the point interpretations having little or nothing to do with the idea of 'recording', which was mainly promoted by the quantum mechanics arising at that time. The interpretations in question where those of Louis de Broglie and Erwin Madelung. They can be easily differentiated from one another if we follow their details mathematically. However there is one essential difference between them that needs to be considered as a matter of principle, *i.e.* for the benefit of human knowledge in general: the difference in the *phase of the wave function* involved in interpretation.

The first one of the two interpretations (de Broglie, 1926) abides by the Schrödinger's *phase linearity with respect to time*, filling in, as it were, for some missing points brought about by the phenomenology of light in the classical theory of light rays. This interpretation, issuing from optics, contains implicitly even a 'recording' manner, inasmuch as it proves that, according to the theory of diffraction of light, the density of the ensemble of particles involved in the interpretation of light *via* a light ray, is given by *the square of the amplitude of the optical signal.* This is an essential quantity according to Schrödinger original interpretation. On the other hand, the second interpretation (Madelung, 1927) *gives up the linearity of phase*, pushing through, as it were, the Hamilton-Jacobi equation considered by Schrödinger as a starting point in his enterprise, but in its classical nonstationary form. This interpretation has, apparently, nothing to do with the idea of recording. While the de Broglie's interpretation makes explicitly use of the mathematical reality condition for the signal representing the light field in a light ray — much to Schrödinger's own satisfaction we should say! — the Madelung interpretation takes the complex wave function defined by the nonstationary Schrödinger equation as a starting point. Being kind of an exotic fact, this last interpretation attracted an overwhelming general attention of theorists, from both the fundamental points of view of the physics and knowledge at large, as well as from the practical point of view, related to the examples of the physics initiated by Schrödinger himself. The story is well-known, and can be entrusted to a, by now, classical literature, so that we do not insist in this work upon Madelung interpretation, but only to the extent that it represents *a matter of fundamental principle* (see Foreword and Chapter 1). The reader can follow the whole story by consulting the literature indicated in our recent work (Mazilu, Agop, and Mercheş, 2019). Here we only point out that a general kind of Madelung interpretation is implicit in the general complex form of the wave function, and that Louis de Broglie showed that the light phenomenon can be interpreted along this idea. But the lessons to be learned from de Broglie's exploits are by far richer, and we had to concentrate here in just a part of them.

Start with the observation that an algebraically complex form of a physical signal, but 'in ordinary space', if it is to use the terminology of Darwin — among the kind of which the wave function is to be considered — involves exhibiting *an amplitude* and *a phase* as functions of a position. As any position in space, this one is assigned by some coordinates x with respect to a physical reference frame at a time moment t, chosen from a time sequence as given by a certain clock. As it happens, this function is by default, if we may say so, the solution of a differential equation that can be identified as what we know today as a *free particle nonstationary Schrödinger equation.* The identification can be carried out, indeed, but only under specific conditions to be revealed mathematically (Schleich, Greenberger, Kobe and Scully, 2013), which, however, can be physically 'interpreted' in the sense of Darwin, thus providing a logical reason for the very concept of interpretation. The first of these conditions is an equation of continuity for a density proportional to the square of modulus of the wave function, and a velocity field given by the gradient of the phase of wave function. The second of conditions is the Hamilton-Jacobi equation, provided one adopts a special definition of the potential (Mazilu, Agop, and Mercheş, 2019): it is determined by the amplitude of the wave function. In other words, in any point in space, reached by the wave described by such a wave function, a potential exists, generated by this wave, and with it a system should exist as described by this potential. This is plainly the holographic property mentioned by us above. It has strong ties with the development of optics, to which the de Broglie's interpretation is explicitly addressed.

Louis de Broglie's theory was, indeed, originally designed to establish that the physical optics does not contradict the wave-mechanical precepts, because these very precepts are coming out of necessities settled by the quantum mechanics (de Broglie, 1926). The main trait of this example of interpretation is, at least in what concerns us, first, of proving the historical continuity of knowledge, and secondly, of providing a striking case of 'taking the consequences such as they are' in the actual facts of modern day theoretical physics. However, there are a few fundamental reasons for which we are taking the example to its intimate details, and follow it in developing a special instrument of theoretical physics: the physical theory of surfaces in Chapter 8.

Let us take notice, once again, that the Schrödinger equation for a free particle is an equation 'in the ordinary space', and that contrary to the usual method, one needs 'to transcribe the ordinary space into coordinate space', as Darwin stipulates in the excerpt provided by us in Chapter 3 here. In setting the problem this way, one cannot but notice that the wave mechanics reproduces in fact a general pattern of classical knowledge: it is actually starting from the ordinary space — the space of positions — that we have to construct a coordinate space — a space occupied by the matter accessible to experiment. This process necessitates, first and foremost, a reference frame to locate the events in space, secondly some clocks to arrange the events in the order necessary to describe the motions, and finally some measurements in order to put together those events. The de Broglie's interpretation is referring to light, and has most of the elements of an interpretation that we need to learn and reproduce in any other of the physical cases. Which, by and large, is what we do in the present work.

The equation of continuity, necessary in order to assure the validity of nonstationary Schrödinger equation, is itself valid provided two essential conditions are secured: (1) *the square of the amplitude of complex function* $\psi(x,t)$ *can be taken as a density* of the interpretative ensemble of free particles, and (2) *the velocity field* of the interpretative ensemble of free particles *is provided by the gradient of the phase* of wave function. It is quite instructive, indeed, for the knowledge at large, and especially for physics, to analyze how this historical interpretation dealt with such requirements. It should be noticed *again*, though, that Louis de Broglie was, and still is, *the only one* in history who ever has undertaken the task of *proving* these conditions, otherwise always taken as axioms. He started from optics, for there was no physical situation involving particles in the physical optics of the day, a field of knowledge which, as a matter of fact, inspired the wave-mechanical ideas. His work goes on, logically we should say, in filling some missing points that mark discontinuities from the epoch's very natural philosophy of light.

The specific problem at that time was, in de Broglie's acceptance at least, that of proving the old idea that the light can be seen as a flux of particles. However, inasmuch as such a proof has had eluded any human attempt before, de Broglie wanted just to add to our experience the one fact turned critical at the time, namely to show that the particulate image of the light contradicts neither the optical, nor the mechanical rules of natural philosophy, for the particular case of *the experiments of diffraction.* Long known for light and, we might add, routinely used in optics even from the times of Augustin Fresnel, such experiments were revived for electrons at the epoch we are in now — and especially illustrated through the work of Davisson and Germer — by de Broglie's own idea of associating a wave with a particle (see Davisson and Germer (1927) for the history and literature on the problem). In hindsight, this idea gives a positive turn to the old belief that the knowledge should be just as continuous as the nature itself: this seems to be the message of the de Broglie's very work on interpretation, anyway, at least as we take it. Indeed, the Fresnel theory of light, the one that instated diffraction as a routine experiment in optics, harmoniously completes, in fact, the classical line of the natural philosophy referring to the light phenomenon at large. Let us summarize the main steps of this side of natural philosophy regarding to the light phenomenon, in order to better understand the position of Louis de Broglie in the work we have under scrutiny at this point.

The first physical image of the light phenomenon was that of Thomas Hobbes, a global concept as it were, involving the idea of 'orb'. It was probably religiously inspired as, in fact, any natural philosophical idea was, by an analogy with the heart. The analogy did not work quite properly: everyone can perceive the light as expanding only, never contracting, like the heart does in completing its job. However, the idea of materiality of light settled this issue in an unexpected way, insofar as Robert Hooke placed a periodic motion where it should belong, just by logic: it takes place, regardless of expansion or contraction, within the 'orb' — read wave surface of light — otherwise the light would be able destroy the transparent materials it penetrates, and such an event has never been observed. The periodic motion is a 'pulse', more precisely an 'orbicular pulse', if it is to use Hooke's own words (Hooke, 1665).

This last concept improves, kinematically, but mostly dynamically, upon Hobbes' purely geometrical 'line of light', naturally completing the fact of expansion by the idea of propagation of light along a direction. This was just about the first case in the Newtonian epoch, whereby the human spirit was starting filling the empty ideal world image provided by geometry, with properties of the world of human experience. The notorious case of this start of the classical natural philosophy in the epoch is, of course, the triad of Kepler laws, facilitating the invention of forces by Isaac Newton.

The phenomenology of light was limited at the time to just two specific phenomena: *reflection* and *refraction* of light. Newton 'fortified', we might say, this phenomenology technologically (see his *Opticks*), in order to avoid the 'invention of hypotheses', by creating an experimental basis which, under different circumstances, of course, works even today. However, based exclusively on that phenomenology, and therefore abundantly still 'inventing hypotheses', as it were, Hooke created a concept of light ray (*loc. cit. ante*, pp. 53–69), in which he incorporated what we think of as the first rational theory of colors. In this concept, the color is controlled by the angle between orbicular pulse and the mathematical rays delimiting a plane construction that can be rightfully called *physical ray.* It is on this concept that the experimental basis created by Newton helped improve, adding one important differentia to it: *the physical ray is not a plane figure*, like that of Hobbes and Hooke, but *a solid one*, a cone or a cylinder, or even a more general *tube*. This was the conclusion of the celebrated detailed experiments with a prism, which proved that the color is a property varying directionally, indeed, but in the cross-section of the light ray, and quite independently of the geometrical form of the ray itself as a physical object.

It should be illuminating, we believe, especially in understanding the concept of physical ray in general, to notice that, eventually, it was discovered that the color is described by a gauge group acting in the cross-section of the ray (Resnikoff, 1974), reproducing a group action of SL(2, R) type, and that Erwin Schrödinger pioneered this very discovery (Schrödinger, 1920). One can rightfully say that, with the theory of colors he created in 1920, Schrödinger was in fact completing an 'apprenticeship' for the physics which he started building six years later (Mazilu, Agop and Mercheş, 2019). Anyway, what we think is worth retaining from this brief history, in the spirit of today's physics, of course, is the fact that the epoch of reflection and refraction phenomenology produced the concept of *a light ray as a tube*, having the color as *a transversal property described by a gauge group*.

Along this historical path, Augustin Fresnel started a new epoch in the phenomenology of light, marked by the introduction of *diffraction* of light as a new phenomenon, in the description of which the periodic properties of light were the usual observables, and thereby the wave nature of light came closer to our understanding. The obvious spatial periodical pattern in the *recordings* of diffraction phenomena could thus be explained physically, as a mechanical interference phenomenon. In so doing, the optics made reference to the harmonic oscillator, in order to understand the intensity of light for instance, to say nothing of some other physically fundamental necessities, like the very definition of the *intensity of light*. However, this reference is, by stretching a little the meaning of words, 'illegal' to say the least, in the case of light, inasmuch as the light is a far cry from exhibiting the inertial properties required by a proper dynamics of the harmonic oscillator. For, as a purely dynamical system, the harmonic oscillator is a system described by forces proportional to displacements (those type of elastic forces, used initially by Hooke to explain the behavior of light), and in the case of physical optics the second principle of dynamics is quite incidental, as it were. It was introduced only by a natural mathematical property of transcendence of the second-order ordinary differential equation: it describes *any type* of periodic processes. And the fact is, that in the foundations of modern physical optics, the periodic processes of diffraction have more to do with the theory of statistics than with the dynamics (Fresnel, 1827).

This is, however, not to say that the harmonic oscillator is to be abandoned altogether, as a model, because as we have shown in Chapter 2, the Berry–Klein theory strongly indicates that this is not the case, either from experimental point of view, or even theoretically. All we want to say is that we need to find its right place and form of expression in the theory, and this is specified, again, through the order imposed by the measure of things, this time as their mass. Indeed, dynamically, the second-order differential equation expressing the principle of inertia, involves a finite mass. On the other hand, for light the mass is virtually inexistent, and if the second-order differential equation is imposed by adding the diffraction to the phenomenology of light, this means that it actually describes *a transcendence between finite and infinitesimal scales of mass* [for a closer description of the concept, one can consult Mazilu, Agop and Mercheş (2019)]. As we shall see later, the mathematics of scale transitions between finite and infinitesimal gauges in a scale relativity, fully respects the rules related to the harmonic oscillator model. In fact, the whole wave mechanics, as a science, can be constructed based on such rules, which appear to be universal.

Now, along with the settling of Fresnel's theory in physical optics, a few changes in the natural philosophy have taken place. First in the order of things, was making the dynamics 'lawful', as it were, in the case of light. The first step was *to identify the phase*, mathematically involved as an independent variable in the trigonometric functions describing the diffraction in the light phenomenon, *with the time of an evolution*: mathematically, the phase had to be linear in time. A condition which brought *the frequency* front and center, and with it *the concept of wavelength*, thus generating right away a whole new experimental technology of the Newtonian kind, but leaving nevertheless behind what the dynamical principle really needed for a sound physical theory: first, the elastic properties of the medium supporting the light and, secondly, the *interpretation of light*, which obviously required the old idea of particle, and therefore the inescapable inertial properties. The Fresnel's ellipsoid of elasticities pretty much fills in for the first aspect of this issue, while the second one was delayed, and left in suspension ever since, being occasionally replaced with *ad hoc* creations of mind, and so is it, actually, even today to a large extent. A proper dynamical use of the second principle of dynamics in the matters of light came in handy only later on, with the advent of the electromagnetic theory of light.

This theory of light has in common with the old Fresnel optics the one equation that models the space and time periodical properties no matter of their physical approach, as long as these properties are described by a frequency: *the D'Alembert equation*, which is the main characteristic of both the optics and the particle theory as Louis de Broglie has proved.

All these historical evidences indicate to us that de Broglie's work harmoniously completes a historical line of thinking, from which we have to learn for the benefit of knowledge in general. Once again, the concept of ray in his acceptance is the instrument of interpretation: it turns out to be essential in the construction of a coordinate system (Coll, 1985), and in this capacity defines a universal coordinate system — the Appell coordinates described by us in Chapter 4 — by a kinematic condition equivalent to old Hooke's explanation. As explained in Chapter 3, the existence of universal Newtonian forces means regularization, and the regularization procedure is a consequence of the condition of analyticity. One thing did not de Broglie's interpretation touch apparently, which seems a consequence of only the Madelung interpretation: the holographic property of matter. And the holography is the fourth of the phenomena that can enter the phenomenology of light, together with reflection, refraction and diffraction.

Nevertheless, as we have shown in Chapter 6, this property may be implicit in the construction of a ray, as designed by de Broglie. If properly constructed, a wave surface involves a Fubini–Pick cubic generating a vector that represents the 'memory' of ray. It is contained in the complete correspondence between time and phase: the classical case of linearity between phase and time is just a particular case of their connection. As we have shown in Chapter 6, there are stable cycles in the space of the parameters dictating the connection between time and phase cycles can be taken as the 'memory' of the relation between phase and time, which is a holographic property in the case of light. When referring the the physical properties, such cycles define the inertia of matter, which thus can be taken as a 'memory' property.

Conclusions: Historical Line for a Theoretical Physics

According to Berry–Klein invariance theory of forces, the central forces of Newtonian type play an essential part in the understanding, through scale transitions, of the construction of the universe we inhabit. The first example of scale transition is implicit to Newton himself, and is referring to the time scales: the validation of central forces, acting permanently, by collision forces, acting instantaneously. Here the action of force is time scaled: permanent action means action in finite or transfinite time intervals, while instantaneous action means action in infrafinite time intervals. The notorious case of transition from finite to infrafinite is represented in mathematics by differential calculus, invented by Newton specifically to serve the purpose of physics. Such a transition is based on continuity [see especially Newton (1974), Volume I, Book II, Lemma II], even though the calculus may not depend on this condition, as in the contemporary fractal calculus.

If it is to build a physics based on the condition of scale transition invariance, we need to take notice of the fact that this invariance places a special importance on the dynamical description of the Kepler problem. This makes out of the planetary model, the structural model of all space and time scales, as shown by us in Chapter 1. Not only this, but considering the forces *per se*, the only forces entitled to existence at any space scale are the central forces of Newtonian character, *i.e.* having a magnitude depending only on distance, namely inversely proportional with the square of this distance. This allows the construction of a differential geometry of physical magnitudes at any scale, based on the mathematical principle of confinement: a Cayleyan or absolute geometry, as we did in Chapter 1. There are three Newtonian forces, whose action is liable to describe an ensemble of Hertz material particles in equilibrium: the gravitational mass and two charges, electric and magnetic. This ensemble is the physical image of the mathematical absolute necessary in constructing the geometry of physical properties of matter. We have to stress once again that, using Poincaré's inspired expression, this ensemble is 'fictitious', more to the point 'logically conjured' if we may. The true, physically verifiable relations, involve non-equilibrium of the Newtonian forces, and characterize the space scale by the domination of forces: at the cosmological scale the gravitation is dominant, while at the microscopic scale the electromagnetic force is dominant. No matter of scale though, the geometry of physical magnitudes turns out to be the same: the hyperbolic geometry. It allows the description of matter by a density, with physical formations represented by solitons. But more importantly this geometry allows us to assert that the scale transitions are described by properties connected with the $SL(2, R)$ group.

In connection with the fictitious equilibrium ensembles of Hertz material particles, there is another connotation of the Berry–Klein theory of transition invariance of the Newtonian forces analyzed by us in Chapter 2. Notice that, from a classical point of view, an ensemble of material points with central forces in equilibrium between them, can be treated as an ensemble of free particles, like the classical ideal gas. Then the radial motion of such a material point is liable to offer us the gauge length of a Berry–Klein gauging, just as the mean path of the gas molecules in the classical case. Only, here, the gauge length describes an acceleration whose fluctuations are, according to the Berry–Klein theory, proportional with the distance between material points. We see this as a generalization of the known Hubble law of contemporary cosmology. A special theorem of H. J. Wagner, given at the end of Chapter 2, then assures us that the motion of material particles of the gauged ensemble is a harmonic oscillator motion.

This makes out of the harmonic oscillator an important structure in describing the physical interactions. It turns out to have been just as important in extending the classical dynamical theory of Kepler problem in the infrafinite space range, by the theory of regularization presented by us in Chapter 3. This theory revealed the importance of the analyticity condition in the definition of forces. A three-dimensional analyticity condition is instrumental in the definition of universal space coordinates, as we did in Chapter 4, and related to such coordinates, physical reference frames in the sense of Bartolomé Coll can be defined (Coll, 1985) as in Chapter 6.

Speaking of equilibrium ensembles we need to touch a special subject connected with the definition of physical surfaces, as given in Chapter 8. This is imposed by what we would like to call the *de Broglie interpretation* of the wave mechanics, whose main instrument is the classical light ray. This interpretation allows us to see the electron in the planetary model as a ray of toroidal shape, and the nucleus in that model as a ray of spherical shape. The theory of Chapter 8 allows one to define physically a wave surface for such a ray by the condition of apolarity between the reference frame and coordinate system. The point we want to make here is that such a construction is part of a historical line followed closely by Louis de Broglie all along his work. And everything commenced with Schrödinger's approach of the description of classical planetary atom.

Schrödinger started his program with a logarithmic transformation of the classical variable of action, which helped transform the Hamilton–Jacobi equation for the planetary hydrogen atom into a functional of Dirichlet type (Schrödinger, 1928, pp. 12). Then, by a classical variational principle applied to the energy thus calculated, he was able to obtain a Helmholtz-type partial differential equation as the eigenvalue equation of energy, thus providing the simplest quantization condition for the hydrogen atom. Now, as an eigenvalue problem, the quantization thus performed can be carried out in any conceivable detail, just by placing the physical system in some specific environments, *via* particular boundary conditions. However, classically speaking, the idea of 'physical system' involves an identification of a system by its physical properties. Therefore, the description of a system like the hydrogen atom should be more of a cosmological type, aiming mostly for a universal description of the physical structure of that system, independently of any specific environment. In the words of Charles Galton Darwin, the details connected to a specific environment, belong to 'coordinate space', while the classical rigor always asks for the description of system 'in ordinary space' (see the end of Chapter 3). In hindsight, one can say that this is actually a general aspect of synthetic knowledge — the only kind of knowledge involving human creativity — implanted in human's mind by the necessity of describing the fundamental 'bricks' in a 'construction' of a universe. Such a universal description was apparently quite handy for the method of approach of physics, which Schrödinger devised, so he wasted no time in adopting, and effectively using it: one simply needs to *get rid of the eigenvalues in the very problem of description of the hydrogen atom*, for the eigenvalues carry the particular details.

Indeed, the problem of energy eigenvalues of the hydrogen atom carry the specific of this physical system by two features. First, we have the potential describing the physical system as such, *viz.* only in 'cosmological' boundary conditions: the potential has to vanish *at infinity*. Secondly, we have particular aspects of our experience with the hydrogen atom — behavior in an electric or magnetic field, in a molecule, in an ensemble of atoms forming a thermodynamical system, etc — all these involving ultimately only the eigenvalues. Thus, getting rid of eigenvalues in the description of a planetary system may be taken as having the connotation of a description which is independent of at least some environments, if not all of them. This last kind of description is, indeed, a 'cosmological' description of the system, satisfying the classical rigor, so we can very well conjecture an equivalence of this kind of rigor with the 'elimination of the eigenvalues'. This can be taken as a clear advantage of the new method over the classical description of a system: it provides the means to extract the essentials about a system from the very particular situations in which it makes itself observable to the mankind.

Fact is that Schrödinger proceeded this way, by using a special periodical time dependence of the wave function he introduced *via* his initial logarithmic transformation. And such a dependence, in the case of a complex function, means an imaginary exponential with the phase linear in time. As Darwin expresses it, "a stationary state is merely a solution of the wave equation that happens to be harmonic in time" (Darwin, 1927, p. 260), so that such a time dependence seems to be the hallmark of the eigenvalue problem altogether. The method of elimination is just as obvious as its well-known and epoch-making result: the partial differential equation resulting from the variational principle is time independent. However, the solution of that equation is defined up to an arbitrary time-dependent factor. If this factor is complex and periodic in time the eigenvalue can be eliminated leading to the conclusion that the universal description of the hydrogen atom stays in a non-stationary partial differential equation, which is the Schrödinger equation as we know it today. In so doing, however, Schrödinger apparently started being embarrassed. Duly reflected in the work cited above is his insistence on the fact that, with consideration of the time exponential factor, the eigenvalue can be eliminated by a time derivative of *any* order. And issuing from this possibility, there are at least a couple of sources of embarrassment to be considered.

In the first place, this raises questions on the very concept of energy as formally represented by a Hamiltonian. So, in order to maintain the 'first derivative elimination', as it were, one should ask for the identity between Hamiltonian and energy as a separate hypothesis of wave mechanics, a condition that came later under special scrutiny (Kennedy and Kerner, 1965). One might suspect that the complex nature of the wave function is the real problem: after all, the classical optics works in real domain with trigonometric functions, and everything goes just fine. Secondly, assuming just real wave functions, in trying to trigonometrically eliminate the eigenvalue, one is forced to multiple time derivatives specifically two of them in view of the duality of the trigonometric functions satisfying the conditions of periodicity. At the time when Schrödinger elaborated his fundamental work, the idea of two-component wave function was a far cry from what it became later through the works of Dirac and Pauli. So it seemed that the complex wave function with the first time derivative still remains the best way to eliminate the eigenvalue, in order to describe the sistem from a classical point of view.

Be it as it may, fact is that the complex character of the wave function became so important for Schrödinger, that he took it as a persisting 'flaw' in the theory, which would irrevocably exclude any idea of 'complete' classical rigor, as long as this rigor is to be satisfied in real mathematical terms. Namely, while the stationary problem of energy eigenvalues could be so conducted as to be accomplished with a function ψ in real terms, when assuming that it has to satisfy the new 'universal' nonstationary Schrödinger equation, *that function had to be essentially complex.* Today this fact would hardly be a problem, and one would virtually accept the condition with not so much concern about it. However, a century ago the things were entirely different when talking about the classical rigor. Thus, in asking why, and how could this happen, Schrödinger found no satisfactory answer in just the obvious mathematical procedure imposed by the particular choice of the time dependence of the wave function: he needed a more profound reason, perhaps even a physical reason. Which he could not find at the time, so he served the eternity only with a footnote joke regarding the human condition in general. This joke carries the unmistaken signature of Wolfgang Pauli, notorious for many such incidents in the epoch we are talking about, incidents that even earned him the nickname *Mephisto* among the fellow physicists. Quoting, then:

> The words $\ll$ essentially complex $\gg$ try to cover here a great difficulty. In his desire to visualize, at any rate, the propagation of the waves ψ as a real phenomenon in the classical acceptance of this term, the autor has refused for himself to recognize frankly that any development of the theory accentuated more and more clearly the essentially complex nature of the wave function. Yet, this function is determined by an equation (*the nonstationary Schrödinger equation, n/a*) whose coefficients are essentially complex!

> But how could $\sqrt{(-1)}$ bring itself into this equation? One answer, which I don't dare indicate here but the general idea, was given to this question by a physicist who once left Austria, but who, despite the long years spent abroad, has not lost completely his biting Viennese humor, and who, besides, is known for his ability to find the right word, all the more fair as it is cruder. Here is that answer: $\ll$ $\sqrt{(-1)}$ slipped into that equation as something we let accidentally escape, providing us however with an invaluable relief, even though we produced it involuntarily.$\gg$ [(Schrödinger, 1933, footnote on pp. 166–167); *our translation, n/a*]

Joke aside, the fact remained unsettled until recently: the complex nature of the wave function seemed just an accidental blunder of the common kind suggested by the joke in this quotation, that we have to accept inasmuch as it provides 'invaluable relief', in this particular instance when using the existing mathematical rules of handling, of course. It took almost a century and a simple observation made quite recently (Schleich, Greenberger, Kobe and Scully, 2013), in order to realize that the wave function, such as Schrödinger wanted it, *must be complex at any rate*, for there is no way around this condition. Besides, the consequences of this condition are staggering, first from historical point of view, but mostly from a gnoseological point of view. To mention just the most important of these consequences, *this condition would mean that a wave-mechanical universe is necessarily holographic*!

It is on this note that the present work was conceived. We had to start, therefore, with an obvious methodological question: what would an *interpretable* complex wave function involve? This means, first and foremost, that everything in elaborating an answer to this question should revolve around the concept of interpretation, and we take it according to Darwin's definition, presented by us in a few important historical details in Chapter 1, but in a simpler, more general form: *there should be an ensemble of particles describing the waves* in general. At the time when Schrödinger published his ideas there were already two wave-mechanical objective interpretations of the wave function, on which, by and large, we have elaborated recently (Mazilu, Agop, and Mercheş, 2019). When we say 'objective' here, we understand interpretations in the very sense of Schrödinger, more to the point interpretations having little or nothing to do with the idea of 'recording', which was mainly promoted by the quantum mechanics arising at that time. The interpretations in question where those of Louis de Broglie and Erwin Madelung. They can be easily differentiated from one another if we follow their details mathematically. However there is one essential difference between them that needs to be considered as a matter of principle, *i.e.* for the benefit of human knowledge in general: the difference in the *phase of the wave function* involved in interpretation.

The first one of the two interpretations (de Broglie, 1926) abides by the Schrödinger's *phase linearity with respect to time*, filling in, as it were, for some missing points brought about by the phenomenology of light in the classical theory of light rays. This interpretation, issuing from optics, contains implicitly even a 'recording' manner, inasmuch as it proves that, according to the theory of diffraction of light, the density of the ensemble of particles involved in the interpretation of light *via* a light ray, is given by *the square of the amplitude of the optical signal.* This is an essential quantity according to Schrödinger original interpretation. On the other hand, the second interpretation (Madelung, 1927) *gives up the linearity of phase*, pushing through, as it were, the Hamilton-Jacobi equation considered by Schrödinger as a starting point in his enterprise, but in its classical non-stationary form. This interpretation has, apparently, nothing to do with the idea of recording. While the de Broglie's interpretation makes explicitly use of the mathematical reality condition for the signal representing the light field in a light ray — much to Schrödinger's own satisfaction we should say! — the Madelung interpretation takes the complex wave function defined by the non-stationary Schrödinger equation as a starting point. Being kind of an exotic fact, this last interpretation attracted an overwhelming general attention of theorists, from both the fundamental points of view of the physics and knowledge at large, as well as from the practical point of view, related to the examples of the physics initiated by Schrödinger himself. The story is well-known, and can be entrusted to a, by now, classical literature, so that we do not insist in this work upon Madelung interpretation, but only to the extent that it represents *a matter of fundamental principle* (see Foreword and Chapter 1). The reader can follow the whole story by consulting the literature indicated in our recent work (Mazilu, Agop, and Mercheş, 2019). Here we only point out that a general kind of Madelung interpretation is implicit in the general complex form of the wave function, and that Louis de Broglie showed that the light phenomenon can be interpreted along this idea. But the lessons to be learned from de Broglie's exploits are by far richer, and we had to concentrate here in just a part of them.

Start with the observation that an algebraically complex form of a physical signal, but 'in ordinary space', if it is to use the terminology of Darwin — among the kind of which the wave function is to be considered — involves exhibiting *an amplitude* and *a phase* as functions of a position. As any position in space, this one is assigned by some coordinates x with respect to a physical reference frame at a time moment t, chosen from a time sequence as given by a certain clock. As it happens, this function is by default, if we may say so, the solution of a differential equation that can be identified as what we know today as a *free particle non-stationary Schrödinger equation.* The identification can be carried out, indeed, but only under specific conditions to be revealed mathematically (Schleich, Greenberger, Kobe and Scully, 2013), which, however, can be physically 'interpreted' in the sense of Darwin, thus providing a logical reason for the very concept of interpretation. The first of these conditions is an equation of continuity for a density proportional to the square of modulus of the wave function, and a velocity field given by the gradient of the phase of wave function. The second of conditions is the Hamilton-Jacobi equation, provided one adopts a special definition of the potential (Mazilu, Agop, and Mercheş, 2019): it is determined by the amplitude of the wave function. In other words, in any point in space, reached by the wave described by such a wave function, a potential exists, generated by this wave, and with it a system should exist as described by this potential. This is plainly the holographic property mentioned by us above. It has strong ties with the development of optics, to which the de Broglie's interpretation is explicitly addressed.

Louis de Broglie's theory was, indeed, originally designed to establish that the physical optics does not contradict the wave-mechanical precepts, because these very precepts are coming out of necessities settled by the quantum mechanics (de Broglie, 1926). The main trait of this example of interpretation is, at least in what concerns us, first, of proving the historical continuity of knowledge, and secondly, of providing a striking case of 'taking the consequences such as they are' in the actual facts of modern day theoretical physics. However, there are a few fundamental reasons for which we are taking the example to its intimate details, and follow it in developing a special instrument of theoretical physics: the physical theory of surfaces in Chapter 8.

Let us take notice, once again, that the Schrödinger equation for a free particle is an equation 'in the ordinary space', and that contrary to the usual method, one needs 'to transcribe the ordinary space into coordinate space', as Darwin stipulates in the excerpt provided by us in Chapter 3 here. In setting the problem this way, one cannot but notice that the wave mechanics reproduces in fact a general pattern of classical knowledge: it is actually starting from the ordinary space — the space of positions — that we have to construct a coordinate space — a space occupied by the matter accessible to experiment. This process necessitates, first and foremost, a reference frame to locate the events in space, secondly some clocks to arrange the events in the order necessary to describe the motions, and finally some measurements in order to put together those events. The de Broglie's interpretation is referring to light, and has most of the elements of an interpretation that we need to learn and reproduce in any other of the physical cases. Which, by and large, is what we do in the present work.

The equation of continuity, necessary in order to assure the validity of non-stationary Schrödinger equation, is itself valid provided two essential conditions are secured: (1) *the square of the amplitude of complex function* $\psi(x,t)$ *can be taken as a density* of the interpretative ensemble of free particles, and (2) *the velocity field* of the interpretative ensemble of free particles *is provided by the gradient of the phase* of wave function. It is quite instructive, indeed, for the knowledge at large, and especially for physics, to analyze how this historical interpretation dealt with such requirements. It should be noticed *again*, though, that Louis de Broglie was, and still is, *the only one* in history who ever has undertaken the task of *proving* these conditions, otherwise always taken as axioms. He started from optics, for there was no physical situation involving particles in the physical optics of the day, a field of knowledge which, as a matter of fact, inspired the wave-mechanical ideas. His work goes on, logically we should say, in filling some missing points that mark discontinuities from the epoch's very natural philosophy of light.

The specific problem at that time was, in de Broglie's acceptance at least, that of proving the old idea that the light can be seen as a flux of particles. However, inasmuch as such a proof has had eluded any human attempt before, de Broglie wanted just to add to our experience the one fact turned critical at the time, namely to show that the particulate image of the light contradicts neither the optical, nor the mechanical rules of natural philosophy, for the particular case of *the experiments of diffraction.* Long known for light and, we might add, routinely used in optics even from the times of Augustin Fresnel, such experiments were revived for electrons at the epoch we are in now — and especially illustrated through the work of Davisson and Germer — by de Broglie's own idea of associating a wave with a particle [see Davisson and Germer (1927) for the history and literature on the problem]. In hindsight, this idea gives a positive turn to the old belief that the knowledge should be just as continuous as the nature itself: this seems to be the message of the de Broglie's very work on interpretation, anyway, at least as we take it. Indeed, the Fresnel theory of light, the one that instated diffraction as a routine experiment in optics, harmoniously completes, in fact, the classical line of the natural philosophy referring to the light phenomenon at large. Let us summarize the main steps of this side of natural philosophy regarding to the light phenomenon, in order to better understand the position of Louis de Broglie in the work we have under scrutiny at this point.

The first physical image of the light phenomenon was that of Thomas Hobbes, a global concept as it were, involving the idea of "orb". It was probably religiously inspired as, in fact, any natural philosophical idea was, by an analogy with the heart. The analogy did not work quite properly: everyone can perceive the light as expanding only, never contracting, like the heart does in completing its job. However, the idea of materiality of light settled this issue in an unexpected way, insofar as Robert Hooke placed a periodic motion where it should belong, just by logic: it takes place, regardless of expansion or contraction, within the 'orb' — read wave surface of light — otherwise the light would be able destroy the transparent materials it penetrates, and such an event has never been observed. The periodic motion is a 'pulse', more precisely an "orbicular pulse", if it is to use Hooke's own words (Hooke, 1665). This last concept improves, kinematically, but mostly dynamically, upon Hobbes' purely geometrical "line of light", naturally completing the fact of expansion by the idea of propagation of light along a direction. This was just about the first case in the Newtonian epoch, whereby the human spirit was starting filling the empty ideal world image provided by geometry, with properties of the world of human experience. The notorious case of this start of the classical natural philosophy in the epoch is, of course, the triad of Kepler laws, facilitating the invention of forces by Isaac Newton.

The phenomenology of light was limited at the time to just two specific phenomena: *reflection* and *refraction* of light. Newton 'fortified', we might say, this phenomenology technologically (see his *Opticks*), in order to avoid the 'invention of hypotheses', by creating an experimental basis which, under different circumstances, of course, works even today. However, based exclusively on that phenomenology, and therefore abundantly still 'inventing hypotheses', as it were, Hooke created a concept of light ray (*loc. cit. ante*, pp. 53–69), in which he incorporated what we think of as the first rational theory of colors. In this concept, the color is controlled by the angle between orbicular pulse and the mathematical rays delimiting a plane construction that can be rightfully called *physical ray*. It is on this concept that the experimental basis created by Newton helped improve, adding one important differentia to it: *the physical ray is not a plane figure*, like that of Hobbes and Hooke, but *a solid one*, a cone or a cylinder, or even a more general *tube*. This was the conclusion of the celebrated detailed experiments with a prism, which proved that the color is a property varying directionally, indeed, but in the cross-section of the light ray, and quite independently of the geometrical form of the ray itself as a physical object.

It should be illuminating, we believe, especially in understanding the concept of physical ray in general, to notice that, eventually, it was discovered that the color is described by a gauge group acting in the cross-section of the ray (Resnikoff, 1974), reproducing a group action of SL(2,R) type, and that Erwin Schrödinger pioneered this very discovery (Schrödinger, 1920). One can rightfully say that, with the theory of colors he created in 1920, Schrödinger was in fact completing an 'apprenticeship' for the physics which he started building six years later (Mazilu, Agop and Mercheş, 2019). Anyway, what we think is worth retaining from this brief history, in the spirit of today's physics, of course, is the fact that the epoch of reflection and refraction phenomenology produced the concept of *a light ray as a tube*, having the color as *a transversal property described by a gauge group*.

Along this historical path, Augustin Fresnel started a new epoch in the phenomenology of light, marked by the introduction of *diffraction* of light as a new phenomenon, in the description of which the periodic properties of light were the usual observables, and thereby the wave nature of light came closer to our understanding. The obvious spatial periodical pattern in the *recordings* of diffraction phenomena could thus be explained physically, as a mechanical interference phenomenon. In so doing, the optics made reference to the harmonic oscillator, in order to understand the intensity of light for instance, to say nothing of some other physically fundamental necessities, like the very definition of the *intensity of light*. However, this reference is, by stretching a little the meaning of words, 'illegal' to say the least, in the case of light, inasmuch as the light is a far cry from exhibiting the inertial properties required by a proper dynamics of the harmonic oscillator. For, as a purely dynamical system, the harmonic oscillator is a system described by forces proportional to displacements (those type of elastic forces, used initially by Hooke to explain the behavior of light), and in the case of physical optics the second principle of dynamics is quite incidental, as it were. It was introduced only by a natural mathematical property of transcendence of the second-order ordinary differential equation: it describes *any type* of periodic processes. And the fact is, that in the foundations of modern physical optics, the periodic processes of diffraction have more to do with the theory of statistics than with the dynamics (Fresnel, 1827).

This is, however, not to say that the harmonic oscillator is to be abandoned altogether, as a model, because as we have shown in Chapter 2, the Berry–Klein theory strongly indicates that this is not the case, either from experimental point of view, or even theoretically. All we want to say is that we need to find its right place and form of expression in the theory, and this is specified, again, through the order imposed by the measure of things, this time as their mass. Indeed, dynamically, the second-order differential equation expressing the principle of inertia, involves a finite mass. On the other hand, for light the mass is virtually inexistent, and if the second order differential equation is imposed by adding the diffraction to the phenomenology of light, this means that it actually describes *a transcendence between finite and infinitesimal scales of mass* [for a closer description of the concept, one can consult Mazilu, Agop and Mercheş (2019)]. As we shall see later, the mathematics of scale transitions between finite and infinitesimal gauges in a scale relativity, fully respects the rules related to the harmonic oscillator model. In fact, the whole wave mechanics, as a science, can be constructed based on such rules, which appear to be universal.

Now, along with the settling of Fresnel's theory in physical optics, a few changes in the natural philosophy have taken place. First in the order of things, was making the dynamics 'lawful', as it were, in the case of light. The first step was *to identify the phase*, mathematically involved as an independent variable in the trigonometric functions describing the diffraction in the light phenomenon, *with the time of an evolution*: mathematically, the phase had to be linear in time. A condition which brought *the frequency* front and center, and with it *the concept of wavelength*, thus generating right away a whole new experimental technology of the Newtonian kind, but leaving nevertheless behind what the dynamical principle really needed for a sound physical theory: first, the elastic properties of the medium supporting the light and, secondly, the *interpretation of light*, which obviously required the old idea of particle, and therefore the inescapable inertial properties. The Fresnel's ellipsoid of elasticities pretty much fills in for the first aspect of this issue, while the second one was delayed, and left in suspension ever since, being occasionally replaced with *ad hoc* creations of mind, and so is it, actually, even today to a large extent. A proper dynamical use of the second principle of dynamics in the matters of light came in handy only later on, with the advent of the electromagnetic theory of light. This theory of light has in common with the old Fresnel optics the one equation that models the space and time periodical properties no matter of their physical approach, as long as these properties are described by a frequency: *the D'Alembert equation*, which is the main characteristic of both the optics and the particle theory as Louis de Broglie has proved.

All these historical evidences indicate to us that de Broglie's work harmoniously completes a historical line of thinking, from which we have to learn for the benefit of knowledge in general. Once again, the concept of ray in his acceptance is the instrument of interpretation: it turns out to be essential in the construction of a coordinate system (Coll, 1985), and in this capacity defines a universal coordinate system — the Appell coordinates described by us in Chapter 4 — by a kinematic condition equivalent to old Hooke's explanation. As explained in Chapter 3, the existence of universal Newtonian forces means regularization, and the regularization procedure is a consequence of the condition of analiticity. One thing did not de Broglie's interpretation touch apparently, which seems a consequence of only the Madelung interpretation: the holographic property of matter. And the holography is the fourth of the phenomena that can enter the phenomenology of light, together with reflection, refraction and diffraction.

Nevertheless, as we have shown in Chapter 6, this property may be implicit in the construction of a ray, as designed by de Broglie. If properly constructed, a wave surface involves a Fubini–Pick cubic generating a vector that represents the 'memory' of ray. It is contained in the complete correspondence between time and phase: the classical case of linearity between phase and time is just a particular case of their connection. As we have shown in Chapter 6, there are stable cycles in the space of the parameters dictating the connection between time and phase cycles can be taken as the 'memory' of the relation between phase and time, which is a holographic property in the case of light. When referring the the physical properties, such cycles define the inertia of matter, which thus can be taken as a 'memory' property.

Bibliography

Aitken, R. G. (1918): *The Binary Stars*, Douglas C. McMurtrie, New York.

Albeverio, S., Hoegh-Krohn, R. (1974): *A Remark on the Connection Between Stochastic Mechanics and the Heat Equation*, Journal of Mathematical Physics, Volume **15**, pp. 1745–1747.

Alfaro, V. de, Fubini, S., Furlan, G. (1976): *Conformal Invariance in Quantum Mechanics*, Il Nuovo Cimento A, Volume **34**, pp. 569–611.

Apostol, T. M., Mnatsakanian, M. A. (2008): *New Descriptions of Conics via Twisted Cylinders, Focal Disks, and Directors*, American Mathematical Monthly, Volume **115**(8), pp. 795–812.

Appell, P. (1889): *De l'Homographie en Mécanique*, American Journal of Mathematics, Volume **12**(1), pp. 103–114.

Appell, P. (1891): *Sur les Lois de Forces Centrales Faisant Decrire a Leur Point d'Application une Conique Quelles Que Soient les Conditions Initiales*, American Journal of Mathematics, Volume **13**(2), pp. 153–158.

Appell, P. (1893): *Traité de Mécanique Rationnelle*, Tome I, Gauthier-Villars, Paris.

Baker, H. F. (1901): *On the Exponential Theorem for a Simply Transitive Continuous Group*, Proceedings London Mathematical Society, Volume **34**, pp. 91–127.

Barbilian, D. (1937): *Die von Einer Quantik Induzierte Riemannsche Metrik*, Comptes Rendus de l'Académie Roumaine des Sciences, Volumul **2**, p. 198.

Barbilian, D. (1938): *Riemannsche Raum Cubischer Binärformen*, Comptes Rendus de l'Académie Roumaine des Sciences, Volumul **2**, p. 345.

Barbilian, D. (1971): *Elementary Algebra*, in Didactic Works of Dan Barbilian, Volume **II**, Editura Tehnica, Bucuresti (in Romanian).

Barbilian, D. (1974): *Geometry and Function Theory*, in Didactic Works of Dan Barbilian, Volume **III**, pp. 5–167, Editura Tehnica, Bucuresti (in Romanian).

Barbour, J. B., Pfister, H. (eds) (1995): *Mach's Principle: From Newton's Bucket to Quantum Gravity*, in *Einstein Studies*, Volume **6**, Birkhäuser, Boston, Basel, Berlin.

Barrow, J. D., Dabrowski, M. P. (1998): *Gödel Universes in String Theory*, Physical Review D, Volume **58**(10), 103502; arxiv: gr-qc/980 3048v2.

Barton, G. (1986): *Quantum Mechanics of the Inverted Oscillator Potential*, Annals of Physics, Volume **166**, pp. 322–363.

Battaglini, G. (1878): *Sul Movimento per una Linea di 2 Ordine*, Giornale di Matematiche (Giornale di Battaglini), Volume **17**, pp. 43–52.

Bekenstein, J., Milgrom, M. (1984): *Does the Missing Mass Problem Signal the Breakdown of Newtonian Gravity*? The Astrophysical Journal, Volume **286**(1), pp. 7–14.

Bellman, R. (1997): *Introduction to Matrix Analysis*, Society for Industrial and Applied Mathematics, Philadelphia, PA.

Berry, M. V. (1984): *Quantal Phase Factors Accompanying Adiabatic Changes*, Proceedings of the Royal Society of London, Series A, Volume **392**, pp. 45–57.

Berry, M. V. (1985): *Classical Adiabatic Angles and Quantal Adiabatic Phase*, Journal of Physics A: Mathematical and General, Volume **18**(1), pp. 15–27.

Berry, M. V. (1989): *The Quantum Phase, Five Years After*, Original introductory contribution to (Shapere & Wilczek, 1989).

Berry, M. V., Klein, G. (1984): *Newtonian Trajectories and Quantum Waves in Expanding Force Fields*, Journal of Physics A: Mathematical and General, Volume **17**(8), pp. 1805–1815.

Bertrand, J. (1873): *Théorème Relatif au Mouvement d'un Point Attiré vers un Centre Fixe*, Comptes Rendus de l'Académie des Sciences Paris, Tome **77**, pp. 849–851.

Bertrand, J. (1877): (a) *Sur la Possibilité de Déduire d'une Seule des Lois de Kepler le Principe de l'Attraction*, Comptes Rendus de l'Académie des Sciences Paris, Tome **84**, pp. 671–674; (b) *Note sur un Problème de Mécanique, ibidem*, pp. 731–732.

Bilaniuk, O. M. P., Deshpande, V. K., Sudarshan, E. C. G. (1962): *"Meta" Relativity*, American Journal of Physics, Volume **30**(10), pp. 718–723.

Bilaniuk, O. M. P., Sudarshan, E. C. G. (1969): *Particles Beyond the Light Barrier*, Physics Today, Volume **22**(5), pp. 43–51.

Bloore, F. J. (1977): *The Shape of Pebbles*, Mathematical Geology, Volume **9**(2), pp. 113–122.

Bogoslovsky, G. Yu. (1973): *On Special Theory of Relativity of Anisotropic Space-Time*, DAN SSSR, Volume **213**(5), pp. 1055–1058 (in Russian).
Bogoslovsky, G. Yu. (2013): *On Relativistic Symmetry of Finsler Spaces with Mutually Opposite Preferred Directions*, Proceedings of the Conference on Physical Interpretations of Relativity Theory, Moscow, pp. 30–39; arXiv:gr-qc/1311.5432v1.
Bohr, N. (1913): *On the Constitution of Atoms and Molecules*, The Philosophical Magazine, Volume **26**, pp. 1–24.
Boltyanskii, V. G. (1974): *Anisotropic Relativism, Differential Equations*, Volume **10**(11), pp. 2101–2110 (in Russian).
Boltyanskii, V. G. (1979): *The Anisotropic Theory of Relativity and Optimization, Differential Equations*, Volume **15**(10), pp. 923–932 (in Russian).
Bond, V. R. (1985): *A Transformation of the Two-Body Problem*, Celestial Mechanics, Volume **35**, pp. 1–7.
Bondi, H. (1947): *Spherically Symmetrical Models in General Relativity*, Monthly Notices of the Royal Astronomical Society, Volume **107**, pp. 410–425.
Bondi, H. (1968): *Cosmology*, Cambridge University Press, Cambridge, UK.
Boyer, T. H. (1969): *Derivation of the Blackbody Radiation Spectrum without Quantum Assumptions*, Physical Review, Volume **182**(4), pp. 1374–1383.
Boyer, T. H. (1975): *Random Electrodynamics: The Theory of Classical Electrodynamics with Classical Electromagnetic Zero-Point Radiation*, Physical Review D, Volume **11**, pp. 790–808.
Brackenridge, J. B. (1995): *The Key to Newton's Dynamics: The Kepler Problem and the Principia*, University of California Press.
Brillouin, M. (1919): *Actions Mécaniques a Hérédité Discontinue par Propagation; Essai de Théorie Dynamique de l'Atome à Quanta*, Comptes Rendus de l'Académie des Sciences Paris, Tome **168**, pp. 1318–1320.
Brillouin, M. (1920): *Actions Héréditaires Discontinues et Équations Différentielles qui en Résultent*, Comptes Rendus du Congrès International des Mathématiciens, Strasbourg, 22–30 Septembre 1920, pp. 526–533.
Brillouin, M. (1922): *Atome de Bohr — Fonction de Lagrange Circumnucléaire*, Le Journal de Physique et le Radium, Tome **3**, pp. 65–73.
Broglie, L. de (1923): *Ondes et Quanta*, Comptes Rendus de l'Académie des Sciences, Paris, Tome **177**, pp. 507–510.
Broglie, L. de (1926): (a) *Sur le Parallélisme entre la Dynamique du Point Matériel et l'Optique Géométrique*, Le Journal de Physique et le Radium, Série VI, Tome **7**(1), pp. 1–6; (b) *Sur la Possibilité de*

Relier les Phénomènes d'Interférence et de Diffraction à la Théorie des Quanta de Lumière, Comptes Rendus de l'Académie des Sciences de Paris, Tome **183**, pp. 447–448; (c) *Interference and Corpuscular Light*, Nature, Volume **118**, pp. 441–442; (d) *Les Principes de la Nouvelle Mécanique Ondulatoire*, Le Journal de Physique et le Radium, Tome **7**(10), pp. 321–337; (e) *Ondes et Mouvements*, Gauthier-Villars, Paris.

Broglie, L. de (1935): *Une Remarque sur l'Interaction entre la Matière et le Champ Électromagnétique*, Comptes Rendus de l'Académie des Sciences de Paris, Tome **200**, pp. 361–363.

Brown, E. W. (1925): *Gravitational Forces in Spiral Nebulae*, The Astrophysical Journal, Volume **61**, pp. 97–113.

Brown, L. M. (ed) (2005): *Feynman's Thesis — A New Approach to Quantum Mechanics*, World Scientific, Singapore.

Burdet, C. A. (1968): *Theory of Kepler Motion: The General Perturbed Two Body Problem*, ZAMP, Volume **19**, pp. 345–368.

Burdet, C. A. (1969): *Le Mouvement Keplerien et les Oscillateurs Harmoniques*, Journal für die Reine und Angewandte Mathematik, Volume **238**, pp. 71–84.

Burns, J. (1966): *Noncentral Forces*, American Journal of Physics, Volume **34**, p. 164.

Burnside, W. S., Panton, A. W. (1960): *The Theory of Equations*, Dover Publications.

Buser, M., Kajari, E., Schleich, W. P. (2013): *Visualization of the Gödel Universe*, New Journal of Physics, Volume **15**, 013063.

Caldirola, P. (1941): *Forze Non Conservative nella Meccanica Quantistica*, Il Nuovo Cimento, Volume **18**(9), pp. 393–400.

Caldirola, P. (1983): *Quantum Theory of Nonconservative Systems*, Il Nuovo Cimento B, Volume **77**(2), pp. 241–262.

Cartan, E. (1930): *Notice Historique sur la Notion de Parallélisme Absolu*, Mathematische Annalen, Tome **102**, pp. 698–706; English translation in (Delphenich, 2011), pp. 130–137.

Cartan, E. (1931): *Le Parallélisme Absolu et la Théorie Unitaire du Champ*, Revue de Métaphysique Et de Morale, Tome **38**(1), pp. 13–28; English translation in (Delphenich, 2011), pp. 202–211.

Cartan, E. (1951): *La Théorie des Groupes Finis et Continus et la Géométrie Différentielle Traitées par la Méthode du Repère Mobile*, Gauthier-Villars, Paris.

Cartan, E. (2001): *Riemannian Geometry in an Orthogonal Frame*, World Scientific Publishing, Singapore.

Casimir, H. B. G. (1948): *On the Attraction Between Two Perfectly Conducting Plates*, Proceedings of the Royal Netherlands Academy of Arts and Sciences at Amsterdam, Volume **51**, pp. 793–795.

Castelnuovo, G. (1931): *De Sitter's Universe and the Motion of Nebulae*, Monthly Notices of the Royal Astronomical Society, Volume **91**, pp. 829–836.

Cayley, A. (1859): *A Sixth Memoir Upon Quantics*, Philosophical Transactions of the Royal Society of London, Volume **149**, pp. 61–90. Reproduced in *The Collected Mathematical Works*, Volume **II**, pp. 561–592, Cambridge University Press.

Cheng, S.-Y., Yau, S.-T. (1986): *Complete Affine Hypersurfaces. Part I: The Completeness of Affine Metrics*, Communications on Pure and Applied Mathematics, Volume **39**(5), pp. 839–866.

Chern, S.-S., Terng, C.-L. (1980): *An Analogue of Bäcklund's Theorem in Affine Geometry*, Rocky Mountain Journal of Mathematics, Volume **10**(1), pp. 105–124.

Clausius, R. (1870): *On a Mechanical Theorem Applicable to Heat*, Philosophical Magazine, Volume **40**, pp. 122–127. Republished in S. G. Brush, *Kinetic Theory*, Volume I, Pergamon Press, 1965.

Coll, B. (1985): *Coordenadas Luz en Relatividad*, In Proceedings of the 1985 Spanish Relativity Meeting, A. Molina (ed), Barcelona, Spain, pp. 29–38. There is an English translation as *Light Coordinates in Relativity*, available on the homepage of Bartolomé Coll at Observatoire de Paris/CNRS.

Coll, B. (2001): *Elements for a Theory of Relativistic Coordinate Systems. Formal and Physical Aspects*, In *Reference Frames and Gravitomagnetism*, Proceedings of the EREs 2000 Spanish Relativity Meeting, J. F. Pascual-Sánchez, L. Flora, A. San Miguel, and F. Vicente (eds), Valladolid, Spain, pp. 53–65.

Coll, B., Morales, J. A. (1988): *Sur les Repères Symmétriques Lorentziens*, Comptes Rendus de l'Académie des Sciences Paris, Tome **306**, Série I, pp. 791–794.

Coll, B., Morales, J. A. (1991): *Symmetric Frames on Lorentzian Spaces*, Journal of Mathematical Physics, Volume **32**(9), pp. 2450–2455.

Compton, A. H. (1915): *The Variation of the Specific Heat of Solids with Temperature*, Physical Review, Volume **6**, pp. 377–389.

Compton, A. H. (1916): *A Physical Study of the Thermal Conductivity of Solids*, Physical Review, Volume **7**, pp. 341–348.

Dainelli, U. (1880): *Sul Movimento per una Linea Qualunque*, Giornale di Matematiche (Giornale di Battaglini), Volume **18**, pp. 271–300.

Darboux, G. (1877): *Recherche de la Loi que Doit Suivre une Force Centrale pour que la Trajectoire qu'elle Détermine Soit Toujours une Conique*, Comptes Rendus de l'Académie des Sciences Paris, Tome **84**, pp. 760–762; 936–938.

Darboux, G. (1884): *Sur les Lois de Kepler*, Note XII en *Cours de Mécanique par Théodore Despeyrous*, Hermann, Paris, Tome **I**, pp. 432–440.

Darboux, G. (1910): *Leçons sur les Systèmes Orthogonaux et les Coordonnées Curvilignes*, Gauthier-Villars, Paris.

Darwin, C. G. (1927): *Free Motion in the Wave Mechanics*, Proceedings of the Royal Society of London Series A, Volume **117**, pp. 258–293.

Davisson, C., Germer, L. H. (1927): *Diffraction of Electrons by a Crystal of Nickel*, The Physical Review, Volume **30**(6), pp. 705–740.

Delphenich, D. H. (2002): *A Geometric Origin of the Madelung Potential*, arXiv:0211065 [gr-qc].

Delphenich, D. H. (Ed) (2011): *Selected Papers on Teleparallelism*, www.neo-classical-physics.info.

Delphenich, D. H. (2013): (a) *A Strain Tensor that Couples to the Madelung Stress Tensor*, arXiv:1303.3582v1 [quant-ph]; (b) *The Use of Teleparallelism Connection in Continuum Mechanics*, arXiv: 1305.3477v1 [gr-qc].

Denman, H. H. (1968): *Time-Translation Invariance for Certain Dissipative Classical Systems*, American Journal of Physics, Volume **36**(6), pp. 516–519.

Devisme, J. (1933): *Sur l'Équation de M. Pierre Humbert*, Annales de la Faculté des Sciences de Toulouse, Volume **25**, pp. 143–238.

Diósi, L., Lukács, B. (1985): *Covariant Evolution Equation for the Thermodynamic Fluctuations*, Physical Review A, Volume **31**, pp. 3415–3418.

Dirac, P. A. M. (1931): *Quantised Singularities in the Electromagnetic Field*, Proceedings of the Royal Society of London, Series A, Volume **133**, pp. 60–72.

Dirac, P. A. M. (1933): *The Lagrangian in Quantum Mechanics*, Physikalische Zeitschrift der Sowietunion, Volume **3**(1), pp. 64–72; reproduced in (Brown, 2005).

Dirac, P. A. M. (1948): *The Theory of Magnetic Poles*, Physical Review, Volume **74**(6), pp. 807–830.

Dirac, P. A. M. (1962): *An Extensible Model for the Electron*, Proceedings of the Royal Society of London, Series A, Volume **268**, pp. 57–67.

Duval, Ch., Ovsienko, V. (2000): *Lorentzian Worldlines and the Schwarzian Derivative*, Functional Analysis and its Applications, Volume **34**(2), pp. 135–137.

Duval, Ch., Guieu, L. (2000): *The Virasoro Group and Lorentzian Surfaces: The Hyperboloid of One Sheet*, Journal of Geometry and Physics, Volume **33**, pp. 103–127; arXiv: mathDG/9806135v1.

Earnshaw, S. (1842): *On the Nature of the Molecular Forces which regulate the Constitution of the Luminiferous Ether*, Transactions of the Cambridge Philosophical Society, Volume **VII**, Part I, pp. 97–112.

Einstein, A. (1905): *On the Electrodynamics of Moving Bodies*, in *The Principle of Relativity, a Collection of Original Memoirs on the Special and General Theory of Relativity*, Dover Publications, 1923, 1952, 2003. English translation of the original from Annalen der Physik, Volume **17** (1905).

Einstein, A. (1907): *Planck's Theory of Radiation and the Theory of Specific Heat*, The Collected Papers of Albert Einstein, Volume **2**, Princeton University Press; English translation of the German original from Annalen der Physik, Volume **22**, pp. 180–190.

Einstein, A. (1909): *Planck's Theory of Radiation and the Theory of Specific Heat*, The Collected Papers of Albert Einstein, Volume 2, Princeton University Press; English translation of the original from Deutsche Physikalische Gesellschaft, Verhandlungen 7, pp. 482–500; also published in Physikalische Zeitschrift, Volume **10**, pp. 817–825.

Einstein, A. (1916): *The Foundation of the General Theory of Relativity*, The Collected Papers of Albert Einstein, Volume **6**, Princeton University Press, pp. 146–200; English translation of the German original from Annalen der Physik, Volume **49**, 1916.

Einstein, A. (1917): *Cosmological Considerations on the General Theory of Relativity*, in *The Principle of Relativity, A Collection of Original Memoirs on the Special and General Theory of Relativity*, Dover Publications, 1923, 1952, 2003. English translation of the German original from Sitzungsberichte der Preussischen Akademie der Wissenschaften, 1917.

Einstein, A. (1918): *Critical Comment on a Solution of the Gravitational Field Equations Given by Mr. De Sitter*, The Collected Papers of Albert Einstein, Volume **8**, Princeton University Press, pp. 36–38; English translation of the German original from Sitzungberichte der Königlich Preussische Akademie der Wissenschaften, Berlin 1918, pp. 270–272.

Einstein, A. (1949): *Remarks Concerning the Essays Brought Together in this Co-Operative Volume*, in *Albert Einstein, Philosopher-Scientist*, P. A. Schilpp (ed); Volume VII, the Library of Living Philosophers, pp. 665–688, MJF Books, NY.

Einstein, A. (1956): *Investigations on the Theory of the Brownian Movement*, edited with notes by R. Fürth, Dover Publications, Inc., New York.

Einstein, A. (1965): *Concerning an Heuristic Point of View toward the Emission and Transformation of Light*, American Journal of Physics, Volume **33**(4), pp. 367–374; English translation of the original from Annalen der Physik, Volume **17** (1905).

Einstein, A. (2004): *The Meaning of Relativity*, Routledge Classics, London and New York.

Einstein, A., Hopf, L. (1910): (a) *On a Theorem of the Probability Calculus and Its Applications in the Theory of Radiation*; (b) *Statistical Investigations of a Resonator's Motion in a Radiation Field*, The Collected Papers of Albert Einstein, Volume 3, Princeton University Press; English translations of the German originals from Annalen der Physik, Volume **33**, pp. 1096–1115.

Einstein, A., Stern, O. (1913): *Some Arguments for the Assumption of Molecular Agitation at Absolute Zero, and Remark Added in Proof*, The Collected Papers of Albert Einstein, Volume 4, Princeton University Press; English translation of the German original: Annalen der Physik, Volume **40**, pp. 551–560.

Eliezer, C. J., Gray, A. (1976): *A Note on the Time-Dependent Harmonic Oscillator*, SIAM Journal of Applied Mathematics, Volume **30**, pp. 463–468.

Emden, R. (1907): *Gaskugeln*, Teubner, Leipzig and Berlin.

Eötvös, L. (2008): *On the Gravitation Produced by the Earth on Different Substances*, The Abraham Zelmanov Journal, Volume **1**, pp. 6–9; English translation of the 1889 original: *Über die Anziehung der Erde auf verschiedene Substanzen*. Mathematische und Naturwissenschaftliche Berichte aus Ungarn, 1890, Band 8, S. 65–68.

Ermakov, V. P. (2008): *Second Order Differential Equations: Conditions of Complete Integrability*, Applicable Analysis and Discrete Mathematics, Volume **2**, pp. 123–145.

Evans, G. C. (1923): *A Bohr-Langmuir Transformation*, Proceedings of the National Academy of the United States, Volume **9**, pp. 230–236.

Fels, M., Olver, P. J. (1998): *Moving Coframes I*, Acta Applicandae Mathematicae, Volume **51**(2), pp. 161–213.

Fels, M., Olver, P. J. (1999): *Moving Coframes II*, Acta Applicandae Mathematicae, Volume **55**(2), pp. 127–208.

Fennelly, A. J. (1974): *Gravitational Charge, Hadron Hydrodynamics and Gödelized Hadrons*, Nature, Volume **248**, pp. 221–223.

Ferguson, T. S. (1978): *Maximum Likelihood Estimates of the Parameters of the Cauchy Distribution for Samples of Size 3 and 4*, Journal of the American Statistical Association, Volume **73**, pp. 211–213.

Fermi, E. (1950): *High Energy Nuclear Events*, Progress of Theoretical Physics (Japan), Volume **5**, pp. 570–583.

Feynman, R. P. (1948): *Space-Time Approach to Non-relativistic Quantum Mechanics*, Reviews of Modern Physics, Volume **20**, pp. 367–387.

Feynman, R. P. (1949): *The Theory of Positrons*, Physical Review, Volume **76**, pp. 749–759.

Feynman, R. P. (1965): *The Development of the Space-Time View of Quantum Electrodynamics*, Nobel Lecture, December 11, 1965.

Feynman, R. P. (1969): *Very High-Energy Collisions of Hadrons*, Physical Review Letters, Volume **23**(23), pp. 1415–1417.

Feynman, R. P. (1995): *Lectures on Gravitation*, F. B. Morinigo, W. G. Wagner & B. Hatfield (eds), Addison-Wesley Publishing Company, Reading, MA.

Feynman, R. P., Hibbs, A. R. (1965): *Quantum Mechanics and Path Integrals*, McGraw-Hill Publishing Company, New York, NY.

Fisher, R. A. (1922): *On the Mathematical Foundations of Theoretical Statistics*, Philosophical Transactions of the Royal Society of London, Series A, Volume **222**, pp. 309–368.

Fisher, R. A. (1925): (a) *Applications of "Student's" Distribution*, Metron, Volume **5**, pp. 90–104; (b) *Theory of Statistical Estimation*, Mathematical Proceedings of the Cambridge Philosophical Society, Volume **22**(4) pp. 700–725.

Fixsen, D. J., Cheng, E. S., Gales, J. M., Mather, J. C., Shafer, R. A., Wright, E. L. (1996): *The Cosmic Microwave Background Spectrum from the Full COBE FIRAS Data Set*, The Astrophysical Journal, Volume **473**, pp. 576–587.

Flamm, L. (2015): *Contributions to Einstein's Theory of Gravitation*, General Relativity and Gravitation, Volume **47**, Article 72; translation of the German original from Physikalische Zeitschrift, Volume **17** (1916), pp. 448–454.

Flanders, H. (1965): *Local Theory of Affine Hypersurfaces*, Journal d'Analyse Mathématique, Volume **15**(1), pp. 353–387.

Flanders, H. (1970): *The Schwarzian as a Curvature*, Journal of Differential Geometry, Volume **4**(4), pp. 515–519.

Flanders, H. (1973): *Differentiation under the Integral Sign*, The American Mathematical Monthly, Volume **80**(5), pp. 615–627.

Flanders, H. (1989): *Differential Forms, with Applications to the Physical Sciences*, Dover Publications, New York.

Fock, V. A. (1959): *The Theory of Space, Time and Gravitation*, Pergamon Press, New York.

Fowles, G. R. (1977): *Self-Inverse Form of the Lorentz Transformation*, American Journal of Physics, Volume **45**(6), pp. 675–676.

Frederick, C. (1976): *Stochastic Space-Time and Quantum Theory*, Physical Review D, Volume **13**(11), pp. 179–196.

Fresnel, A. (1821): *Considérations Mécaniques sur la Polarisation de la Lumière*, Annales de Chimie et de Physique, Tome **17**, pp. 179–196; Reprinted in Oeuvres Complètes, Imprimerie Impériale, Paris 1866, Tome Premier, pp. 629–645 (one can find it at Gallica.bnf.fr).

Fresnel, A. (1827): *Mémoire sur la Double Réfraction*, Mémoirs de l'Académie des Sciences de l'Institute de France, Tome **7**, pp. 45–176; reproduced in Oeuvres Complètes, Imprimerie Impériale, Paris 1866, (can be found at Gallica.bnf.fr).

Fulton, D. G., Rainich, G. Y. (1932): *Generalizations to Higher Dimensions of the Cauchy Integral Formula*, American Journal of Mathematics, Volume **54**, pp. 235–241.

Gallavotti, G. (2001): *Quasi Periodic Motions from Hipparchus to Kolmogorov*, Atti della Accademia Nazionale dei Lincei. Classe di Scienze Fisiche, Matematiche e Naturali, Volume **12**, pp. 125–152; arXiv:chao-dyn/9907004v1.

Georgescu-Roegen, N. (1971): *On the Texture of the Arithmetical Continuum*, Appendix A in *The Entropy Law and the Economic Process*, Harvard University Press, Cambridge MA, USA.

Gibbons, G. W., Warnick, C. M. (2011): *The Geometry of Sound Rays in a Wind*, Contemporary Physics, Volume **52**(3), pp. 197–209; arXiv: 1102.2409v1.

Gibbs, J. W. (1883): *On the General Equations of Monochromatic Light in Media of Every Degree of Transparency*, American Journal of Science, Volume **25**, pp. 107–118; reproduced in *The Scientific Papers of J. Willard Gibbs*, Longmans, Green, & Co, New York 1906, Volume II, pp. 211–222.

Glaisher, J. W. L. (1878): *On the Law of Force to Any Point in the Plane of Motion, in Order that the Orbit May be Always a Conic*, Monthly Notices of the Royal Astronomical Society, Volume **39**, pp. 77–91.

Gödel, K. (1949): (a) *Lecture on Rotating Universes*, in *Collected Works*, edited by S. Feferman, J. W. Dawson Jr., W. Goldfarb, C. Parsons and R. N. Solovay, Oxford University Press, Oxford UK, 1995, Volume III, pp. 269–287; (b) *An Example of a New Type of Cosmological Solutions of Einstein Field Equations of Gravitation*, Reviews of Modern Physics, Volume **21**, pp. 447–450; reproduced in General Relativity and Gravitation, Volume **32**, (2000) pp. 1409–1417; (c) *A Remark about the Relationship Between Relativity Theory and Idealistic Philosophy*, in *Albert Einstein, Philosopher-Scientist*, P. A. Schilpp editor; Volume VII, the Library of Living Philosophers, pp. 557–562.

Gödel, K. (1952): *Rotating Universes in the General Relativity*, in *Proceedings of the International Congress of Mathematicians*, Cambridge, MA 1950; American Mathematical Society, 1952, Volume **1**, pp. 175–181;

reproduced in General Relativity and Gravitation, Volume **32**(6), 2000, pp. 1419–1427.

Gonera, J. (2013): *Conformal Mechanics*, arXiv: hep-th/1211.4403 v4.

Gradshteyn, I. S., Ryzhik, I. M. (2007): *Table of Integrals, Series and Products*, Seventh Edition, A. Jeffrey & D. Zwillinger (eds), Academic Press–Elsevier.

Guggenheimer, H. W. (1977): *Differential Geometry*, Dover Publications, New York.

Güngör, F., Torres, P. J. (2017): *Lie Point Symmetry Analysis of a Second Order Differential Equation with Singularity*, Journal of Mathematical Analysis and Applications, Volume **451**(2), pp. 976–989.

Halphen, G-H. (1877): *Sur les Lois de Kepler*, Comptes Rendus de l'Académie des Sciences, Paris, Tome **84**, pp. 939–941.

Halphen, G. (1879): *Sur l'Équation Différentielle des Coniques*, Bulletin de la Societé Mathématique de France, Tome **7**, pp. 83–85.

Hamilton, W. R. (1841): *On a Mode of Deducing the Equation of Fresnel's Wave*, Philosophical Magazine, Volume **19**, pp. 381–383; reproduced in "The Mathematical Papers of Sir William Rowan Hamilton", A. W. Conway & J. L. Synge (eds), Cambridge University Press 1931, Volume **I**, pp. 341–343.

Hamilton, W. R. (1847): *On Theorems of Central Forces*, Proceedings of the Royal Irish Academy, Volume **3**, pp. 308–309; reproduced in "The Mathematical Papers of Sir William Rowan Hamilton", A. W. Conway & J. L. Synge (eds), Cambridge University Press 1931, Volume **II**, p. 286.

Hannay, J. H. (1985): *Angle Variable Holonomy in Adiabatic Excursion of an Integrable Hamiltonian*, Journal of Physics A: Mathematical and General Volume **18**(2), pp. 221–230.

Harrison, E. R. (1965): (a) *Microcosmic Model of the Universe*, Astronomical Journal, Volume **70**, p. 677; (b) *Cosmology without General Relativity*, Annals of Physics, Volume **35**, pp. 437–446.

Harrison, E. R. (2000): *Cosmology, the Science of the Universe*, Cambridge University Press, Cambridge UK.

Heckmann, O. (1942): *Theorien der Kosmologie*, Springer-Verlag, Berlin.

Heckmann, O., Schücking, E. (1955): *Bemerkungen zur Newtonschen Kosmologie. I*, Zeitschrift für Astrophysik, Band **38**, S. 95–109.

Heckmann, O., Schücking, E. (1956): *Bemerkungen zur Newtonschen Kosmologie. II*, Zeitschrift für Astrophysik, Band **40**, S. 81–92.

Heisenberg, W. (1925): *Über quantentheoretische Umdeutung kinematischer und mechanischer Beziehungen*, Zeitschrift für Physik, Volume **33**, pp. 879–893.

Hertz, H. (1893): *Electric Waves*, Dover Publications, Inc. 1962.

Hertz, H. (2003): *The Principles of Mechanics, Presented in a New Form*, Dover Phoenix Editions.

Hobbes, T. (1644): *Tractatus Opticus*, in *Thomae Hobbes Malmesburiensis Opera Philosophica Quae Latine Scripsit Omnia*, Volume **5**, pp. 215–248, J. Bohn, London, 1839.

Hoffman, W. C. (1966): *The Lie Algebra of Visual Perception*, Journal of Mathematical Psychology, Volume **3**, pp. 65–98.

Hooke, R. (1665): *Micrographia, or Some Physiological Descriptions of Minute Bodies Made by Magnifying Glasses*, Martyn & Allestry, London.

Hooke, R. (1705): *The Posthumous Works of Robert Hooke*, Johnson Reprint Corporation, New York 1969.

Hu, Z.-J., Zhao, G.-S. (1997): *Isometric Immersions of the Hyperbolic Space* $H^2(-1)$ *Into* $H^3(-1)$, Proceedings of the American Mathematical Society, Volume **125**, pp. 2693–2697.

Hubble, E. P. (1929): *A Relation Between Distance and Radial Velocity Among Extra-Galactic Nebulae*, Proceedings of the National Academy of Sciences USA, Volume **15**, pp. 168–173.

Humbert, P. (1929): (a) *Potentiel Correspondant à une Attraction Proportionnelle à* $\rho\,\exp(?^2/2)$, Mathematica, Cluj-Napoca, Volume **1**, pp. 117–121; (b) *Sur une Généralisation de l'Équation de Laplace*, Journal de Mathématiques Pures et Appliquées, Tome **8**, pp. 145–159.

Hurwitz, A. (1898): *Über die Komposition quadratischer Formen von beliebig vielen Variablen*, Nachrichten der Königlichen Gesellschaft der Wissenschaften zu Göttingen. Mathematisch-physikalische Klasse, pp. 309–316.

Jackiw, R., Nair, V. P., Pi, S.-Y., Polychronakos, A. P. (2004): *Perfect Fluid Theory and its Extensions*, Journal of Physics A: Mathematical and General, Volume **37**, pp. R327–R432.

Jammalamadaka, S. R., Sengupta, A. (2001): *Topics in Circular Statistics*, World Scientific, Singapore.

James, D. F. V., Agarwal, G. S. (1996): *The Generalized Fresnel Transform and its Applications to Optics*, Optics Communications, Volume **126**, pp. 207–212.

Jeans, J. H. (1913): *Discussions on Radiations, with the Participation of H. A. Lorentz, E. Pringsheim, A. E. H. Love, and J. Larmor*; in British Association Report on Birmingham Meeting from, September 1913, pp. 376–386.

Jeans, J. H. (1923): *Internal Motions in Spiral Nebulae*, Monthly Notices of the Royal Astronomical Society, Volume **84**, pp. 60–76.

Jeans, J. H. (1954): *The Dynamical Theory of Gases*, Dover Publications, New York.

Jessop, C. M. (1903): *A Treatise on the Line Complex*, Cambridge University Press, Cambridge, UK.

Kanai, E. (1948): *On the Quantization of Dissipative Systems*, Progress of Theoretical Physics (Japan), Volume **3**(4), pp. 440–442.

Katz, E. (1965): *Concerning the Number of Independent Variables of the Classical Electromagnetic Field*, American Journal of Physics, Volume **33**, pp. 306–312.

Kelvin, William Thomson, Lord (1847): *Extraits de Deux Lettres Adressées a M. Liouville*, Journal des Mathématiques Pures et Appliquées (Liouville), Tome **12**, pp. 256–264, and Liouville's own following commentaries, ibidem, pp. 265–290.

Kennedy, F. J., Kerner, E. H. (1965): *Note on the Inequivalence of Classical and Quantum Hamiltonians*, American Journal of Physics, Volume **33**(5), pp. 463–466; *Adendum*, Ibidem, Volume **34**(3), p. 271.

Kindlmann, G. (2004): *Visualization and Analysis of Diffusion Tensor Fields*, PhD Dissertation, School of Computing, University of Utah, Salt Lake City, UT, USA.

Klein, F. (1918): *Über die Integralform der Erhaltungssätze und die Theorie der räumlich-geschlossenen Welt*, Nachrichten der Königlichen Gesellschaft der Wissenschaften zu Göttingen. Mathematisch-physikalische Klasse, 1918, pp. 394–423; *On the Integral Form of Conservation Laws and the Theory of Spatially-Closed World*, English translation by D. H. Delphenich, of the German original; to be found at www.neo-classical-physics.info.

Klein, F. (1919): *Bemerkungen über die Beziehungen des DE SITTER 'schen Koordinatensystems B zu der allgemeinen Welt konstanter positiver Krummung*, Proceedings of the Royal Netherlands Academy of Arts and Sciences at Amsterdam, Volume **21 I**, pp. 614–615.

Klein, G. (1984): *An Invariant for Repeated Reflections in an Expanding Ellipsoid*, Proceedings of the Royal Society of London, Series A, Volume **396**, pp. 217–226.

Klein, G., Mulholland, H. P. (1978): *Repeated Elastic Reflections of a Particle in an Expanding Sphere*, Proceedings of the Royal Society of London A, Volume **361**, pp. 447–461.

Kundt, W. (1956): *Trägheitsbahnen in einem von Gödel angegebenen Kosmologischen Modell*, Zeitschrift für Physik, Volume **145**, pp. 611–620.

Kustaanheimo, P. E., Stiefel, E. (1965): *Perturbation Theory of Kepler Motion Based on Spinor Regularization*, Journal für die Reine und Angewandte Mathematik, Volume **218**, pp. 204–219.

Lalan, V. (1937): *Sur les Postulats qui sont à la Base des Cinématiques*, Bulletin de la Societé Mathématique de France, Tome **65**, pp. 83–99.

Landau, L. D. (1965): *On Multiple Production of Particles during Collisions of Fast Particles*, in Collected Papers, D. Ter Haar editor, Gordon & Breach, Ltd., New York, pp. 569–585.

Langevin, P. (1905): *Sur l'Origine des Radiations et l'Inertie Électromagnétique*, Journal de Physique Théorique et Appliquée, Tome **4**(1), pp. 165–183.

Langmuir, I. (1921): *Forces Within a Static Atom*, Physical Review, Volume **18**, p. 104.

Larmor, J. (1893): *The Singularities of the Optical Wave-Surface, Electric Stability, and Magnetic Rotary Polarization*, Proceedings of the London Mathematical Society, Volume **24**, pp. 272–290.

Larmor, J. (1893): *A Dynamical Theory of the Electric and Luminiferous Medium*, Proceedings of the Royal Society of London, Series A, Volume **54**, pp. 438–461.

Larmor, J. (1894): *A Dynamical Theory of the Electric and Luminiferous Medium*, Philosophical Transactions of the Royal Society of London, Series A, Volume **185**, pp. 719–822.

Larmor, J. (1895): *A Dynamical Theory of the Electric and Luminiferous Medium*, Philosophical Transactions of the Royal Society of London, Series A, Volume **186**, pp. 695–743.

Larmor, J. (1897): *A Dynamical Theory of the Electric and Luminiferous Medium*, Philosophical Transactions of the Royal Society of London, Series A, Volume **190**, pp. 205–300.

Larmor, J. W. (1900): *On the Statistical Dynamics of Gas Theory as Illustrated by Meteor Swarms and Optical Rays*, Nature, Volume **63**(1624), pp. 168–169; British Association Report, September 1900, pp. 632–634.

Lavenda, B. H. (1990): *Underlying Probability Distributions of Planck's Radiation Law*, International Journal of Theoretical Physics, Volume **29**(12), pp. 1379–1392.

Lavenda, B. H. (1992): *Statistical Physics, A Probabilistic Approach*, Wiley & Sons, New York.

Lavenda, B. H. (2012): *A New Perspective on Relativity — An Odyssey in Non-Euclidean Geometries*, World Scientific, New Jersey.

Layzer, D. (1954): *On the Significance of Newtonian Cosmology*, The Astronomical Journal, Volume **59**, pp. 268–270.

Layzer, D., Heckmann, O., Schücking, E. (1956): *Newtonian Cosmology*, The Observatory, Volume **76**, pp. 73–75.

Lebedeff, P. (1900): *Les Forces de Maxwell-Bartoli dues a la Pression de la Lumière*, Rapports Présentés au Congrès International de Physique, Ch-Éd. Guillaume & L. Poincaré (eds), Volume **II**, pp. 133–140, Gauthier-Villars, Paris.

Lebedew, P. (1902): *Experimental Investigations on the Pressure of Light*, Astrophysical Journal, Volume **15**, pp. 60–61.

Lemons, D. S. (1988): *A Newtonian Cosmology Newton would Understand*, American Journal of Physics, Volume **56**, pp. 502–504.

Letac, G. (1989): *Le Probléme de la Classification des Familles Exponentielles Naturelles de R^d ayant une Fonction-Variance Quadratique, n Probability Measures on Group IX*. Lecture Notes in Mathematics, Volume **1379**, pp. 194–215, Springer, Berlin.

Letac, G. (2016): *Associated Natural Exponential Families and Elliptic Functions*, in *The Fascination of Probability, Statistics and their Applications*, in *Honour of Ole E. Barndorff-Nielsen*, M. Podolskij & Al, (eds), pp. 53–83, Springer International Publishing Switzerland, 2016.

Levi-Civita, T. (1920): *Sur la Régularisation du Problème des Trois Corps*, Acta Mathematica, Volume **42**, pp. 99–144.

Lévy, P. (1943): *Un Théorème d'Invariance Projective Relatif au Mouvement Brownien*, Commentarii Mathematici Helvetici, Tome **16**, pp. 242–248.

Lévy, P. (1944): *Une Propriété d'Invariance Projective dans le Mouvement Brownien*, Comptes Rendus de l'Académie des Sciences Paris, Tome **219**, pp. 377–379.

Lévy, P. (1954): *Le Mouvement Brownien*, Gauthier-Villars, Paris.

Lévy, P. (1965): *Processus Stochastiques et Mouvement Brownien*, Gauthier-Villars, Paris.

Lewis, G. N. (1926): *The Conservation of Photons*, Nature, Volume **118**, pp. 874–875.

Lewis, H. R. (1968): *Class of Exact Invariants for Classical and Quantum Time-Dependent Harmonic Oscillators*, Journal of Mathematical Physics, Volume **9**(11), pp. 1976–1986.

Lorentz, H. A. (1892): *La Théorie Électromagnétique de Maxwell et son Application aux Corps Mouvant*, Archives Néerlandaises des Sciences Exactes et Naturelles, Tome **XXV**, pp. 363–552.

Lorentz, H. A. (1904): *Electromagnetic Phenomena in a System Moving with any Velocity Smaller than that of Light*, Proceedings of the Royal Netherlands Academy of Arts and Sciences at Amsterdam, Volume **6**, pp. 809–831.

Lorentz, H. A. (1916): *Theory of Electrons*, Teubner, Leipzig, New York.

Lorentz, H. A. (1921): *Deux Mémoires de Henri Poincaré sur la Physique Mathématique*, Acta Mathematica, Tome **38**, pp. 293–308.

Lorenz, L. (1867): *On the Identity of the Vibrations of Light with Electrical Currents*, Philosophical Magazine, Volume **84**, Series 4, pp. 287–301 (English translation of the German original from Poggendorff's *Annalen*, June 1867).

Ludford, G. S. S., Martinek, J., Yeh, G. C. K. (1955): *The Sphere Theorem in Potential Theory*, Mathematical Proceedings of the Cambridge Philosophical Society, Volume **51**, pp. 389–393.

Lummer, O., Pringsheim. E. (1899): *Die Vertheilung der Energie im Spectrum des schwarzen Körpers*, Verhandlungen der Deutschen Physikalischen Gesellschaft im Jahre 1899, pp. 23–41.

Lutzky, M. (1978): (a) *Symmetry Groups and Conserved Quantities for the Harmonic Oscillator*, Journal of Physics A: Mathematical and General, Volume **11**(2), pp. 249–258; (b) *Noether's Theorem and the Time-Dependent Harmonic Oscillator*, Physics Letters A, Volume **68**, pp. 3–4.

Lynden-Bell, D. (2006): *Hamilton Eccentricity Vector Generalized to Newton Wonders*, The Observatory, Volume **126**, pp. 176–182; arXiv: astro-ph/0604428v1.

Mach, E. (1902): *The Science of Mechanics*, The Open Court Publishing Company, Chicago.

Madelung, E. (1927): *Quantentheorie in Hydrodynamischer Form*, Zeitschrift für Physik, Volume **40**, pp. 322–326; *Quantum Theory in Hydrodynamical Form*, English translation by D. H. Delphenich; www.neo-classical-physics.info.

Malament, D. B. (1995): *Introductory Notes to Gödel's Lecture*, in (Gödel, 1949a).

Mandel, L. (1974): *Interpretation of Instantaneous Frequencies*, American Journal of Physics, Volume **42**, pp. 840–848.

Mandelbrot, B. (1956): (a) *An Outline of a Purely Phenomenological Theory of Statistical Thermodynamics: I. Canonical Ensembles*, I. R. E. Transactions on Information Theory, Volume **2**(3) pp. 190–203; (b) *Exhaustivité de l'Énergie Totale d'un Systeme en Équilibre, pour l'Estimation de sa Température*, Comptes Rendus de l'Académie des Sciences Paris, Tome **243**, pp. 1835–1838.

Mandelbrot, B. (1982): *The Fractal Geometry of Nature*, W. H. Freeman & Company, New York.

Mariwalla, K. H. (1982): *Integrals and Symmetries: The Bernoulli-Laplace-Lenz Vector*, Journal of Physics A: Mathematical and General, Volume **15**(8), pp. L467–L471.

Marshall, T. W. (1963): *Random Electrodynamics*, Proceedings of the Royal Society of London, Series A, Volume **276**, pp. 475–491.

Marshall, T. W. (1965): *Statistical Electrodynamics*, Proceedings of the Cambridge Philosophical Society, Volume **61**, pp. 537–546.

Mashhoon, B., Hehl, F. W., Theiss, D. S. (1984): *On the Gravitational Effects of Rotating Masses*, General Relativity and Gravitation, Volume **16**(8), pp. 711–750.

Maxwell, J. C. (1861): *On Physical Lines of Force*, Philosophical Magazine, Volume **21**, pp. 161–175; 281–291; 338–348.

Maxwell, J. C. (1862): *On Physical Lines of Force*, Philosophical Magazine, Volume **23**, pp. 12–24; 85–95.

Maxwell, J. C. (1904): *Theory of Heat*, Longmans, Green & Co., London.

Maxwell, J. C. (1965): *Scientific Papers*, edited by W. D. Niven, Dover Publications.

Mazilu, N. (2006): *The Stoka Theorem, a Side Story of Physics in Gravitation Field*, Supplemento ai Rendiconti del Circolo Matematico di Palermo **77**, pp. 415–440.

Mazilu, N. (2010): *Black-Body Radiation Once More*, Bulletin of the Polytechnic Institute of Jassy, Volume **54**, pp. 69–97.

Mazilu, N. (2019): *Physical Principles in Revealing the Working Mechanisms of Brain, Part I*, Bulletin of the Polytechnic Institute of Jassy, Volume **65**(69), Issue 4.

Mazilu, N., Agop, M. (2012): *Skyrmions — A Great Finishing Touch to Classical Newtonian Philosophy*, Nova Publishers, New York.

Mazilu, N., Ioannou, P. D., Diakonos, F. K., Maintas, X. N., Agop, M (2013): *A Quark-Independent Description of Confinement*, Modern Physics Letters A, Volume **28**(30), 1350126.

Mazilu, N., Agop, M., Axinte, C. I., Radu, E., Jarcău, M., Gârţu, M., Răuţ, M., Pricop, M., Boicu, M., Mihăileanu, D., Vrăjitoriu, L. (2014): *A Newtonian Message for Quantization*, Physics Essays, Volume **27**(2), pp. 204–214.

Mazilu, N., Agop, M (2015): *The Concept of Physical Surface in Nuclear Matter*, Modern Physics Letters A, Volume **30**(6), 1550026.

Mazilu, N., Porumbreanu, M. (2018): *Devenirea Mecanicii Ondulatorii*, Editura Limes, Cluj-Napoca (soon to appear in English as *The Coming to Being of Wave Mechanics*).

Mazilu, N., Agop, M., Mercheş, I. (2019): *The Mathematical Principles of Scale Relativity Physics; The Concept of interpretation*, CISP & CRC Press (Taylor & Francis Group).

McCrea, W. H. (1955): *On the Significance of Newtonian Cosmology*, The Astronomical Journal, Volume **60**, pp. 271–274.

McCrea, W. H., Milne, E. A. (1934): *Newtonian Universes and the Curvature of Space*, The Quarterly Journal of Mathematics, Volume **5**(1), pp. 73–80; reprinted in the General Relativity and Gravitation, Volume **32**(9), pp. 1949–1958 (2000).

McCullagh, P. (1996): *Möbius Transformation and Cauchy Parameter Estimation*, The Annals of Statistics, Volume **24**(2), pp. 787–808.

Mihaileanu, N. (1972): *Analytic, Projective and Differential Geometry Complements*, Editura Didactica si Pedagogica, Bucharest (in Romanian).

Miles, E. P. (1954): *Three Dimensional Harmonic Functions Generated by Analytic Functions of a Hypervariable*, The American Mathematical Monthly, Volume **61**(10), pp. 694–697,

Milgrom, M. (1983): *A Modification of the Newtonian Dynamics as a Possible Alternative to the Hidden Mass Hypothesis*, The Astrophysical Journal, Volume **270**, pp. 365–389.

Milgrom, M. (1984): *Isothermal Spheres in the Modified Dynamics*, The Astrophysical Journal, Volume **287**, pp. 571–576.

Milnor, J. (1983): *On the Geometry of the Kepler Problem*, The American Mathematical Monthly, Volume **90**(6), pp. 353–365.

Milonni, P. W., Shih, M.-L. (1991): *Zero-Point Energy in Early Quantum Theory*, American Journal of Physics, Volume **59**(8), pp. 684–698.

Moigno, l'Abbé (1868): *Leçons de Mécanique Analytique: Statique*, Gauthier-Villars, Paris.

Morris, C. N. (1982): *Natural Exponential Families with Quadratic Variance Functions*, Annals of Statistics, Volume **10**, pp. 65–80.

Nedeff, V., Lazăr, G., Agop, M., Eva, L., Ochiuz, L., Dimitriu, D., Vrăjitoriu, L., Popa, C. (2015): *Solid Components Separation from Heterogeneous Mixtures through Turbulence Control*, Powder Technology, Volume **284**, pp. 170–186.

Needham, T. (2001): *Visual Complex Analysis*, Clarendon Press, Oxford.

Nelson, E. (1966): *Derivation of the Schrödinger Equation from Newtonian Mechanics*, Physical Review, Volume **150**, pp. 1079–1085.

Németi, I., Madarász, J. X., Andréka, H., Andai, A. (2008): *Visualizing Some Ideas about Gödel-type Rotating Universes*, arXiv:0811.2910v1 [gr-qc].

Newton, Sir Isaac (1952): *Opticks, or a Treatise of the Reflections, Refractions, Inflections & Colours of Light*, Dover Publications, Inc., New York.

Newton, I. (1974): *The Mathematical Principles of Natural Philosophy*, Translated into English by Andrew Motte in 1729. The Translation Revised, and Supplied with an Historical and Explanatory Appendix, by Florian Cajori, University of California Press, Berkeley and Los Angeles, CA.

Niall, K. K. (2017): *Erwin Schrödinger's Color Theory, Translated with Modern Comentary*, Springer International Publishing AG.

Niederer, U. (1972): *The Maximal Kinematical Invariance Group of the Free Schrödinger Equation*, Helvetica Physica Acta, Volume **45**, pp. 802–810.

Norton, J. D. (1995): *Mach's Principle before Einstein*, in (Barbour & Pfister, 1995), pp. 19–57.

Nottale, L. (1992): *The Theory of Scale Relativity*, International Journal of Modern Physics A, Volume **7**(20), pp. 4899–4936; the www version complemented by notes and errata in 2003.

Nottale, L. (2011): *Scale Relativity and Fractal Space-Time*, Imperial College Press, London, UK.

Novozhilov, V. V. (1952): *On the Physical Meaning of the Invariants Used in the Theory of Plasticity* (in Russian), Prikladnaya Matematika i Mekhanika, Volume **16**, pp. 617–619.

Olariu, S. (2002): *Complex Numbers in n Dimensions*, North-Holland Mathematics Studies 190, Elsevier, Amsterdam.

Olver, P. J. (1998): *Applications of Lie Groups to Differential Equations*, Springer-Verlag, New York.

Ozsváth, I., Schücking, E. (1962): *Finite Rotating Universe*, Nature, Volume **193**, pp. 1168–1169.

Ozsváth, I., Schücking, E. (1969): *The Finite Rotating Universe*, Annals of Physics, Volume **55**, pp. 166–204.

Ozsváth, I., Schücking, E. (1998): *The World Viewed from Outside*, Journal of Geometry and Physics, Volume **24**, pp. 303–330.

Ozsváth, I., Schücking, E. (2001): *Approaches to Gödel's Rotating Universe*, Classical and Quantum Gravity, Volume **18**, pp. 2243–2252.

Penrose, R. (2002): *Gravitational Collapse: The Role of General Relativity*, General Relativity and Gravitation, Volume **34**, pp. 1141–1165.

Penzias, A. A., Wilson, R. W. (1965): *A Measurement of Excess Antenna Temperature at 4080 Mc/s*, The Astrophysical Journal, Volume **142**, pp. 419–421.

Perjés, Z. (1970): *Spinor Treatment of Stationary Space-Times*, Journal of Mathematical Physics, Volume **11**, pp. 3383–3391.

Pinney, E. (1950): *The Nonlinear Differential Equation* $y'' + p(x)y + cy^{-3} = 0$, Proceedings of the American Mathematical Society, Volume **1**(5), p. 681.

Pitman, E. J. G., Williams, E. J. (1967): *Cauchy-Distributed Functions of Cauchy Variates*, The Annals of Mathematical Statistics, Volume **38**(3), pp. 916–918.

Planck, M (1900): *Planck's Original Papers in Quantum Physics*, translated by D. Ter-Haar and S. G. Brush, and Annotated by H. Kangro, Wiley & Sons, NY, 1972.

Poincaré, H. (1889): *Leçons sur la Théorie Mathématique de la Lumière*, rédigées par J. Blondin, Georges Carré, Paris.

Poincaré, H. (1895): *Capillarité*, Georges Carré, Paris.

Poincaré, H. (1895): *A Propos de la Théorie de M. Larmor*, L'Éclairage Électrique, Tome **3**, pp. 5–13; 289–295; Tome **5**, pp. 5–14; 385–392. Reproduced in Œeuvres de Henri Poincaré, Tome IX, pp. 369–426, Gauthier-Villars 1951.

Poincaré, H. (1896): *Remarques sur une Expérience de M. Birkeland*, Comptes Rendus de l'Académie des Sciences de Paris, Tome **123**, pp. 530–533.

Poincaré, H. (1897): *Les Idées de Hertz sur la Mécanique*, Revue Générale des Sciences Pures et Appliquées, Tome **8**(17), pp. 734–743 (there is an English translation of this work at vixra.org).

Poincaré, H. (1900): *La Théorie de Lorentz et le Principe de Réaction*, Archives Néerlandaises des Sciences Exactes et Naturelles, Série II, Tome V, pp. 252–278 (there is an English translation of this work at physicsinsights.org).

Poincaré, H. (1906): *Sur la Dynamique de l'Électron*, Rendiconti del Circolo Matematico di Palermo, Tome **21**, pp. 129–176 [there is an English translation of this work at Wikisource; there are also other translations as indicated in a rendition of this work in modern terms by H. M. Schwartz; see (Schwartz, 1971, 1972)].

Poisson, S. D. (1813): *Remarques sur une Équation qui se Présente dans la Théorie des Attractions des Sphéroïdes*, Nouveau Bulletin des Sciences, Société Philomatique de Paris, Tome **3**(12), pp. 388–392.

Popescu, I. N. (1982): *Gravitation: Pleading for a New Theory of Gravitation*, Editura Stiintifica si Enciclopedica, Bucuresti (in Romanian).

Popescu, I. N. (1988): *Gravitation: Pleading for a New Unified Theory of Motion and Fields*, Nagard, Rome.

Poynting, J. H. (1884): *On the Transfer of Energy in the Electromagnetic Field*, Philos. Trans. of the Royal Society of London, Volume **175**, pp. 343–361.

Price, R. H., Thorne, K. S. (1986): *Membrane Viewpoint on Black Holes: Properties and Evolution of the Stretched Horizon*, Physical Review D, Volume **33**(4), pp. 915–941.

Price, R. H., Thorne, K. S. (1988): *The Membrane Paradigm for Black Holes*, Nature, Volume **258**(4), pp. 69–77.

Ray, J. R. (1973): *Modified Hamilton's Principle*, American Journal of Physics, Volume **41**(10), pp. 1188–1190.

Resnikoff, H. L. (1974): *Differential Geometry of Color Perception*, Journal of Mathematical Biology, Volume **1**, pp. 97–131.

Reyes, E. G. (2003): *On Generalized Bäcklund Transformation for Equations Describing Pseudo-spherical Surfaces*, Journal of Geometry and Physics, Volume **45**(3–4), pp. 368–392.

Riemann, B. (1867): *Ueber die Hypothesen, welche der Geometrie zu Grunde liegen*, Abhandlungen der Königlichen Gesellschaft der Wissenschaften zu Göttingen, Volume **13**, pp. 2082–2089.

Rindler, W. (1960): (a) *Hyperbolic Motion in Curved Space Time*, Physical Review, Volume **119**, pp. 2082-2089; (b) *Remarks on Schrödinger's Model of de Sitter Space*, Physical Review, Volume **120**, pp. 1041–1044.

Rindler, W. (1966): *Kruskal Space and the Uniformly Accelerated Frame*, American Journal of Physics, Volume **34**, pp. 1174–1178.

Rindler, W. (2009): *Gödel, Einstein, Mach, Gamow and Lanczos: Gödel's Remarkable Excursion into Cosmology*, American Journal of Physics, Volume **77**(6), pp. 498–510.

Rogers, C., Ramgulam, U. (1989): *A Non-Linear Superposition Principle and Lie Group Invariance: Application in Rotating Shallow Water Theory*, International Journal of Non-Linear Mechanics, Volume **24**(3), pp. 229–236.

Rogers, C., Schief, W. K. (2002): *Bäcklund and Darboux Transformations*, Cambridge University Press, Cambridge, UK.

Römer, O. (1676): *Demonstration Touchant le Mouvement de la Lumi-ère*, Le Journal des Sçavans, Lundi 7 Décembre 1676, pp. 233–236.

Römer, O. (1677): *A Demonstration Concerning the Motion of Light*, Philosophical Transactions of the Royal Society of London, Volume **12**, pp. 893–894.

Salmon, R. (1982): *Hamilton's Principle and Ertel's Theorem*, American Institute of Physics Conference Proceedings, Volume **88**, pp. 127–135.

Salmon, R. (1988): *Hamiltonian Fluid Mechanics*, Annual Reviews in Fluid Mechanics, Volume **20**, pp. 225–256.

Salmon, R. (1998): *Lectures on Geophysical Fluid Dynamics*, Oxford University Press, Oxford, UK.

Sambursky, S. (1937): *Static Universe and Nebular Red Shift I*, Physical Review, Volume **52**, pp. 335–338.

Sambursky, S., Schiffer, M. (1938): *Static Universe and Nebular Red Shift II*, Physical Review, Volume **53**, pp. 256–263.

Sasaki, R. (1979): *Soliton Equations and Pseudospherical Surfaces*, Nuclear Physics B, Volume **154**, pp. 343–357.

Schiff, L. I. (1966): *Nonrelativistic Quark Model*, Physical Review Letters, Volume **17**(10), pp. 612–613; *Quarks and Magnetic Poles*, Ibidem, Volume **17**(14), pp. 714–716.

Schiff, L. I. (1967): *Quarks and Magnetic Poles*, Physical Review, Volume **160**, pp. 1257–1262.

Schleich, W. P., Dahl, J. P. (2002): *Dimensional Enhancement of Kinetic Energies*, Physical Review A, Volume **65**(5), 052109.

Schleich, W. P., Greenberger, D. M., Kobe, D, H., Scully, M. O. (2013): *Schrödinger Equation Revisited*, Proceedings of the National Academy of Sciences USA, Volume **110**(14), pp. 5374–5379.

Schrödinger, E. (1920): *Grundlinien einer Theorie der Farbenmetrik im Tagessehen*, Annalen der Physik, Band **63**(21), pp. 397–426; 427–456; **63**(22), pp. 481–520; English translations by Keith K Niall, in the collection (Niall, 2017).

Schrödinger, E. (1933): *Mémoires sur la Mécanique Ondulatoire*, Librairie Félix Alcan, Paris.

Schrödinger, E. (1956): *Expanding Universes*, Cambridge University Press, Cambridge, UK.

Schwarzschild, K. (1916): *Über das Gravitationsfeld einer Kugel aus inkompressiebler Flüssigkeit nach der Einsteinschen Theorie*, Sitzungsberichte der Königlich Preussischen Akademie der Wissenschaften zu Berlin, Mathematisch-Physikalischen Klasse, pp. 424–435; English translations: S. Antoci, arXiv:physics/9912033; L. Borissova, D. Rabounski, The Abraham Zelmanov Journal, Volume **1**, pp. 20–32, 2008.

Schwartz, M. H. (1971): *Poincaré's Rendiconti Paper on Relativity I*, American Journal of Physics, Volume **39**(11), pp. 1287–1294.

Schwartz, M. H. (1972): *Poincaré's Rendiconti Paper on Relativity II*, American Journal of Physics, Volume **40**(6), pp. 862–872; *Poincaré's Rendiconti Paper on Relativity III, ibidem* Volume **40**(9), pp. 1282–1287.

Sciama, D. W. (1953): *On the Origin of Inertia*, Monthly Notices of the Royal Astronomical Society, Volume **113**(1), pp. 34–42.

Sciama, D. W. (1969): *The Physical Foundations of General Relativity*, Doubleday, New York.

Scott, A. C., Chu, F. Y. F., McLaughlin, D. W. (1973): *The Soliton: A New Concept in Applied Science*, Proceedings of the IEEE, Volume **61**(10), pp. 1443–1483.

Segev, R. (2017): *Geometric Analysis of Hyper-Stresses*, International Journal of Engineering Science, Volume **120**(1), pp. 100–118; arXiv: 1705.10080v1 [math-ph].

Serret, J.-A. (1886): *Cours de Calcul Différentiel et Intégral*, Gauthier-Villars, Paris.

Shapere, A., Wilczek, F. (1989): *Geometric Phases in Physics*, World Scientific Publishing Co., Singapore.

Shapiro, A. E. (1973): *Kinematic Optics: A Study of the Wave Theory of Light in the Seventeenth Century*, Archives for the History of Exact Sciences, Volume **11**, pp. 134–266.

Shapiro, A. E. (1975): *Newton's Definition of a Light Ray and the Diffusion Theories of Chromatic Dispersion*, Isis, Volume **66**, pp. 194–210.

Shchepetilov, A. V. (2003): *The Geometric Sense of the Sasaki Connection*, Journal of Physics A: Mathematical and General, Volume **36**(12), pp. 3893–3898.

Shirokov, P. A., Shirokov, A. P. (1959): *Affine Differential Geometry*, GIFML, Moscow (in Russian); (there is no English translation of this work, only a German translation).

Shirokov, A. P. (1988): *On the Geometry of Tangent Bundles of Homogeneous Spaces*, The Works of the Geometry Seminar of the Kazan University, Volume **18**, pp. 153–163 (in Russian).

Shivamoggi, B. K., Muilenburg, L. (1991): *On Lewis' Exact Invariant for the Linear Harmonic Oscillator with Time-Dependent Frequency*, Physics Letters A, Volume **154**(1,2), pp. 24–28.

Sitter, W. de (1916): *On Einstein's Theory of Gravitation and its Astronomical Consequences*, Monthly Notices of the Royal Astronomical Society: *First Paper*, Volume **76**, pp. 699–728; *Second Paper*, Volume **77**, pp. 155–184; *Third Paper*, Volume **78**, pp. 3–28.

Sitter, W. de (1917): *On the Relativity of Inertia. Remarks Concerning Einstein's Latest Hypothesis*, Proceedings of the Royal Netherlands Academy of Arts and Sciences at Amsterdam, Volume **19 II**, pp. 1217–1225.

Sitter, W. de (1918): *On the Curvature of Space*, Proceedings of the Royal Netherlands Academy of Arts and Sciences at Amsterdam, Volume **20 I**, pp. 229–243.

Skinner, D. (2016): *Mathematical Methods*, Course Notes at the University of Cambridge, UK.

Sofue, Y., Rubin, V. (2001): *Rotation Curves of Spiral Galaxies*, Annual Review of Astronomy and Astrophysics, Volume **39**, pp. 139–174.

Sparnaay, M. J. (1958): *Measurement of Attractive Forces Between Flat Plates*, Physica, Volume **24**, pp. 751–764.

Spivak, M. (1995): *Calculus on Manifolds*, Addison-Wesley, Reading, Massachusetts.

Spivak, M. (1999): *A Comprehensive Introduction to Differential Geometry*, Volume III, Publish or Perish, Inc., Houston, Tx.

Stiefel, E. (1952): *On Cauchy-Riemann Equations in Higher Dimensions*, Journal of Research of the National Bureau of Standards, Volume **48**(5), pp. 395–398.

Stiefel, E. L., Scheifele, G. (1975): *Linear and Regular Celestial Mechanics*, Nauka, Moscow (Russian translation of the 1971 English original).

Stoka, M. I. (1968): *Géométrie Intégrale*, Mémorial des Sciences Mathématiques, Fascicule **165**, pp. 1–65, Gauthier–Villars, Paris.

Stratton, J. A. (1941): *Electromagnetic Theory*, McGraw-Hill Book Company, Inc., NY.

Struik, D. J. (1988): *Lectures on Classical Differential Geometry*, Dover NY.

Synge, J. L. (1972): *A Special Class of Solutions of the Schrödinger Equation for a Free Particle*, Foundations of Physics, Volume **2**(1), pp. 35–40.

Szilard, L. (1925): *On the Extension of Phenomenological Thermodynamics to Fluctuation Phenomena*, n *The Collected Works of Leo Szilard. I. Scientific Papers*, The MIT Press, London, UK & Cambridge, MA, 1972, pp. 70–102; English translation of the German original from Zeitschrift für Physik, Volume **32**(1), pp. 753–788.

Szilard, L. (1929): *On the Decrease of Entropy in a Thermodynamic System by the Intervention of Intelligent Beings*, in *The Collected Works of Leo Szilard. I. Scientific Papers*, The MIT Press, London, UK & Cambridge, MA, 1972, pp. 120–129; English translation of the German original from Zeitschrift für Physik, Volume **53**(10), pp. 840–856.

Takabayasi, T (1952): *On the Formulation of Quantum Mechanics Associated with Classical Pictures*, Progress of Theoretical Physics, Volume **8**(2), pp. 143–182.

Ţiţeica, G., (1908): *Sur une Nouvelle Classe de Surfaces I*, Rendiconti del Circolo Matematico di Palermo, Tome **25**, pp. 180–187.

Ţiţeica, G., (1909): *Sur une Nouvelle Classe de Surfaces II*, Rendiconti del Circolo Matematico di Palermo, Tome **28**, pp. 210–216.

Tolman, R. C. (1929): *On the Astronomical Implications of the de Sitter Line Element for the Universe*, The Astrophysical Journal, Volume **69**(4), pp. 245–274.

Tolman, R. C. (1934): *Relativity, Thermodynamics and Cosmology*, Clarendon Press, Oxford.

Toomre, A. (1977): *Theories of Spiral Structure*, Annual Review of Astronomy and Astrophysics, Volume **15**, pp. 437–478.

Uy, Z. E. S. (1976): *A Solution to the SU(2) Classical Sourceless Gauge Field Equation*, Nuclear Physics B, Volume **110**, pp. 389–396.

Villarceau, Y. (1850): *Memoires et Notes sur les Étoiles Doubles*, Paris, Bachelier.

Vinti, J. P. (1966): *Invariant Properties of the Spheroidal Potential of an Oblate Planet*, Journal of the National Bureau of Standards, Series B, Volume **70**(1), pp. 1–16.

Vivarelli, M. D. (2015): *The Kepler Problem Primigenial Sphere*, Meccanica, Volume **50**, pp. 915–925.

Volk, O. (1973): *Concerning the Derivation of the KS-Transformation*, Celestial Mechanics, Volume **8**, pp. 297–305.

Volk, O. (1976): *Miscellanea from the History of Celestial Mechanics*, Celestial Mechanics, Volume **14**, pp. 365–382.

Vrânceanu, G. (1962): *Lessons of Differential Geometry*, Volume I, Editura Academiei, Bucureşti (in Romanian; there is a French edition of this book).

Wagner, H. J. (1991): *Comment on "On Lewis' Exact Invariant for the Linear Harmonic Oscillator with Time-Dependent Frequency"*, Physics Letters A, Volume **158**(3,4), pp. 181–182.

Weiss, P. (1944): *On Hydrodynamical Images: Arbitrary Irrotational Flow Disturbed by a Sphere*, Proceedings of the Cambridge Philosophical Society, Volume **40**, pp. 259–261.

Weyl, H. (1923): *Raum. Zeit. Materie*, Vorlesungen über allgemeine Relativitätstheorie, Sechste unveränderte Auflage, Springer, Berlin.

Weyl, H. (1952): *Space-Time-Matter*, Dover Publications, Inc., New York, NY.

Whittaker, E. T. (1903): *On the Partial Differential Equations of Mathematical Physics*, Mathematische Annalen, Volume **57**, pp. 333–355.

Whittaker, E. T. (1917): *Analytical Dynamics*, Cambridge University Press.

Widder, D. V. (1979): *The Airy Transform*, The American Mathematical Monthly, Volume **86**, pp. 271–277.

Wien, W. (1896): *Über die Energievertheilung im Emissionsspectrum eines schwarzen Körpers*, Annalen der Physik, Volume **58**, pp. 662–669.

Wien, W. (1900): *Les Lois Théoriques du Rayonnement*, Rapports Présentés au Congrès International de Physique, editori Ch-Éd. Guillaume & L. Poincaré, Volume **II**, pp. 23–40, Gauthier-Villars, Paris.

Wien, W., Lummer, O. (1895): *Methode zur Prüfung des Strahlüngsgesetzes absolut schwarzer Körper*, Annalen der Physik (Wiedemann), Band **56**, pp. 451–456.

Wigner, E. P. (1954): *Conservation Laws in Classical and Quantum Physics*, Progress of Theoretical Physics (Japan), Volume **11**(4/5), pp. 437–440.

Wilhelm, H. E. (1971): *Hydrodynamic Model of Quantum Mechanics*, Physical Review D, Volume **1**, pp. 2278–2285.

Williamson, J. G., van der Mark, M. B. (1997): *Is the Electron a Photon with Toroidal Topology?* Annales de la Fondation Louis de Broglie, Volume **22**(2), pp. 133–157.

Wilson, E. B. (1919): *Radiationless Orbits*, Proceedings of the National Academy of the United States, Volume **5**, pp. 588–591.

Wilson, E. B. (1924): *Coulomb's Law and the Hydrogen Spectrum*, Proceedings of the National Academy of the United States, Volume **10**, pp. 346–348.

Winkelmolen, A. M. (1982): *Critical Remarks on Grain Parameters, with Special Emphasis on Shape*, Sedimentology, Volume **29**(2), pp. 255–265.

Wu, T. T., Yang, C. N. (1969): *Some Solutions of the Classical Isotopic Gauge Field Equations*, in 'Properties of Matter Under Unusual Conditions', H. Mark & S. Fernbach (eds), Wiley-Interscience, New York, pp. 349–354; reprinted in 'Selected Papers of C. N. Yang, with Commentaries', World Scientific, Singapore, 2005, pp. 400–405.

Yamamoto, T. (1952): *The Analytic Representation of Spin*, Progress of Theoretical Physics (Japan), Volume **8**, p. 258.

Yang, C. N., Mills, R. L. (1954): *Conservation of Isotopic Spin and Isotopic Gauge Invariance*, Physical Review, Volume **96**, pp. 191–195.

Yau, S-T. (1989): *Affine Conormal of Convex Hypersurfaces*, Proceedings of the American Mathematical Society, Volume **106**(2), pp. 465–470.

Zelikin, M. I. (2000): *Control Theory and Optimization*, Encyclopaedia of Mathematical Sciences, Volume **86**, Springer.

Printed in the United States
by Baker & Taylor Publisher Services

Printed in the United States
by Baker & Taylor Publisher Services